I0032468

Anonymous

Weekly World's Fair Supplement

Electrical Industries

Anonymous

Weekly World's Fair Supplement
Electrical Industries

ISBN/EAN: 9783743477841

Manufactured in Europe, USA, Canada, Australia, Japa

Cover: Foto ©berggeist007 / pixelio.de

Manufactured and distributed by brebook publishing software (www.brebook.com)

Anonymous

Weekly World's Fair Supplement

Weekly World's Fair Supplement.

ELECTRICAL INDUSTRIES

DEVOTED TO THE ELECTRICAL AND ALLIED INTERESTS OF THE WORLD'S FAIR, ITS VISITORS AND EXHIBITORS.

Vol. I, No. I.   CHICAGO, JUNE 15, 1893.   TEN CENTS A COPY. FIVE MONTHS $1.00.

The Western Electric Company's Exhibit in the Electricity Building.

Among the large number of exhibits in the Electricity hind glass of rare color and design together with the rapid and mysterious changes in form and color has fascinated the beholder. The large column covered with hundreds of incandescent lamps of red, white and blue from the top

WESTERN ELECTRIC COMPANY'S EXHIBIT—THE EGYPTIAN TEMPLE.

building none has attracted more attention from all classes of visitors than that of the Western Electric Company. The arrangements of lights, many of them concealed be of which reach out four zig zag arms of colored lamps with globes at their ends also covered with colored lights, and constantly revolving, attracts constantly a crowd of visit ors who watch the band of light rise from the bottom of the column to the top and then spread out over the arms to the globes turning first red, then white, then blue. The large automatic switch or commutator in the basement is especially of interest to those engaged in electrical pursuits. Its construction, the manner of making and breaking the contacts, the arrangement of the air blast which extinguishes the sparks, etc., are all very ingenious

reaches it. This effect is produced by two cams at right angles and of suitable shape for producing the motion through racks and gears. Another ingenious device operated on the same principle is that in which an incandescent lamp on the end of a pointer is made to spell out the words Western Electric Co.

The Egyptian temple which has been erected for the display of telegraph, testing, and other small instruments

WESTERN ELECTRIC CO.'S EXHIBIT—COMBINATION COVERING AND SILK MACHINES.

WESTERN ELECTRICAL CO.'S EXHIBIT—NUT AND MULTIPLE DRILL MACHINES.

and interesting. A window has been placed in the floor over this machine, and it is always surrounded by a crowd of curious people trying to understand its operation.

The large illuminated sign on which is the name of the company, and the cities at which are located its different factories is another attractive object. The letters are composed of bits of colored glass set in the darker stained glass background, and the fourteen are lamps which are

and brass work is lighted entirely by lights concealed behind translucent glass of many colors. A large number of lights are arranged above the false ceiling of colored glass, and are so manipulated as to vary the light and render it mild and more pleasing to the eye. Columns of heavy green

WESTERN ELECTRICAL CO.'S EXHIBIT—AUTOMATIC SCREW MACHINE.

WESTERN ELECTRIC CO.'S EXHIBIT—BATTERIES.

concealed within the sign are so moved by means of a motor that the letters on the sign have a sparkling appearance which seems mysterious until explained. This sign is especially noticeable at night. Some unique features have been devised by the use of cams and gears in connection with electric lights. A cabinet surmounted with the large letters W. E. Co. in incandescent lamps of red, white and blue is made especially attractive by the lamps being lighted in rotation commencing with W, and a pointer on the top of the box at the back of the letters follows their outline and seemingly lights each lamp as it

glass illuminated by incandescent lamps placed within them add to the effect produced by the other lamps. Plush lined cases sunk in the walls and lighted by incandescent lamps concealed above are filled with the various articles displayed. For the lighting of this temple between 1,100 and 1,200 incandescent lamps are used, 700 of which are constantly burning.

An examination of the interior and exterior frescoing of the temple show forms and figures of Egyptian design illustrating the various departments of electricity. The lineman and wireman are shown at work, the operation of the fire and police patrol systems as well as the operation of the telephone and telegraph are illustrated in these interesting decorations. The combination of the ancient figures with the various applications of electricity, the most modern force, is certainly unique. From the upper part of the walls light is thrown out through ground glass globes which add to the striking appearance of this temple. The wire for this building is laid in interior conduit tubing and is easily accessible through suitable traps in the upper part of the structure, and a switchboard is placed in the top for controlling the different circuits of

with no extra foundation above or below the floor boards and they are so well balanced that there is no vibration whatever.

Back of the motor-generator exhibit toward the main aisle is another enclosure with some representative machines designed and in use by this company to save labor. Among them is an automatic nut-making machine, which from strips of brass fed into it, cuts out a blank, stamps the company's name on it, faces it up on both sides, threads it and casts it out for inspection. Next comes a multiple drill press so arranged that any number of small holes up

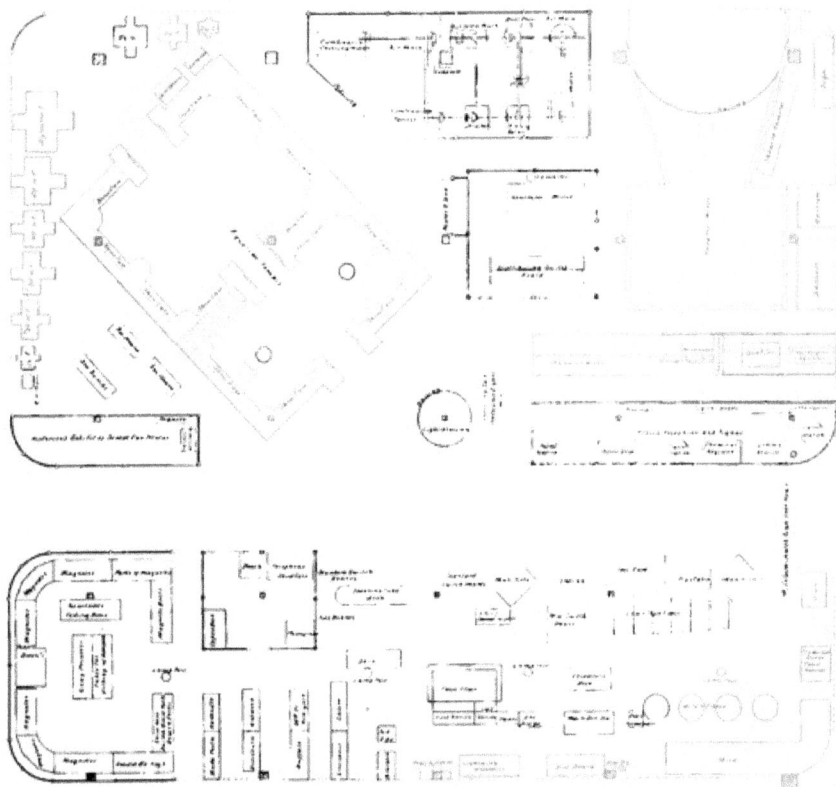

WESTERN ELECTRIC CO'S EXHIBIT- FLOOR PLAN

lamps. Leaving the temple by a northerly doorway the combination switchboard of marble attracts the attention of the visitor. This board is equipped with the necessary switches and apparatus for controlling arc lamps, incandescent lamps, and power or motor circuits and is a finely finished piece of work; in fact, the back of the board containing all the electrical connections is, contrary to the usual case, fully as well finished as the front side. In the same enclosure with the switchboard is one of the company's 120 ampere 220 volt motors directly connected by a shaft coupling to a 240 ampere 110 volt generator of the same type. The two machines are set on a common floor,

to 32 in any position may be drilled all at one motion, by setting the drills once by jig. Another is a very ingenious screw machine for turning small brass work. It is almost human in its actions, and is completely automatic going through the various motions, cutting and even transferring a partly finished piece from one set of jaws to the other with an accuracy which defies criticism.

Other machines are for insulating small wires with silk, paper and cotton, and for twisting and forming different flexible cords. A machine for winding the armature for magneto bells is also shown. A miniature theatre has been built to the north of the machinery exhibit in which are shown in connection with a standing scene, the various effects of lights, changing from daylight to night and moonlight, making the complete cycle of the day. The operation of an electric lighting plant is neatly illustrated in miniature. On a stand of polished wood and enclosed in a glass case is shown a fine model of a three cylinder engine complete with steam pipes, etc. Belted to this is a small motor, and on a bracket in the center is a group of six incandescent lamps, giving the appearance of a dynamo run by an engine and lamps burning from the current generated. The deception is quite complete as no wires leading to or from this stand are seen. The various dry and wet batteries made by this company are shown with all the parts going to make up the completed cells, carbons, zincs, porous cups, etc.

A large number of samples are shown of the hotel calls and annunciators with all the accompanying equipment. A complete system of police patrol booths with call boxes of various descriptions is also shown. Reels of wire large and small, for all classes and kinds of work, from the finest green silk covered magnet wire up to the many circuit lead

to-day. Statistical tables showing the growth of various exchanges and multiple switchboards are submitted for inspection. Samples of telephones made abroad at the Antwerp factory of the company are shown also, while magneto bells and telephones unassembled with all parts laid out for inspection make a fine exhibit with reference to the amount of detail necessary for such construction.

A Hollerith electrical tabulating machine, as used by the United States and other governments, is exhibited, and it is expected that a sample of Lieutenant Bradley Fiske's range finders will soon be; both of these machines being made for the inventors by this company. Among the historical exhibits is a sample set of the Gray printing telegraph which was just coming into use when the telephone came out and supplanted it. A model of an electric street car designed by the late Prof. Moses G. Farmer, in 1847, is also shown. It has batteries under the seat and is equipped with proper levers for controlling the motor. There is also one of the original bi-polar motors, built at a very early date by Prof. Farmer. In a conspicuous place over this exhibit is a very handsome portrait in oils of the Pro-

WESTERN ELECTRIC CO.'S EXHIBIT—COMBINATION SWITCH BOARD, BACK AND FRONT VIEW.

covered underground cable, are on exhibition. Two brick manholes for subways are shown with cables placed as in actual use, and the methods of drawing in and splicing are indicated in an excellent manner. The Western Electric Company, it may be said in passing, was the first to adopt the underground system and to develop it in proper form. Various types of telephone transmitters and receivers as made for the American Bell Telephone Company are shown, including the regular instruments, long distance sets and the various forms of desk sets now coming largely into use in large office buildings.

There is a fine historical exhibit of telephone switchboards, commencing with one made in 1884 and taking in all the improvements, up to and including a section of the latest board now made for and used by nearly all the local telephone exchanges in the country. These boards are all connected up just as they would be in use, and where improvements have been made in the connections which were not radical enough to require a new board to demonstrate it, the diagram is on the back of the board showing the circuits in outline. This permits of following each and every step of this branch of the business from the beginning up

fessor made in recent years. One unique feature is a portal over the entrance of one section constructed wholly of speaking tube, showing the straight tubing and various curves; another is a coil of coppered iron rheostat wire spirally coiled and the spiral, over 200 feet in length, is festooned from post to post.

In the arc lighting exhibit is shown a case containing the various small supplies used for that work, such as side brackets, insulators, switches, hanger-boards, etc.; an ornamented rack suspended from which are the various styles of arc lamps made by the company; one of its standard arc switchboards with marble face, and samples of each of the company's high tension arc dynamos of 20, 30, 40, 50 and 60 light capacity. In connection with the large service plant of arc dynamos in Machinery Hall is shown the Rudd ground alarm, used for detecting grounds in arc circuits while the current is on. All the dynamos shown are set up on the regular wooden base and show fine workmanship in all parts. Samples of the Chicago City lamp post as designed and constructed by this company are on exhibition in its section, but are also to be found in actual use about the galleries of the Electricity building. In the line of incandescent dynamos, are samples of 110-volt machines of 30, 100, 165, 230, 400 and 650-ampere capacity, and of 220 volt motors, aside from the 40-horse power four pole motor used for driving the machinery exhibit, the company has samples of seven, 17 and 35-ampere four pole slow speed motors and the previously mentioned bi-polar motor directly connected to the lighting generator.

Taken altogether this exhibit covers more ground in detail and classes of work shown than any other in the department. Comment on the nature of the work is unnecessary in connection with the well-known reputation of this company and it only need be said that it is finished up to the high standard set by it. The comprehensiveness of the exhibit can be judged by the fact that outside of the time taken in designing and building the apparatus it has taken six months to complete its erection and installation. Hard and persistent work has been required by all hands

WESTERN ELECTRIC CO'S. EXHIBIT—MODEL OF FARMER'S RAILWAY MOTOR.

to overcome the difficulties that have beset work at the Fair during the winter, the elements and all else seeming to combine to prevent any but the very slowest progress.

Search Lights at the Fair.

The field in which the search light is used being limited but few manufacturers of electrical goods have given any attention to its construction. At the present time the principal use to which it has been put is on river and excursion steamers, on the large ocean steamers and in the navy, and in a few cases large business and office buildings have been furnished with them for advertising purposes. Of the three exhibitors of search lights but one is American, the General Electric Co., the others are Schuckert & Co., of Nuremberg, and Sautter-Harle & Co., of Paris.

Messrs. Schuckert & Co. exhibit four search lights with lenses respectively five, four, three and two feet in diameter. These are placed at the four corners of the main roof of the Manufacturers and Liberal Arts building. Only two of them have been used so far, but the wiring is nearly completed for the other two, then all four of them will be seen the three nights of each week, on which the Fair is open. The largest of these lamps uses a carbon one and one-half inches in diameter for the positive and one inch for the negative. In the center of the positive carbon is a glass core, three-sixteenths of an inch in diameter for the

better focusing of the arc. This light requires 150 amperes of current at 60 volts, and represents the lights designed for coast defense. In the manufacture of these lights great precision and exactness is required as well as excellence in the quality of the material used. The reflector must be of a certain shape and finely finished. The lense must be of a particularly clear glass carefully ground to the required shape to give the rays of light a parallel or slightly diverging direction. This company also exhibits in the Electricity Building a three-foot marine search light, a standard search light intended for use in war on land. All the lamps manufactured by Schuckert & Co. are automatically adjusted and controlled. Its lamps are all designed for war purposes. The beam from the largest lamp is estimated at 194 million candle power and has been seen in Milwaukee.

The General Electric Company shows a variety of search lights, some controlled automatically, others by hand, the adjustment being automatic in all. The lights controlled automatically can by setting suitable stops on the frame, be made to swing automatically through any desired arc without any attention. On the roof of Music hall is a light with a 30 inch lens. In the equipment of the model battle ship Illinois, which was done by the General Electric Company are two 30 inch automatically controlled lamps. There are also two lamps of this type but smaller in the galleries of the Electricity Building for lighting the tower. An interesting part of the exhibit is a search light recovered from the U. S. Trenton that was sunk in the harbor at Samoa and was the first U. S. Man-of-War to be equipped with an electrical plant. The company also shows a 65,000 candle power focusing lamp for making blue prints or photographic printing.

Sautter-Harle & Co. show a 24-inch Mangin projector automatically adjusted and controlled electrically. The controlling apparatus is very simple and ingenious. Nearly all the naval search lights are provided with a double lens, the outer one being movable so that the direction of the rays of light may be changed so as to cover a small or large area as desired.

Models from the Patent Office.

The Government Building contains a large number of articles from the patent office including a number of models of early electrical inventions, among which is a model of a magnetic machine invented by Pixii in 1832, an engine by M. H. Jacobis in 1834, magnetic motor by Joseph Henry, an electric motor by Thos. Davenport in 1837, etc.; then the more modern models of Gramme, Sims, Farmer, Fuller, Weston, Brush, Edison, Thompson-Houston and others. Besides these models of motors and dynamos there are also models of arc lamps, meters, welding machines, converters, telephone and telegraph apparatus. There is also shown in this building samples of the apparatus used by the U. S. Signal Corps, both the stationary and portable sets. Among them is a light iron wagon holding all the apparatus used in the field, with the wire wound on reels so that it may be quickly run out, all the batteries, instruments, etc., placed where they can be got hold of to enable the complete line to be quickly put in working order.

The Standard Electric Company has at present twenty 50 light dynamos in operation daily in Machinery Hall. The company is also putting in two additional machines for the exclusive purpose of lighting the power plant.

ELECTRICAL INDUSTRIES.

Entered as Chicago Postoffice as second-class mail matter.

PUBLISHED EVERY THURSDAY BY THE

ELECTRICAL INDUSTRIES PUBLISHING COMPANY,
INCORPORATED 1888.
MONADNOCK BLOCK, CHICAGO.
TELEPHONE HARRISON 185.

E. L. POWERS, Pres. and Treas.   E. E. WOOD, Secretary.

E. L. POWERS,   Editor

H. A. FOSTER, |
                       }   Associate Editors
W. A. REMINGTON, |

E. E. WOOD,   Eastern Manager

FLOYD E. SHORT,   Advertising Department

EASTERN OFFICE, WORLD BUILDING, NEW YORK.
World's Fair Headquarters, Y 27 Electricity Building.

SUBSCRIPTION:

FIVE MONTHS $2.00
SINGLE COPY 10

Advertising Rates Upon Application.

News items, notes or communications of interest to World's Fair
Visitors are earnestly desired for publication in these columns and will
be heartily appreciated. We especially invite all visitors to call upon us
or send address at once upon their arrival in the city or at the grounds.
ELECTRICAL INDUSTRIES PUBLISHING CO.,
Monadnock Block, Chicago.

Salutatory.

The World's Columbian Exposition is recognized throughout the world as the most important event of the day. Volumes have already been written upon the beauty of its architecture and its educational advantages to every one interested in the progress of the times and still the Fair is only now just fairly begun. Visitors are filled with wonder at what they see and go away satisfied that they have beheld the most marvelous exhibition of the kind ever held or is likely to be held for many years to come. The electrical and mechanical features of the Exposition, it is agreed, however, are among the most attractive and important of all, and the field of research for the electrician, electrical engineer and manufacturer is almost beyond comprehension. The growth, extent, and importance of the electrical industries is here fully shown and the phenomenal development of these interests is a matter of which we may well be proud. Indeed without electricity the Exposition would fall far short of being the great success that it is. To gather in a comprehensive manner and put in acceptable form for the reader the vast amount of data here found is the function of the trade journal, and the task to say the least is not an easy one. The pressure upon the reading columns of our regular monthly issue has become so great that we have found it necessary to publish this weekly supplemental sheet of modest proportions. The mission of this paper will be to advance the electrical and allied mechanical interests of the Fair and at the same time be of interest and value to the visitor. With this object in view we present on another page a map of the city of Chicago, giving the name and location of the electrical interests in the city. This map is to be in each issue, thus making it a permanent feature. On another page will also be found diagrams of the Electricity Building, giving the name and location of each exhibitor. This map likewise is to be a permanent feature and will, we believe, be invaluable to every one visiting that building. The paper, in fact, is to cover an important field now existing and we have every assurance that our enterprise in thus serving our readers will be appreciated.

WORLD'S FAIR NOTES.

The electric pianos in the west gallery of the Electricity building furnish very enjoyable music.

In the French section of the Electricity building there is shown a model of the Letrange process for the reduction of zinc by electricity.

On the evenings of August 24, 25 and 26, during the International Electrical Congress, public lectures of a popular character will be given by eminent electricians.

A. Groetzinger & Sons, of Allegheny, Pa., show samples of gears and pinions for street car work of dermaghdine. This process of preparing raw hide for this work has been very successful.

The exhibit of the Excelsior Electric Company was nearly completed when Mr. Fuller decided to send a large amount of additional material which would make the exhibit better represent the factory.

Prof. Barrett is now preparing some very attractive electrical features which will be placed in the galleries of the Electricity building, which will, without doubt, draw a crowd to those parts of the building.

The basins of the electric fountains are being boarded over on a level with the bottom of the spray caps. This covering will conceal the pipes and other material, giving the fountains a more finished appearance.

The entire south wall of the Electricity building above the gallery has been for some days in the hands of the decorators, who are preparing some very fine decorations under the direction of the Westinghouse company.

Both the General Electric and Westinghouse companies have made elaborate preparations for exhibiting the triphase systems and also the Allgemeine Electricitats Gesellschaft have several machines installed for that purpose.

The Page Belting Company, of Concord, N. H., has a very neat and appropriate booth erected on the main floor of the Electricity building. It is made up entirely of the well known link belting and makes a very attractive display.

A number of changes are being made in the Bell Telephone Company's building, which will cause the building to be partly closed to visitors for a number of days, but when it is again opened the change will undoubtedly be appreciated.

Lieut. Spencer reports that when the exhibit of the General Electric Company is completed, there will be exhibits of Prof. Thomson's experiments with currents of very high tension. A regular program will be issued so people may know when the experiments are to take place.

The General Electric Company has just issued for the World's Fair a very handsome folder describing and illustrating its works and the most noteworthy central stations furnished with its system. The application of electricity to its many uses is also shown by numerous cuts, and the company's exhibits at various expositions are described. It is a very neat and interesting souvenir of the company.

When completed, the Tesla Multiphase exhibit of the Westinghouse Electric and Manufacturing Co., will be one of the handsomest on the grounds. The structure in the center of the space, painted cream color, touched out with gold leaf and lighted brilliantly with incandescent bulbs and short ornamental arc lamps is one of the finest yet built. Many of the features in this exhibit are entirely new and will prove of very great interest to electrical engineers.

Although some of the exhibits in the various buildings are covered at night, and almost all the exhibitors leave the grounds at 6 o'clock, enough remains open for inspection to consume much time. It is a mistake that people are apt to assume, that little is to be seen at night. Many of the exhibits are seen to advantage only at night, and this is especially true of many of those in the Electricity Building, the tower, the various illuminated signs and other curious features being well worth a special visit.

It is reported that Nikola Tesla will show some entirely new experiments in high tension currents in a room provided for the purpose in the Westinghouse section. The Westinghouse company are preparing to exhibit various phenomena produced by currents of this nature. They will also show the working of the Wurtz non-arcing lightning arrester for various voltages and will endeavor to produce by means of a Holtz machine all the phenomena and effects accompanying lightning on alternating currents.

In the southwestern part of the Transportation Building, there is being installed a portable electric lighting plant, which is designed for use in war or for temporary lighting for any other purpose away from a central station. It consists of a heavy wagon enclosing a gasoline engine with tanks holding a supply of fuel, a dynamo of the Manchester type, and ten arc lamps hung on a portable frame with a spool of well insulated flexible conduit for each lamp. When completed this exhibit will undoubtedly be very interesting.

The cable for feeding the lights on the buoys for marking a safe way between Van Buren street pier and the pier at the World's Fair grounds for boats will be laid this week. The course is now marked out with white spar buoys, where the lamp buoys will be located. Commander J. J. Brice is pushing the work as fast as circumstances will permit. Mr. Ira W. Henry, the electrical engineer, and Capt. Marks, superintendent of the Bishop Gutta Percha Co., are to lay the cable. Mr. Frank Marks will make the connections and look after the details.

A large and very comprehensive exhibit of the University of Illinois in the Illinois Building at the Fair includes many excellent and interesting features, especially from the department of electrical engineering and physics, in which are shown apparatus and instruments with photographs of the machinery used in its laboratories. The exhibit is in charge of Prof. Crawford who invites inspection of the school's work and facilities. It is his intention at the close of the school year to have students do practical testing with the apparatus during the day to more fully illustrate the instruction given in the University.

PERSONAL.

Mr. Gisbert Kapp arrived in Chicago June 1st.

Mr. Waddell, of the Waddell Entz company, is in the city.

Prof. Elihu Thompson registered at the Hyde Park Hotel last week.

Mr. E. Wilbur Rice of the General Electric Company was in Chicago visiting the Fair last week.

Mr. H. D. Fuller, president of the Excelsior Electric Company, was registered at the Auditorium last week.

Mr. W. H. Knight, superintendent of the railway department of the Lynn factories was in Chicago last week.

Mr. J. R. Lovejoy of the Boston office of the General Electric

Company has just returned home after a hasty trip to the Fair. He will make a more extended visit later in the season.

Mr. H. McL. Harding lately with the Westinghouse Company, is registered at the Auditorium, and visiting the Fair.

Mr. George Cutter has gone east for a brief trip in the interest of some of his specialties which he is exhibiting at the Fair.

Captain W. L. Candler of the Okonite Company, Ltd., is spending a few days at the Fair looking after his exhibit which has been delayed by the press of work at the factory.

BUSINESS NOTES.

The Mather Electric Company is enlarging its factory at Manchester, Conn., by the addition of a new 100 ft. building.

The Chicago Insulated Wire Co., Chicago, has recently added to its power plant in the factory at Sycamore, Ill., a 100-horse power engine and boilers. The factory is now in shape to fill all orders promptly.

The Marion Street Railway Company has commenced work on the line from Marion, Ind., to Gas City, via the Soldiers Home. The overhead fixtures and attachments are being furnished by the Ansonia Electric Company.

The J. B. & D. Lane Company, manufacturer of all kinds of friction clutch pulleys, has just removed from its old quarters at Hornellsville, N. Y., to new and larger shops and foundry just completed at Massillion, Ohio, where the company will have better facilities for handling its increasing trade.

The Great Western Manufacturing Company, 204-207 South Canal St., in its monthly bulletin for June gives lists with prices of the standard supplies and specialties, for which this company is so well known. The new fixture catalogue will be ready for distribution within a very few days.

Bartholomew, Stow & Co., Chicago, selling agents for the McNutt Incandescent Lamp have had some difficulty in getting a sufficient number of lamps from the factory to fill orders promptly, but they now state that arrangements have been perfected so that all orders will receive immediate attention.

Mr. J. H. Gates is now manager of the western department of the Waddell-Entz Company, of Bridgeport, Conn., with a temporary office at 1122, until the offices 1108 and 1110 are ready for occupancy, Monadnock building, Chicago. Direct coupled lighting and railway generators are made a specialty of by this company.

The Buckeye Engine Company, Salem, Ohio, has a very fine exhibit in the large 1,500-horse power engine in Machinery hall running one of the 10,000-light Westinghouse dynamos. This engine has run very smoothly since it was started, and has given good satisfaction. There are also five other smaller Buckeye engines exhibited running dynamos.

The Ansonia Electric Company, Chicago, reports that the old Sunbeam incandescent lamp finds a ready sale, as the former buyers gave it the preference as soon as they learned it was in the market; that favorable reports are coming in from the users of the W-W lightning arrester, and that it has just shipped a large order of shield brand wire to Iowa.

The Central Electric Company, 115 Franklin St., Chicago, is furnishing the Sentinel building, Milwaukee, Wis., with conduit and fittings, manufactured by the Interior Conduit & Insulation Company, New York. The Lundell fan motors, manufactured by the same company, for whom the Central Company is western agent, are being placed in a number of places. The demand for Okonite wires and tapes continues as lively as ever.

The Ansonia Electric Company, Chicago, has just completed the elegant club rooms, it has had the enterprise to fit up for the special accommodation of its friends and World's Fair visitors. These rooms are handsomely furnished throughout and will be formally opened this week. Great credit is due Mr. George B. Shaw, who has charge of the World's Fair Bureau of the company and under whose personal supervision the fitting up of the rooms has been.

The Electric Appliance Company has made some novel additions in its incandescent lamp department. The first is a six candle power lamp made to fit any regular socket and burn on regular circuits. In the direction of large lamps the stock includes lamps up to 1,000-candle power and of a high efficiency. One of these lamps will soon be placed on exhibition at the World's Fair. These lamps are the Packard lamp manufactured by the well known New York & Ohio Company for whom the Electric Appliance Company is western agent.

The Detroit Dynamo Company, Detroit, Mich., has been very successful in its construction work which was started less than a year ago. The new hotel Ste. Claire, Detroit, opened to the public on June 7th was equipped by it with motors for ventilation and power, the Herzog telesame system, fire alarms, annunciators, speaking tubes, etc., and it shows the excellent quality of the work done by this company. A large number of orders has been received for dynamos among which was one for the yacht of Mr. Cudahy of Chicago.

ELECTRICITY BUILDING.

DIAGRAMS OF FLOOR SPACE GIVING EXHIBITORS AND THEIR LOCATION.

GALLERY.

MAIN FLOOR.

WE ARE PLEASED TO EXTEND A HEARTY

WELCOME

to our customers and the trade in general to visit us at our World's Fair headquarters

ELECTRICITY "U 16" BUILDING

in the Southwestern part of gallery. Make our space your headquarters — you will find a place to rest yourself and

MESTON FAN MOTORS....to keep you cool.

ELECTRIC APPLIANCE COMPANY,
ELECTRICAL SUPPLIES.

242 Madison Street, - - - CHICAGO.

The "BUCKEYE" Lamp

A KNOWN Quantity	Life 1000 to 4000 Hours
No EXPERIMENTING	Western Deliveries can be
Low Operating Cost	made from Chicago Stock
Improved Lighting Effects	

"The BUCKEYE sets the pace"

THE BUCKEYE ELECTRIC CO., Cleveland, O.

Chicago
437 THE
ROOKERY

New York
49 DEY STREET.

THE MATHER ELECTRIC CO,
MANCHESTER, CONN.

Dynamos, Motors, Generators,

Offices, 116 Bedford St., BOSTON.
—AND—
1002 Chamber of Commerce Bldg., CHICAGO.

THE "NOVAK" LAMP.

CLAFLIN & KIMBALL (Inc.)

General Selling Agents.

116 Bedford Street, BOSTON.

1002 Chamber of Commerce Bldg., CHICAGO.

Enterprise
Electric
Company

307 Dearborn Street,
Chicago ....

Manufacturers' Agents and Mill Representatives for

Electric Railway,
Telegraph, Telephone and
Electric Light

SUPPLIES OF EVERY DESCRIPTION

Agents for Cedar Poles,
Cypress Poles, Oak Pins,
Locust Pins, Cross Arms, Glass
———Feeder Wire, Insulators,
WIRES, CABLES, TAPE and TUBING

ELECTRICAL INDUSTRIES.

Map of Chicago.

Showing Location of its Electrical and Allied Business Interests, Principal Hotels, Theatres, Depots and Transportation Lines to the World's Fair Grounds. (Index numbers refer to the black squares.)

...New...
Lundell Ceiling Fan Motor

NOISELESS   **EFFICIENT**

FINISHED IN BLACK JAPAN AND ANTIQUE POLISHED BRASS
FAN BLADES BLACK WALNUT—PENDANT SWITCH

**Okonite Wires and Cables   Okonite and Manson Tapes
Interior Conduit   Lundell Power and Exhaust Fan Motors.**

GENERAL SUPPLIES OF ALL KINDS

SEND IN YOUR ORDERS

CENTRAL ELECTRIC CO.
118 FRANKLIN ST., CHICAGO, ILL.

GATE CITY ELECTRIC CO.   WESTERN ELECTRICAL SUPPLY CO.
KANSAS CITY, MO.   OMAHA, NEB.

SOUTHERN ELECTRICAL SUPPLY CO.
ST. LOUIS, MO.

CLARK ELECTRIC COMPANY, NEW YORK.

192 Broadway and 11 John Street.

MANUFACTURERS OF ARC LIGHTING APPARATUS FOR EVERY PURPOSE A SPECIALTY.
The CLARK ARC LAMPS for use on EVERY CURRENT, have the reputation of being the best and most durable of any ever made in the United States.

Consolidated Electric Co.

Manufacturers and Dealers in all kinds of

ELECTRICAL . SUPPLIES,

115 Franklin Street,
CHICAGO.

BEAR IN MIND

that the regular monthly issue of ELECTRICAL IN-
DUSTRIES contains the most complete and correct
directories published of the electric light central stations
and the electric railways in North America.

World's Fair Headquarters Y 27 Electricity Building.
CITY OFFICES, Monadnock Block.

RAWHIDE PINIONS FOR ELECTRIC MOTORS
A SPECIALTY.
RAWHIDE DYNAMO BELTING

Greatest Adhesive Qualities.   A Non-Conductor of Electricity
Causes Less Friction than any other Belt.

THE CHICAGO RAWHIDE MANUFACTURING CO.
THE ONLY MANUFACTURERS IN THE COUNTRY.

LACE LEATHER ROPE
and OTHER RAWHIDE

GOODS
OF ALL KINDS
BY KRUEGER'S PATENT

This Belting and Lace Leather is
not affected by steam or dampness;
never becomes hard; is stronger,
more durable and the most economical Belting made. The Rawhide Rope for Round Belting
Transmission is superior to all
others . . . .

75 Ohio Street,   CHICAGO, ILL.

RESERVED FOR THE

STANDARD ELECTRIC CO.

CHICAGO.

Western Electric Company,

CHICAGO.   NEW YORK.

Arc Lighting Apparatus
High and Low Tension,
Double and Single Service Lamps,
All Night Single Lamps,
Theater and Focusing Lamps.

SLOW SPEED

INCANDESCENT DYNAMOS AND POWER GENERATORS.

SLOW SPEED MOTORS.

THE GEORGE CUTTER 1893 ROOKERY CHICAGO

See Our
Exhibits:
East Gallery
Electricity Bldg.

And another:
851 853 855
The Rookery.

George Cutter, Chicago.

SIMPLEX WIRES

SIMPLEX
Ever Onward and Upward!

**INSURE
HICH
INSULATION**

Simplex Electrical Co.
620 Atlantic Ave.,
BOSTON, MASS.

XNTRIC

"That's the Switch"

And we control that movement.

H. T. PAISTE,
10 South 1th St.,
PHILADELPHIA,
PA.

Made 5 amp. S. P.
10 amp. S. P.
5 amp. 3 way.
10 amp. 3 way.

P. & S.
CHINA
SWITCHES

Simplest
Neatest

Made only by

PASS & SEYMOUR,
Syracuse, N. Y.

George Cutter, Chicago.

See Our Exhibit of **ELECTRICAL FIXTURES**

IN SECTION "N", BETWEEN COLUMNS 62 AND 64, MANUFACTURES BUILDING,

GLOBE LIGHT & HEAT CO., 52 & 54 Lake St., CHICAGO.

. . . SEE AD . . .

Western Electric Co.,

PAGE 13.

CHAS. A. SCHIEREN & CO.

MANUFACTURERS OF

Genuine Perforated Electric Leather Belting.

46 So. Canal Street,   -   CHICAGO.

Section 15, Dpt. F, Clm. 27.
MACHINERY HALL.

Section D, Space 3.
ELECTRICITY BUILDING.

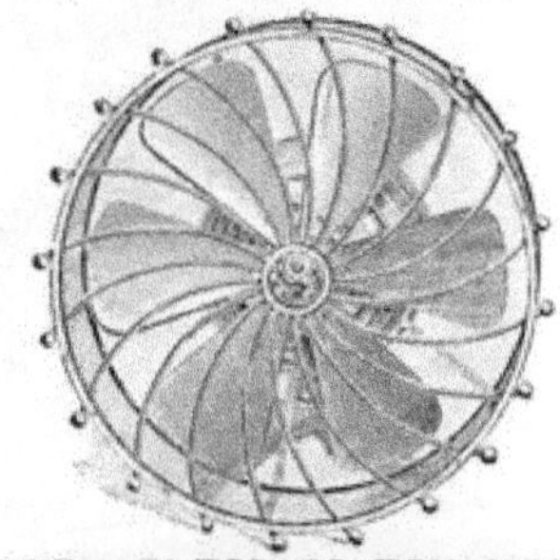

WAGNER ELECTRIC FAN MOTORS

For Direct or Alternating Currents.

These motors give a steady light with less consumption of current than any other fan motor on the market. They are full 18 horse power. Six bladed fan wheel. Self oiling. Furnished with or without guards.

IT WILL PAY YOU TO SEE THE WAGNER BEFORE BUYING ELSEWHERE.

TAYLOR, COODHUE & AMES,
348 Dearborn Street, CHICAGO.

See Map of Chicago, page 10, showing location of Electrical Business Houses.

Weekly World's Fair Supplement.

ELECTRICAL INDUSTRIES

DEVOTED TO THE ELECTRICAL AND ALLIED INTERESTS OF THE WORLD'S FAIR, ITS VISITORS AND EXHIBITORS.

Vol. I, No. 2.　　　　CHICAGO, JUNE 22, 1893.　　　　FIVE MONTHS $1.00. TEN CENTS A COPY.

The Exhibit of the Ft. Wayne Electric Company.

It is now nearly 15 years since the first Wood dynamo was placed on the market, and since then these machines have been installed all over the country, especially through the south and west. The exhibit of the Ft. Wayne

some entirely new ideas. These dynamos and generators are fully up with the times and the alternators shown are especially interesting.

The exhibit is on the left main aisle as one faces north just beyond the Bell Telephone building. In connection with the display of machines there has been fitted up a very

THE EXHIBIT OF THE FT. WAYNE ELECTRIC COMPANY.

Electric Company, the manufacturers of the Wood machines, has received a large number of visitors which is undoubtedly due to these machines being so well known. This exhibit was one of the first arranged. It is confined almost entirely to the standard machines for commercial use. It also contains a number of machines embodying

elaborate office, furnished with everything necessary to make it a comfortable resting place. The carpets, rugs, portieres and polished oak furniture tastily arranged are an exhibit in themselves. Next to the office, an ample storeroom has been erected for keeping supplies, etc.

For the convenient distribution of power a shaft was

placed, running the length of the space equipped with sufficient friction clutch pulleys so that the power may be distributed to the different machines as desired. Two Wood constant potential motors one of 120 horse power, and the other of 80 horse power are belted to the pulley shaft and furnish power for the different machines. The current for the motors is furnished from the Machinery Hall plant. The motors and incandescent lighting dynamos are placed on the west side of the shaft, while the arc dynamos are on the east side. Among the machines shown are a 50 and a 45 K.W. incandescent 110 volt Wood machine, while near the office are placed 20, 40, 65 and 125 light Wood dynamos, and near them a 1,200 light Slattery alternating current dynamo, and a 25 light, 6.8 amperes Wood arc. On the opposite side are a 75 K.W. Wood alternator and exciter, a 1,200 light composite Slattery alternator with exciter, a 40 light nine and a half ampere Wood arc, and also Wood arc dynamos of 60, 80, and 120 light capacity. A 3,000 light alternator is on the way and will soon be in place in the exhibit. This dynamo is a new departure in armature construction, it being made up of sheet iron links put together like a link belt. It is said to make a very efficient machine producing one volt for each 3.6 inches of conductor. A one light arc machine, coupled direct to a

EXHIBIT OF THE FT. WAYNE ELECTRIC CO. ALTERNATOR.

Wood motor is used to run a large locomotive head light which occasions considerable comment when running. There is also the historical dynamo and lamp described and illustrated in our June issue. A sample of a plain marine projector is also shown. Various sizes and styles of armatures are on exhibition; samples of the Fort Wayne recording meter and a lamp bank for testing them, a hanging rack with 80 arc lamps comprising 26 different styles and supplied with the different currents in order to test them; a bank of transformers of all the styles made by the company and of a capacity for 1,500 lights, samples of various forms and sizes of bearings, commutators and a new style hangerboard rheostat, are all arranged in good shape for close inspection. The hangerboard rheostat deserves more than passing mention as it apparently solves a problem heretofore often very awkwardly handled. A skeleton iron frame no larger than the ordinary wooden hangerboard, has a coiled iron wire rheostat covered with insulating material located on the back side, and when secured against a ceiling nothing shows but the ornamental front, and it also allows sufficient space for thorough ventilation. The new alternating arc lamp shown is remarkable in its simplicity, there being nothing but a couple of coarse wire magnet spools, and a few gear wheels for the whole

mechanism; it is of the constant potential type, and run two in series from a 100 volt converter with the above described hangerboard for a resistance. Samples of the regular standard rack lamp both single and double, and of the newer single and double clutch lamps are shown.

In switch boards there is a combination arc and alternating board with marble face, on a wooden framework, the arc board being on the front and the alternating instruments and devices on the back; it is very compact and exceedingly simple. A large low tension switch board made of old Tennessee marble, and provided with all the necessary ammeters, volt meter, switches and fuse cut-outs is shown. Two large rheostats are placed on the floor beneath it. The alternating current switch board is after plans by Mr. A. E. Barnes, who has charge of the exhibit and is very complete in all details. It is of the skeleton wooden frame type, and has a capacity for six dynamos and 42 circuits, the necessary ammeters, volt meters, pilot lamps, and fuse blocks being also provided for. The dynamo switches are located at either end of the board, and the circuit switches

EXHIBIT OF THE FT. WAYNE ELECTRIC CO.—MOTOR.

in the center space. The dynamo switches are so arranged that circuits can be changed from one medium or source of current to another without change of light and without danger of coupling two dynamos together. The whole scheme has been very cleverly worked out, and Mr Barnes states that he has installed a number of them about the country that are reported as giving excellent satisfaction.

This plant furnishes current for lighting part of the building with arc lamps and supplies incandescent current for the lamps in the booths in newspaper row.

The Fort Wayne arc dynamos are carefully examined by all comers. It is a common test to drop a pair of pliers across the two binding posts thus shortcircuiting the dynamo; the regulator takes care of the armature perfectly, and will immediately return to place on removal of the pliers. This was recently done for an eminent English electrician, and he was loath to believe it until the action of the regulator was explained to him.

The exhibit of H. T. Paiste together with the other exhibits in the center of the south gallery have been practically closed since the erection of the scaffolding by the decorators at the south end of the building.

The Brixey Exhibit of Day's Kerite Wires.

This exhibit occupies a prominent position in the gallery fronting the main eastern stairway of the Electricity Building. The whole end of the section is occupied, a space over 1,000 square feet in area. The center is taken up by a very handsome pavilion, serving as an office and assembling place for the company's numerous visitors. This pavilion is 10 by 12 feet, finished outside in white, touched out with gold leaf, while the interior is prettily finished in sycamore with a number of incandescent lamps for light and ornamentation. Desks, chairs, a comfortable lounge and a well stocked book case furnish comfort for the friends of the company.

A handsome sign is suspended above the roof of the office calling attention to the exhibit. Numerous portraits of the proprietors and managers and several fine photographs of cable laying and different samples of cable are hung on the outside of the pavilion. The diplomas received by the firm in commendation of its wires are also suspended on the office walls. Coils of wire for interior use and with ous sizes from the two wire submarine to the 100 conductor underground cable.

This company received medals of the highest award from the Centennial exhibition and has been doing business for a quarter of a century, being among the oldest insulated wire companies in America. Its display is large, complete and very comprehensive, and is a credit to the owner, Mr. W. R. Brixey.

Mr. Eckert in charge gladly shows the wires of the company, explains their uses and is untiring in his entertainment of visitors.

J. C. Vetter & Company's Exhibit.

This company exhibits in the west gallery of the Electricity building the various batteries and apparatus for physician's use, it manufactures. They are displayed in a finely designed and constructed cabinet of cherry, ornamented by numerous miniature incandescent lamps. It is

EXHIBIT OF DAY'S KERITE WIRES.

J. C. VETTER & CO'S EXHIBIT.

covering of different colors are displayed on a table in front of the office; a pyramid of wire on reels, and short lengths of 154 conductor cable on the top add to the finish of the figure.

The column in front of the building is draped in the national colors, combined with those of several foreign nations. Show cases surround the space and contain samples of cables, the Kerite tape, feeder wires, medals and other smaller products of the company.

The nature of the insulating material used is shown from its very origin up through the various stages of manufacture. The Kerite compound is a combination of pure para rubber and crude kerite, the nature of the latter being a secret known only to Mr. W. R. Brixey, the proprietor of the business. A rubber plant, blocks of pure para rubber as gathered for shipment, blocks of crude kerite and samples of the kerite compound are shown, illustrating the various stages in the development of the material. The compound here exhibited has the appearance of long skins of black yarn matted together.

Reels of wire of various sizes and kinds are shown, wire for electric lighting, numerous samples of telegraph and telephone cables, both underground and submarine of vari-

near one of the central stairways on the west gallery of the Electricity building.

The exhibit consists more particularly of the Vetter moist and dry Leclanche disque batteries, several styles of an improved form of portable dry Leclanche faradic batteries for physicians' use, and numerous pieces of apparatus for use in electro-therapeutics, among which are switches, standard milliamperemeters, carbon current controllers, and the Vetter current adapter. Cases of apparatus combining almost everything required by the physician are also shown, constructed and finished with the finest of workmanship.

The current adapter accompanying these outfits is an ingenious device consisting of a receptacle plug with three binding posts. Where a physician has incandescent lighting current handy he can unscrew a lamp and insert this adapter in its place, replace the lamp in the receptacle and turn on the light again. When a slight current is desired, flexible terminals from the medical apparatus are attached to the first two binding posts. This cuts the lamp and the apparatus in series and therefore furnishes a very small amount of current which can be varied by changing lamps. In case a current of the full potential is required the terminals are attached to the two outer binding posts and the current is governed by the carbon current controller. This controller is manufactured on the principle that the resistance of carbon increases as it is compressed. The exhibit although not large is nicely arranged, and makes an attractive object in that part of the gallery.

A Large Holtz Machine.

The Waite & Bartlett Manufacturing Company, has on exhibition among several important features, a Holtz induction machine which Dr. Waite says, as far as he is able to learn, is the largest one in existence, the next largest being one constructed by Wimshurst and but 36 inches in diameter. This machine has six revolving plates of plate glass, one-quarter inch thick and 40 inches in diameter; they are covered with a lacquer which was discovered by the makers. The shaft of the machine is of two inch steel, covered with a hard rubber bushing, four inches in diameter. The plates are secured on this bushing, separated

WAITE AND BARTLETT HOLTZ MACHINE.

by hard rubber collars, drawn into place and fastened by a large brass nut on the outside.

The stationary plates are common glass of the ordinary thickness and are in halves, one part being supported on hard rubber insulators from below, and the other half from above. The stationary plates have the usual paper covering and a small piece of tinsel on the edges. As the machine receives its initial charge from a 16 inch Wimshurst machine the cat-skin and other similar apparatus are not used. The combs are of one half inch brass rod with numerous small points. These combs and the stationary plates are all easily adjustable, all parts of the machine being easily and quickly reached for adjustment. The pole pieces project two feet from the combs and the balls of the prime conductors are of hollow brass, six inches in diameter. These pole pieces are insulated for 19 inches by fine hard rubber made by the Butler Hard Rubber Co., two and one quarter inches diameter. Two leyden jars are used with this machine, 14 inches high by six inches in diameter of glass, one-quarter inch thick.

The case is of old oak and glass finished with brass ornaments and balls. It is four and a half feet high, six feet long and 30 inches wide, setting on a table of the same wood, two and one-half feet high. All joints being packed

the case is air tight and by the use of lime kept dry so that no trouble is found in starting. Various electrodes, chains and chain holders are provided, also an insulated table for seating or for patients to stand on for treatment or experiment.

The machine is run by a Perret motor of one half horse power and in dry cold weather will produce a 20-inch spark. Owing to the state of the atmosphere at the Fair it has been impossible to get a spark of over 12 inches. With the leyden jars in contact and short-circuited terminals this spark is quite heavy and makes a noise like the sharp crack of a whip. For such people as think they are helped by electrical treatment the Doctor places them on the insulated stool, gives them an electrode chain and with rubber holders passes the other electrode over the surface of the body or affected parts at a distance of three or four inches; bright sparks are seen and the sensation is not exactly pleasant. One very queer effect is that made by seating a person on the insulated stool, with one electrode and with the other attached to a light brass frame suspended over the head at a distance of a foot or more, the hair rises and a sensation as of a cool breeze seems to sweep up all around the head towards the terminal. The feeling is entirely novel and pleasant and is said to be of use in insomnia and some other nervous diseases.

The Exhibit of Hardtmuth Carbons.

In connection with their American representative, the International Thomson-Houston Company, Messrs. F. Hardtmuth & Company of Vienna have made a very creditable exhibit of carbons. The demand for a high grade of carbon has gradually increased, and although more expen-

HARDTMUTH CARBONS.

sive, large sales are reported in this country. In the exhibit a general view of which is given in the accompanying cut carbons of from an eighth of an inch to 12 inches are shown, of different shapes according to the various uses for which they are intended.

The display is arranged in a very tasty open booth of

ebonized wood. It is located in section P at the right of the north entrance of Electricity Building near the French exhibit. The manufacture of carbons is an industry that has grown up with that of electric lighting and the improvement in the quality and grade of carbons manufactured has kept pace with the general advancement. The Exposition company has purchased 100,000 of these carbons for use on the grounds so that they may be seen in use.

Cutter's Boulevard Streethood.

Among the novelties exhibited by George Cutter is the handsome form of outdoor incandescent reflector which has been nicknamed Cutter's Boulevard Streethood. In designing it the many little improvements which are found only in Mr. Cutter's streethoods have been retained, the

CUTTER'S BOULEVARD STREETHOOD.

departure lying in the special bracket arm which takes the place of the ordinary gooseneck. This is as strong and well braced as it is neat, and holds the lamp five or six feet from the pole so as to give a good effect on thickly shaded streets. The appearance of these hoods when seen on well kept streets is said to be unusually fine; and the maker is already pleased with the calls for them.

The Lake Cable Laid.

For some two weeks past the U. S. Light House tender, Dahlia, commander J. J. Brice with Mr. Ira W. Henry, the electric engineer of the Bishop Gutta Percha Co., and a party of assistants, has been engaged in the laying of the cable and the anchoring of the light buoys between the Van Buren St. pier and the pier at the Fair grounds. The ends of the cables were brought up and tied to the pier on Tuesday night, and as soon as the tests are completed, the current will be turned on.

The work has been done without accident or trouble except on the occasion when the new whaleback steamer Christopher Columbus backed into the end of the cable at the pier. One of the propellers caught the cable and after twisting it up badly one of the blades broke and it was thus released. Thus about 100 feet of the cable was torn off, but the end was easily recovered by the grapple and the cable carefully spliced out to the required length and the end again made fast, but in a permanent manner to prevent a repetition of the mishap.

The ends of the cable are connected with the switch board in the house erected on the end of the pier, which is surmounted by a flag staff and pennant. The use of this system of lighting for this purpose is entirely new and as soon as completed and in working order a careful examination will be made of it by Capt. W. S. Schley of the United States Light House Establishment at New York. It will also be inspected by other officers of the Light House Establishment and harbor masters from different points. Great credit is due Commander J. J. Brice for his persistence in carrying out the scheme and to Mr. Henry for this application of the system and his boldness in carrying it out.

De Laval's Turbine-Dynamo.

In a very neatly arranged space on one of the main aisles of Machinery Hall, is a dynamo which is decidedly novel in having two armatures in the place of one. The machine is of the bi-polar type of the direct current dynamos, and in appearance resembles very much some of the ordinary dynamos with the exception of the double armatures. The armatures are placed close together, and revolve in the same direction, being geared direct to the shaft of the steam turbine. The whole combination takes up but little room. The motor dynamo and all attachments and connections are placed on one foundation. It is fitted with modern oil cups, etc. The connections and general workmanship seems to be of an excellent kind.

It is supplied with a governor, and the necessary regulating apparatus, and is said to be entirely safe. The coverings and boxes are so constructed as to be easily opened permitting the inspection of any and all parts.

These machines are manufactured at the G. de Laval's Steam Turbine Works, Stockholm, Sweden.

Exhibit of the Williams Engines.

The Williams engines are being shown in Machinery Hall in various uses and types. Two are used for driving line shafts and one for driving a 200 K. W. Siemens Brothers' direct current inverted field type of dynamo. This engine made and used so largely in England is about to be built in Chicago by the M. C. Bullock M'f'g. Company. While they are somewhat after the Westinghouse type, being wholly inclosed, and single acting, they are compound, have a central valve and in place of cushioning on a back pressure of steam are supplied with an air chamber which takes its place and saves that amount of power. The engines are comparatively high speed, the 500 horse power machines running 350 revolutions per minute. The governor is the old form of weights in a horizontal position on the end of the main shaft which extends out through the engine casing into a box provided for the purpose. A bell crank lever changes the motion of the axis of the balls from a horizontal to a perpendicular position and through a long rod connects directly with a throttling valve in the steam pipe controlling the admission of steam to the high pressure cylinders. About half way up on this rod is a lever the fulcrum of which may be changed at any time, thus controlling the throw of the throttle valve in comparison with the action of the governor balls in order to vary the speed of the engine. The balls are held together by a very light spring and are apparently very sensitive to any change. A steam separator is used in connection with the main supply pipe to prevent as far as possible the entrance of water to the cylinders.

The agents and engineers claim to be able to compete in economy of steam consumption with any of the Corliss type of engines. The machine requires very little space, runs remarkably free from jar and noise and is of fine design and appearance.

ELECTRICAL INDUSTRIES.

Entered at Chicago Postoffice as second class mail matter.

PUBLISHED EVERY THURSDAY BY THE

ELECTRICAL INDUSTRIES PUBLISHING COMPANY,

INCORPORATED 1893.

MONADNOCK BLOCK, CHICAGO.

TELEPHONE HARRISON 366.

R. L. POWERS, PRES. AND TREAS.   E. E. WOOD, SECRETARY.

L. L. POWERS, -   EDITOR.

B. A. FOSTER,   }
W. A. REMINGTON,   }   ASSOCIATE EDITORS.

E. E. WOOD,   EASTERN MANAGER.

FLOYD T. SHORT,   ADVERTISING DEPARTMENT.

EASTERN OFFICE, WORLD BUILDING, NEW YORK.

World's Fair Headquarters, Y 27 Electricity Building.

SUBSCRIPTION.

FIVE MONTHS - $1.00
SINGLE COPY -   10

Advertising Rates Upon Application.

News items, notes or communications of interest to World's Fair
Visitors are earnestly desired for publication in these columns and will
be heartily appreciated. We especially invite all visitors to call upon us
or send subjects at once upon their arrival in any one of the grounds.
ELECTRICAL INDUSTRIES PUBLISHING CO.,
Monadnock Block, Chicago.

On a number of occasions the crowd has been so great at the north and south entrances of the Electricity Building as to cause a dangerous jam at these places. Cannot the crowds be directed so as to go in at one entrance and out at another or be separated so as to use the side entrances more? A number of complaints have come to our notice of this fact.

A few chairs or settees might be placed in the gallery of the Electricity building in some of the open space along the railing on which persons fatigued by walking might rest. It is not an uncommon sight to see large numbers resting on the stairs or on the platforms among the exhibits. The chairs or settees so placed would not be in the way and would be greatly appreciated by the visitors.

That the World's Columbian Exposition is an assured success there can now be no doubt. The attendance has increased gradually but steadily each week since the Fair was opened, and as the season advances when schools are closed, the increase will unquestionably continue. The attendance the first week was large on account of the opening exercises. Since the second week in May between the 7th and 13th, when the number of paid admissions was 131,000, the attendance has increased to 724,000 for the week ending June 17. The Department of Electricity certainly has no reason to complain with regard to the number of its visitors. No building on the grounds shows the crowds present that are to be found in the Electricity Building and great interest is shown by the visitors in the various exhibits. Later in the season as the time approaches for the assembling of the Congress of Electricians, the prominent electricians from all parts of the world will be found among the visitors to this building.

The historical features of the exhibits in the Electricity Building are worthy of especial attention. A number of the older companies have gone to a considerable expense and trouble to collect instruments and apparatus which are of historical interest. In nearly all the fields of electricity are shown the apparatus used in its earlier applications to commercial use. The first instrument with the subsequent ones arranged as one succeeded another, as

shown in a number of cases. Where it was impossible to secure the original instruments, models have been constructed from diagrams or memory to make the list complete. A number of articles are also shown interesting from their association with important events. The gradual advancement of the Mechanical Art, the drawing together of the scientist and the mechanic, or rather the growth of a highly intelligent class of mechanics and the utilization in every day life of what was only a chemist's toy, were never exemplified as in the exhibits in the Electricity Building.

Some of our foreign friends have had many troublesome and vexing delays in the receipt of their exhibits. Mr. J. G. Lorrain, of London, England, reports that some of his exhibits are still on the way and what he has received are in very bad shape. He hopes to receive it all soon and will then have his display arranged for exhibition.

The Government exhibit from the naval observatory occupies a position near the lake just east of the Government building and the camp of the U. S. regulars. The various instruments, transits, chronographs, etc., used for ascertaining and signalling the correct time to all parts of the United States by means of the telegraph are exhibited.

The British Government Postal Telegraph exhibit has just arrived and is being rapidly arranged under the direction of Mr. Chapman. It includes a number of instruments of great historical value. The first telegraph instruments made by Cooke and Wheatstone, a very cumbersome affair, having five needles and requiring five wires between each instrument. A relic of the line between Euston and Camden Town is shown in the shape of a short piece of wood into which the five bare copper wires were sunk separately and covered with strips of wood. Instruments are displayed showing the gradual improvement and perfection of the instruments, and the system from the first instrument requiring five wires to send one message to the latest Delaney multiplex system, by which six messages are sent over one wire at the same time. A sample of the first submarine cable laid is shown, having been laid between Dover and Calais in 1850. It was insulated with gutta percha with no protection against mechanical injury, consequently its useful life was short. It was recovered in 1875. The instrument used in the House of Commons, a two needle machine with case finished in the same style as the interior is shown, with many other interesting articles. Most of the historical part of this exhibit is from the British Post Office section of the South Kensington Museum and is worthy of a careful inspection.

THE OPENING OF THE FAIR EVENINGS.

On Monday of this week the council of Administration decided to open the Fair evenings, commencing with Thursday night. This decision has given universal satisfaction, especially to those interested in the Electricity Building. Many of the exhibits in this building are seen to advantage only during the evening, and under the new arrangement every visitor will have an opportunity of seeing them. The exhibitors will undoubtedly give more attention to the lighting of their exhibits, and should, as far as possible, have a representative in charge of their display at night as well as during the day. The buildings will be closed at 10 o'clock, and it is expected that all visitors will leave the grounds by 11 o'clock. The work of completing the lighting of the grounds is being pushed as rapidly as possible.

WORLD'S FAIR NOTES.

The changes in the electric fountains are nearly completed, and great improvement has been noticed when in operation.

The exhibit of L. J. Wing & Company has been increased by the addition of several electrically driven exhaust fans and other apparatus.

The efficiency of the electric launch was shown in rather a disastrous manner a few evenings since in the collision between a launch and a gondola.

The Western Electric Company's miniature theatre is practically finished and the apparatus is being adjusted. It will be opened to the public in a few days.

The Jewell Belting Company, of Hartford, have an exhibit of their well-known oak tanned leather belting in the northeastern alcove of the Electricity building.

The Brush Electric Company are expecting to open their exhibit to the public this week. The temple erected in the center of the exhibit is very artistic and is much admired.

The General Electric Company started last week several electric percussion drills in which a considerable interest has been aroused. Huge blocks of stone have been placed for the purpose of testing them.

Some very beautiful negatives of lightning flashes are shown in the case exhibited by E. Ducretet and L. Lejeune, of 75 Rue Claude Bernard, Paris. They also exhibit many fine electrical testing instruments.

The Electricity building is the great evening attraction; when no special feature is to be seen elsewhere the building is crowded. As soon as the fire works cease, or the play of of the electric fountains, the crowd moves back to the building. The crowd is especially noticeable at the times set for the illumination of the tower.

The Otto Gas Engine Company of Cologne have in their exhibit in Machinery Hall a vertical high speed gas engine coupled direct by shaft connection to a Schukert dynamo. A combination is thus presented that has been successfully used in a number of places in Europe, but which has not been used to any great extent in America, probably on account of the greater cost of gas.

One of the most valued historical relics at the Fair is a magneto electric machine loaned to the Hanson & Van Winkle Company by the corporation of Birmingham, England. It is said to have been built by Woolrych in 1844 for Thos. Prince & Son and was the first machine successfully used. It was in constant use until 1877 when it was superseded by a more modern machine.

On the 15th inst. Mr. George Ferris, the engineer and designer of the Ferris wheel, with a party of friends and stockholders of the company made a trial trip around the great wheel. When completed it is to be lighted by electric lights and will be undoubtedly a very attractive object at night. The current is to be conveyed to the lights through sliding contacts on the axis of the wheel.

The Conz electric night signal apparatus recently installed in one corner of the German exhibit and near the central tower, is not large, but deserves attention. It seems remarkably simple, but is said to have been adopted by the German and Dutch navies. It consists of three double lanterns, one half the globes of which are colored red and the other half white. Two incandescent lamps corresponding to the colors are placed in the lantern. The lanterns are suspended by cables about seven feet apart. Each light is controlled from a controlling box, which is about eight inches square. This box has 14 points on the top, representing different combinations of the lights. A central handle with pointer is the arrangement by which the combinations are made. Mr. Chas. H. S. Schultz is the Chicago representative.

Tests were made on Saturday night of the high tension apparatus of Prof. Thomson in the General Electric Company's exhibit. The apparatus is capable of producing a spark 64 inches in length with a potential estimated at 2,000,000 volts. Some little difficulty was experienced in manipulating the apparatus and the full capacity of the machine was not shown. With some adjustments it will undoubtedly work better at the next test.

F. A. Ringler & Company, 21 Barclay St., New York, have a beautiful exhibit of electroplated material and electrotypes in their space just east of the department offices in the gallery of the Electricity building. The electrotypes of engraved steel plates are especially fine. A unique feature is the plating of any article which one desires to preserve of any material. They are first covered with copper and then plated with any metal desired.

It is expected that the headquarters of the American Institute of Electrical Engineers will be an exhibit in themselves. Mr. Hammer has shipped six large frames of photographs, Mr. William Wallace contributes some interesting apparatus, Prof. Feraris, of Turin, Italy, has shipped the experimental apparatus used in his discovery of the principle of the rotating field and Van Nostrand has promised a complete library of electrical books.

The Charles Munson Belting Company, of 22 36 Canal St., Chicago, have a unique exhibit, not only of their well-known leather belting, but also of leather ornaments and furniture, leather chairs, tables, carpet, fence and railing, gate posts, etc. A sample of an 84 inch double leather belt shows the capacity of their hydraulic press and the method of jointing. Mr. H. B. Morgan is in charge and will be found a genial companion for a talk on belting.

Advisory Council, Chicago World's Congress of Electricians.

President, Dr. Elisha Gray, Highland Park, Ill.

EXECUTIVE COMMITTEE.

Prof. Elihu Thomson, Chairman, Lynn, Mass.
F. S. Terry, Chicago.
George M. Phelps, New York City.
Prof. B. F. Thomas, Columbus, O.
E. N. Barton, Chicago.

COMMITTEE ON INVITATIONS:

T. C. Martin, Chairman, New York.
Dr. Louis Duncan, Baltimore.
Prof. C. R. Cross, Boston.
T. D. Lockwood, Boston.
C. H. Wilmerding, Chicago.

COMMITTEE ON PROGRAMME:

Prof. T. C. Mendenhall, Chairman, Washington, D. C.
Carl Hering, Philadelphia, Pa.
Prof. W. A. Anthony, Manchester, Conn.
Prof. H. A. Rowland, Baltimore, Md.
A. E. Kennelly, Orange, N. J.
Prof. F. B. Crocker, New York City.
Prof. E. L. Nichols, Ithaca, N. Y.

FINANCE COMMITTEE.

B. E. Sunny, Chairman, Chicago.
Prof. E. J. Houston, Philadelphia.
W. A. Kreidler, Chicago.

Hon Sec : Prof. H. S. Carhart, Ann Arbor, Mich. (Member Prog'me Com.)

PERSONAL.

Mr. Frank B. Rae, of Detroit, made a hasty visit to Chicago on the 14th.

Thos. D. Lockwood, of the American Bell Telephone Co., Boston, is in the city.

Mr. Chas. B. Fairchilds, editor of the Street Railway Journal, was in the city on the 16th.

Mr. W. F. Richardson, of the Enterprise Electric Company, returned a few days ago from a ten days' trip in the east.

Mr. Ludwig Gutmann, of Pittsburgh, lately with the Westinghouse Electric & Manufacturing Co., is in town for a few days.

Mr. W. H. Preece, chief of the British Postal Telegraph service, sails for America August 5th on the Paris, and will pay quite an extended visit to the Fair.

Hon. Jacob Hess, commissioner of the New York Board of Electrical Control, was in town for a few days last week at the Auditorium. The commissioner will return later and make an extensive examination of the electrical features of the Fair.

The Cummings & Engelman Conduit Company.

The Cummings & Engelman Conduit Company has just been organized in Detroit, with a paid up capital stock of $300,000. The president of the new company is Mr. James F. Cummings, the treasurer and general manager being Mr. E. M. Engelman. Mr. Cummings is well known to the electrical fraternity from his former connection with the Edison company whose interests he served for several years in the laying of underground conduit. As a result of his former experience he has brought out a new conduit system which will be known as the Cummings and Engelman system, recently illustrated in these columns. Mr. Engelman is a thorough mechanical engineer, having had a practical experience covering the past five years. The new company has patents for the United States, Canada, England, Germany, France and Austria for high and low tension long distance transmissions of power, and for electric railways.

The low tension system which it will promote, will, it is claimed, take care of as high as 5,000 volts, using bare copper wires, while the high tension system will take care of 100,000 volts using bare copper wires. The company is at present putting in a conduit for the Co-operative Electric Light & Power Company, of Chicago, which is to use the Siemens & Halske system of lighting. In this plant bare copper wires will be used to carry a 2,000-volt alternating current. In addition to this large contract the company is also putting in a low tension system for the Edison Company, of St. Paul, the Edison Company, of Detroit, and a number of others. The company has established an extensive plant to manufacture its patents in Detroit, at the corner of Fourth and West Congress sts., with main offices in the new telephone building of that city. The new company invites all interested in electrical work to call and visit them. At present Messrs. Cummings and Engelman are in Chicago overseeing the putting in of their system for the Co-operative company, and besides are just now establishing a company in Canada for the manufacture of their patents.

BUSINESS NOTES.

The souvenir of the World's Fair issued by the Eureka Tempered Copper Company, of North East, Pa., is very tastily gotten up.

The Stow Manufacturing Company, Binghamton, N. Y., has recently placed on the market a portable electric motor for driving the well-known Stow flexible shaft.

The Enterprise Electric Company, Chicago, reports very brisk trade in fan and power motors. N. C. R. wire, which the company is agent for, is in good demand, and the capacity of the factory is now being doubled in order to take care of its rapidly increasing business.

The Hoppes Manufacturing Company, Springfield, Ohio, have recently made a number of large sales of live steam feed water heaters to street railway companies. This company is able, with its excellent shop facilities, to manufacture feed water heaters of any size and capacity.

The Tice-Lininger Electric Company, Minneapolis, Minn., are among the pushing electrical houses of the northwest. The new potential indicator recently placed on the market by this company has received many favorable comments from the trade and will no doubt find a ready sale.

Queen & Co., Incorporated, Philadelphia, Pa., carry a large stock of magnetic vane ammeters and voltmeters, which they can ship promptly. These instruments are specially valuable for isolated plants, combining qualities essential for switch board work. A number are in use by various exhibitors at the World's Fair.

Chas. A. Schieren & Co., 46 Canal St., Chicago, make a specialty of perforated leather belting for central stations and power plants. Their exhibits in Machinery Hall and the Electricity Building show the goods for which they are well known and will well repay a visit.

Chas. Munde & Sons, 88, 90 Walker St., New York, report large sales of perforated metal to makers of motors and dynamos. The kind and quality of work done by this firm is well known to all branches of the electrical trade. All kinds of perforated metal are made by them, but a specialty is made of perforated metal for street railway work.

The Ansonia Electric Co., Michigan Ave. and Randolph St., Chicago, have recently issued a circular descriptive of the Wirt dimming switch, which is a great convenience and luxury in sick rooms. This company has recently closed a contract for upwards of 40 miles of shield brand wires, and during the past month has sold large quantities of Habirshaw wire. The sale of Stanley transformers is active, largely due to the good grade of the goods.

R. Thomas & Sons Company, East Liverpool, Ohio, has established western headquarters at 115 Dearborn St., Chicago, with Mr. J. E. May, manager. A full stock of porcelain electrical supplies will be kept there which will enable it to supply its western customers promptly. The company has made a number of improvements recently at its factories, including new buildings and new machinery.

The Electric Appliance Co., Chicago, promises some marked improvements in the Elkhart transformers in the way of a new fusing arrangement, particularly adapted for circuits of 2,000 volts primary. It is also improved by the introduction of a system of removable coils, making it a very simple matter to repair an occasional burn out. Since the Electric Appliance Co. took the western agency, the sales of the Elkhart converters have been very large, and are now used in all parts of the country.

General Manager Eckert, of the Telautograph, reports considerable progress in the business, and is expecting to fit out a number of central stations, now under contract, as soon as the material can be gotten under way. Prof. Gray is constantly improving the apparatus as well as getting out new material. The exhibit at the Fair will be connected this week with the city of Chicago, so that a practical demonstration of its usefulness can be had by all. Prospects in this field promise very well.

The managers of the Columbian Intramural Railway treated the denizens of Midway Plaisance to an excursion around the grounds of the Exposition recently. Representatives from the different nations exhibited turned out for the ride. A brass band from the Indian school furnished the music for the party. Numerous photographs of the novel sight were taken and will be used to advertise the company. The sight of representatives from the older nations in their odd costumes riding on one of the most modern roads was certainly unique.

The Western Electric Company, Chicago, has just closed a contract for putting in an arc and incandescent central station plant at St. Clair, Mich. The company has also received a contract for lighting the new Hanneman Hospital, Chicago, with incandescent lights, including the complete electric equipment throughout of annunciators, call bells, etc. The new St. Nicholas Hotel of St. Louis which is to be one of the finest in the country is to be lighted by the Western Electric Company's system of incandescent lighting, the contract for which was closed recently.

Electrical heating apparatus is receiving considerable attention of late, but we are conceited enough to think that we have something more to the point and appropriate to the season in our

ELECTRICAL COOLING
MESTON MOTORS

apparatus. We are having a great run on

which have undisputed control of the alternating current fan motor field.
Are you pushing them and getting your share of the profits that are to be made out of the fan motor business this year?

ELECTRIC APPLIANCE COMPANY,
ELECTRICAL SUPPLIES,
242 Madison Street,　　　　　　　　　　　　　　　CHICAGO.

The "BUCKEYE" Lamp

A KNOWN Quantity
No EXPERIMENTING
Low Operating Cost
Improved Lighting Effects

Life 1000 to 4000 Hours
Western Deliveries can be made from Chicago Stock

"The BUCKEYE sets the pace"
THE BUCKEYE ELECTRIC CO., Cleveland, O.

Chicago
437 THE
ROOKERY

New York
49 DEY STREET.

THE MATHER ELECTRIC CO,
MANCHESTER, CONN.

Dynamos, Motors, Generators,

Offices, 116 Bedford St., BOSTON.
—AND—
1002 Chamber of Commerce Bldg., CHICAGO.

THE "NOVAK" LAMP.

CLAFLIN & KIMBALL (Inc.)
General Selling Agents.

116 Bedford Street, BOSTON.
1002 Chamber of Commerce Bldg., CHICAGO.

Enterprise
Electric
Company

307 Dearborn Street,
Chicago . . . .

Manufacturers' Agents and Mill Representatives for

Electric Railway,
Telegraph, Telephone and
Electric Light

SUPPLIES OF EVERY DESCRIPTION

Agents for　Cedar Poles,
Cypress Poles, Oak Pins,
Locust Pins, Cross Arms, Glass
Feeder Wire,　Insulators,
WIRES, CABLES, TAPE and TUBING

ELECTRICITY BUILDING—EXHIBITORS AND THEIR LOCATION.

GALLERY.

MAIN FLOOR.

Exhibitor.	Section.	Exhibitor.	Section.	Exhibitor.	Section.	Exhibitor.	Section.
Ansonia		Electrical Review		Jaeger, Chas. L.	T	Reliance Gauge Co.	N
Ansonia Electric Co.	T	Electricity	Y	Johns Mfg Co., H. W.	T	Rossell & Hunbacher Chem. Co.	N
Am. Inst. of Elec. Eng.	N	Electric Gas Co.	M	Jewell Belting Co.	F	Street Railway Journal	
American Battery Co.	T	Electrical Engineer		Jenney Elec. Motor Co.	T	Schrager Ant. Teleph. Co.	T
Axtond, H. M		Electrical World	Y	Knapp Electric Works	T	Standard Paint Co.	T
Alleg. Elec. Co.	D	Eddy Electric Motor Co.	B	K. A. P. Elec. Novelty Co.	V	Sparhole, C. F.	
Bates Mfg. Co.		Excelsior Electric Co.	B	Knapp & Buckley		Star Iron Tower Co.	W
Bryant Electric Co.	B	Eastern Forging Co.	D	Kennedy Electric Co.		Spith	
Billings & Spencer	B	Republic Dynamo Co.	Q	Lawson, B.	V	Schenten, Chas. A. & Co.	
Brixey, W. R		Ericsson Mfg Co.		LeClanche Battery Co.	F	Schomacker & Sohne	
B Snap Motor Co.		Electrical Conduit Co.	P	McNeil Tinfoil Elec. Co.		Siemens & Halske	E
Bell Telephone Co.	E-I	England		Marine, W. N	Y	Schoenert & Co.	
Brush Electric Co.	L	Empire Chem Works		Meeker, Dr. G.		Short Electric Co.	
Caldwell El. Cloth Cut. Mch. Co.	Y	Franklin Elec. Appliances		McIntosh B & Opt. Co.	W	Sperry Elec. Railway Co.	L
Conrad, Ele. Storage Co.	T	French Fusee Exhibit		Man on, C., Belting Co	D	Standard Underg. Cable Co.	
Collet, George	T	Fellows & Goldhamer	Y	Mather, A. C.		Standard Electric Co.	P
Canton Elec. Co.		France	K-P-V	Mather Electric Co.	M	Samson Battery Co.	
Chicago Elec Wire Co.		Ft. Wayne Elec. Co.	M	Newman Clock Co.		Tne. Ant. El. Signal Co.	
Copenhagen Fire Alarm. Co.	N	Gamla Co., N. E		Non M Quote Watch Co		Todd, Applegate Co.	
Central Electric Co.	A	Gamewell Fire Alarm Co		N. Y. Insulated Wire Co.		Taylor, Goodhue & Ames	A
Consolidated Cable Co.		General Electric Co.	H-N-C-A	National Carbon Co.		Thomson Elec. Welding Co.	
A. C. Ele. Motor Co.	A	General Insulated Wire LT Co.		Norwich Ins. Wire Co.		Twistperough, Elvira Gray	
Cleveland Elec. & Mfg. Co.		Gridley, E. S. & Co.	A	North Am. Phonograph Co.	S-F	Union Electric Co.	
Chicago Belting Co.		Germania		N. Y. A. L. E. A.		Vetter J. C. & Co.	W
Diamond Clock Co.		Habbard, Wm. A. & Co.	X	Nat. Ant Fire Alarm Co.	A	Wells, G.	
Department of Electricity	E	Hofmann, C. J		Nat. Express no Machine Co.	A	Weston El. Instrum nt Co.	E
Everson at Insernats		Hart & Hegeman Mfg. Co.	V	Owen, Dr. A.		Washburn & Moen Mfg Co.	V
Bler Lausch & Sue Co.		Hal-Eye Appliance Co.		Pinless Clasp Co.		Western Union Tel. Co.	
Electric Separator		Hall, Chas.		Polaie, H. P		Waite & Bartlett Mfg. Co.	
Edgerton, E. M		Holtzer, W. J	W	Pollermacher Batt. Co.		White S. S., Dental Mfg. Co.	
Rhein Telephone Co.		Hartman & Braun		Pampelly, E. K	T	Western Electrician	
Edison Elec. Mfg. Co.		Hemen & A sa Wohl		Pratt El. Med. Sup. Co		White Ant Burglar Mfg. Co.	
Interpaose Elec. Co.		Hard, J. M		Powell, Wm. & Co.		Weeden Electric Co.	
Eureka Temp. Copper Co.		Hardmann, F. A & Co.	P	Phelps, A. B		Westinghouse El. & Mfg. Co.	B-H-J
Electric Appliance Co.		Hemin Ming Co.		Page Belting Co.	D	Weber & Stenbold	
Elec. Sel. & Sig'l Co.		Infernal Ind LT A Pr Co.	T	Queen & Co.		Wing, L. J. & Co.	
Electric Heat Alarm Co.		India Rubber Comb Co.	N	Rougier, F. A.	R	Zucker & Levett Chem. Co.	F

CLARK ELECTRIC COMPANY, NEW YORK.

192 Broadway and 11 John Street.

MANUFACTURERS OF ARC LIGHTING APPARATUS FOR EVERY PURPOSE A SPECIALTY.
The CLARK ARC LAMPS for use on EVERY CURRENT, have the reputation of being the best and most durable of any ever made in the United States.

RAWHIDE PINIONS FOR ELECTRIC MOTORS
A SPECIALTY.
RAWHIDE DYNAMO BELTING

Greatest Adhesive Qualities. A Non-Conductor of Electricity.
Causes Less Friction than any other Belt

THE CHICAGO RAWHIDE MANUFACTURING CO.
THE ONLY MANUFACTURERS IN THE COUNTRY

LACE LEATHER ROPE
and OTHER RAWHIDE

GOODS
OF ALL KINDS
BY KRUEGER'S PATENT

The Belting and Lace Leather is not affected by steam or dampness, never becomes hard; is stronger, more durable and the most economical Belting made. The Rawhide Rope for Round Belting Transmission is superior to all others

75 Ohio Street, CHICAGO, ILL.

THE REGULAR MONTHLY EDITION

.... OF ....

Electrical Industries

Is read more widely and used more constantly for reference by actual buyers in the electrical field than any other similar publication. WHY? Because it has in every issue handsomely illustrated special articles and descriptions of everything of interest in the electrical world besides a complete

BUYERS DIRECTORY giving the names of all the manufacturers and dealers in the trade; a complete directory of

ELECTRIC LIGHTING CENTRAL STATIONS and a complete directory of the

ELECTRIC RAILWAYS of North America corrected to the date of going to press, features found in no other electrical journal in the world.

The Weekly World's Fair issue contains the most novel and unique features yet undertaken by an electrical journal.

Sample copies free on application. Every electric lighting company should write for our special terms. We are now making the most liberal offer to subscribers for both publications during the next few months ever made by an electrical paper.

ELECTRICAL INDUSTRIES PUB. CO.

Monadnock Block, CHICAGO.

...New...
Lundell Ceiling Fan Motor

NOISELESS   **EFFICIENT**

**FINISHED IN BLACK JAPAN AND ANTIQUE POLISHED BRASS
FAN BLADES BLACK WALNUT—PENDANT SWITCH**

Okonite Wires and Cables   Okonite and Manson Tapes
Interior Conduit   Lundell Power and Exhaust Fan Motors.

GENERAL SUPPLIES OF ALL KINDS

SEND IN YOUR ORDERS

CENTRAL ELECTRIC CO.

118 FRANKLIN ST., CHICAGO, ILL.

GATE CITY ELECTRIC CO.   **WESTERN ELECTRICAL SUPPLY CO.**
KANSAS CITY, MO.   **OMAHA, NEB.**

SOUTHERN ELECTRICAL SUPPLY CO.
ST. LOUIS, MO.

Map of Chicago.

Showing Location of its Electrical and Allied Business Interests, Principal Hotels, Theatres, Depots and Transportation Lines to the World's Fair Grounds. (Index numbers refer to the black squares.)

Western Electric Company,

CHICAGO. NEW YORK.

Arc Lighting Apparatus
High and Low Tension,
Double and Single Service Lamps,
All Night Single Lamps,
Theater and Focusing Lamps.

SLOW SPEED

INCANDESCENT DYNAMOS AND POWER GENERATORS.

SLOW SPEED MOTORS.

Fort Wayne Electric Co.

FORT WAYNE, IND.

Manufacturers of Apparatus for Arc Lighting; Alternating Current Lighting; Direct Current Incandescent Lighting; Generators for Railway and other Motor Circuits; Accurate Measuring Instruments, Meters and General Supplies. This Apparatus possesses the highest Electrical Efficiency, best Mechanical Construction and the most Artistic Design.

The Armature of the new "Wood" Iron Clad Slow Speed Alternator is indestructible. These dynamos are being manufactured in the following sizes. 750 light, 1500 light, 3000 light and 6000 light capacity.

Twenty-six styles "Wood" Arc Lamps for Constant Current Circuits, Constant Potential Circuits and Alternating Circuits. Noiseless, no extra Rheostat.

The largest stations in existence are of the Fort Wayne Company's System.

SEE OUR APPARATUS IN OPERATION AT OUR WORLD'S FAIR EXHIBIT.

BRANCH OFFICES:

New York, 42 & 44 Broad St.

Chicago, 185 Dearborn St.

Philadelphia, 907 Filbert St.

Pittsburgh, 405 Times Building.

Syracuse, Kirk Building.

Columbus, Ohio, 57 East State St.

San Francisco, 35 New Montgomery St.

City of Mexico.

Dallas, Tex.

New Orleans, 52 Union St.

16   ELECTRICAL INDUSTRIES.

See Our
Exhibits:
East Gallery
Electricity Bldg.

And another:
851-853-855
The Rookery.

SIMPLEX WIRES

**INSURE
HIGH
INSULATION**

SIMPLEX

Ever Onward and Upward!

George Cutter, Chicago.

Simplex Electrical Co.
620 Atlantic Ave.,
BOSTON, MASS.

XNTRIC

"That's the Switch"

And we control that movement.

H. T. PAISTE,
10 South 16th St.,
PHILADELPHIA,
PA.

Made   5 amp. S. P.
10 amp. S. P.
5 amp. 3 way.
10 amp. 3 way.

**P. & S.
CHINA
SWITCHES**

**Simplest
Neatest**

Made only by

PASS & SEYMOUR,
Syracuse, N. Y.

George Cutter, Chicago.

Consolidated Electric Co.

Manufacturers and Dealers in all kinds of

ELECTRICAL . SUPPLIES,

115 Franklin Street,
CHICAGO.

BEAR IN MIND

that the regular monthly issue of ELECTRICAL IN-
DUSTRIES contains the most complete and correct
directories published of the electric light central stations
and the electric railways in North America.

World's Fair Headquarters Y 27 Electricity Building.
CITY OFFICES, Monadnock Block.

. . . SEE AD . . .

Western Electric Co.,

PAGE 14.

CHAS. A. SCHIEREN & CO.

MANUFACTURERS OF

Genuine Perforated Electric Leather Belting.

46 So. Canal Street,   -   CHICAGO.

Section 15, Dpt. F Clm. 27.   Section D, Space 3.
MACHINERY HALL.   ELECTRICITY BUILDING

WAGNER ELECTRIC FAN MOTORS

For Direct or Alternating Currents.

These motors give a ratio per horse—with less consumption of current than
any other fan motor on the market. They will fall 1½ horse-power. Six listed
15-inch tap, belt-joining. Furnished with or without guards.

IT WILL PAY YOU TO SEE THE WAGNER BEFORE BUYING ELSEWHERE

TAYLOR, GOODHUE & AMES,
346 Dearborn Street, CHICAGO.

See Our Exhibit of **ELECTRICAL FIXTURES**

IN SECTION "N", BETWEEN COLUMNS 62 AND 64, MANUFACTURES BUILDING.

GLOBE LIGHT & HEAT CO., 52 & 54 Lake St., CHICAGO.

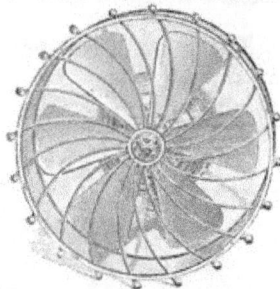

Weekly World's Fair Supplement.

ELECTRICAL INDUSTRIES

DEVOTED TO THE ELECTRICAL AND ALLIED INTERESTS OF THE WORLD'S FAIR, ITS VISITORS AND EXHIBITORS.

Vol. I, No. 3.　　　　　CHICAGO, JUNE 29, 1893.　　　　　FIVE MONTHS $1.00
TEN CENTS A COPY

The Ferris Wheel.

What the Eiffel Tower was to the Paris Exposition the modern feat of engineering skill rises far above the different villages in Midway Plaisance in marked contrast to the simple structures of the natives from the eastern hemisphere.

THE FERRIS WHEEL.

Ferris Wheel is to the World's Columbian Exposition. It is one of the most prominent objects and is seen by the visitor a long time before he reaches the entrance to the grounds. This Although the idea of carrying people in seats attached to the circumference of a wheel is not new the construction of a wheel of such mammoth size and the manner of erecting

It is decidedly novel. The engineering problems involved demanded the best of engineering skill and the workmen and material employed were necessarily of the best. As it is the most prominent object during the day it will also be one of the most prominent at night when it is lighted as it is now intended. Although the plans are not yet fully completed, the cars and offices as well as the framework of the wheel itself will be lighted by thousands of incandescent lamps, and also numerous arc lamps.

The wheel was designed by Mr. George W. G. Ferris, of G. W. G. Ferris & Co., the well known engineers of Pittsburgh. It is built entirely of steel. The wheel is of the bicycle pattern 250 feet in diameter and 28 feet wide. The two outer bands are 40 feet apart. The total height of the wheel is 264 feet while the weight is three million pounds. It is built on a sub-foundation made of 30 to 40 foot piles

Steam, for motor power, is furnished, at 125 pounds pressure, by three Heine safety boilers of 400-horse power each located 700 feet from the structure. The main steam supply pipe is of wrought iron with screw joints and 10 inches in diameter.

The engines built by Wm. Tod & Co., of Youngstown, Ohio, are double, reversible and have piston valves. The cylinder dimensions are 36-inch diameter by 48-inch stroke running at 18 revolutions per minute. They have, in all, 1,000 nominal horse power. The gear on the engine shaft is about three feet in diameter with 14-inch face, the teeth being V shaped across the face. The transmission shaft 12 inches in diameter is geared into the engine gear and extends each way to the nests of gears running to the sprocket wheels. There is a counter shaft gearing into the transmission on one hand and into the sprocket shaft on

THE EXHIBIT OF CHAS. A. SCHIEREN & CO. IN MACHINERY HALL.

above which is a second foundation of concrete. The towers that support the wheel are pyramidal in form 135 feet high to the center of the shaft with a base 40 by 50 feet. The tops of the towers are six feet square and support the bearings which are of cast iron lined with carbon bronze bushings. The four main legs of the towers are made of 24½ inch channel beams with latticed bracings. The shaft is of forged steel 45 feet long, 32 inches in diameter and weighs, with the hubs, 90 tons. This shaft was raised to its place by four engines, each with a four sheave block and falls. The piece was raised, resting against the west side of the towers, until it reached the top when it swung over and dropped into place. There are 36 cars, 26 feet long, 13 feet wide over all and seven feet high in the clear inside, which will seat 38 persons each. They are suspended from the periphery of the double wheel by heavy steel pins in the top of the car so that the weight of the car always keeps its floor very nearly level.

the other. The ratio of sprocket to engine revolutions is about one to 16. The gears are very heavy, being made of cast iron, and having a flange on each face rising as high as one-half the depth of tooth.

The two sprocket wheels on each side are nine feet in diameter and are placed 22 feet apart center to center. There are seven teeth on each wheel and on the face of the wheels are raised studs which catch the pin as the chain drops off the large wheels. The shaft of these wheels is 18 inches in diameter.

On the periphery of the two sides of the wheel are cast iron lugs having gear spaces sunk in them 24 inches apart the pitch diameter of which is 126 feet and the depth six inches.

The chain is formed like a huge bicycle chain of double iron links, 24.08 inches long from center to center of pins. The eyes are 10 inches in diameter outside, a five-inch pin being used, secured in place by a nut and cotter pin. The web of the links is five inches in width by one inch in thickness. In order that the links may wear as little as possible rolls are provided on the pins with the eyes and adjustable platforms under both upper and lower sides of the chain are provided for keeping the weight of the chain from the sprocket wheels.

To stop the motion of the wheel a band brake wheel nine feet in diameter, with a wood lined sheet iron band, eight inches wide, is used on the counter shaft. This band is tightened from the engine by a common 12-inch air cylinder attached to a lever on the band. A five-eighths inch pipe supplies air pressure, which is under the control of the engineer. Each loading gate, of which there are six, is provided with a push button connecting with an annunciator placed directly in front of the engineer so that he may know when to start up.

Mr. Ferris is quite a young man, but 36 years of age, a graduate of the Rensselaer Polytechnic Institute, Troy, N. Y.

The fact that the steel work was not commenced until March, 1893, speaks well for the plant and ability of the Detroit Bridge Company, which built all the steel work. The steel was furnished by Jones & Laughlin, the Carbon Iron Company and Oliver of Pittsburgh. The castings for gears were furnished by the Griffin Car Wheel Company, and the sprockets were made by the Walker Manufacturing Company, of Cleveland, O.

A large force is necessary for operating the wheel. Three shifts of engineers and firemen are required, a guard in each car and one on each platform, a total of 42, being re-

EXHIBIT OF CHAS. A. SCHIEREN & CO. IN ELECTRICITY BUILDING.

ator placed directly in front of the engineer so that he may know when to start up.

Six cars are loaded at each stop from the six platforms provided. It takes about 15 minutes to load all the cars and ten minutes to go around and unload.

The huge structure is owned by the Pittsburgh Construction Co., which is incorporated under the laws of the State of Illinois. The officers are: Robt. W. Hunt, president; George W. G. Ferris, engineer; L. V. Rice, general manager; H. R. Thornton, secretary. The contractors were McCain Bros., the Rookery, Chicago; the inspector of construction being Geo. R. Buchan.

The wheel was designed by Mr. Ferris, the calculations made by Wm. F. Gronan and the draughting done by Paul Laur, both the latter being with Mr. Ferris in Pittsburgh.

quired for the wheel itself, together with six ticket sellers and six turnstile guards stationed at the entrances.

Exhibits of Chas. A. Schieren & Co.

To visitors to the Centennial and Paris Expositions the name of Chas. A. Schieren & Company must be familiar. Their exhibits at the various expositions have attracted the attention of visitors and have been appreciated by the juries of award as shown by the various diplomas and medals received.

Their exhibits at the World's Columbian Exposition are unique in design and the construction of the booth is worthy of a careful inspection. The method of constructing booths and pavilions of link belting has been studied with care and some very artistic combinations have been produced in these exhibits which display to good advantage the construction and general qualities of the leather and the finished belts. The firm has quite outdone itself in this respect in the booth on the east side of the Electricity Building. A classic structure has been erected with four square supporting columns made of the link belt, using black links for the ornamental figures and the Greek scroll around the cornice. The columns are surmounted by small pyramids of belting and adorned with the Spanish colors and American flags. The huge bull's head with graceful and wide spreading horns, which serves also as a trade mark, is erected directly over the entrance to the pavilion and as the horns are each tipped with an incandescent bulb, the effect, joined with that of the light from the other lamps in each panel is pleasing in the extreme.

The interior of the booth is provided with chairs, tables, and a show case containing numerous curiosities in belting, among which might be mentioned several old styles of belt joints; one made in 1830 is laced and riveted without cement, another of 1840 riveted and sewed, and various old style English joints both sewed and riveted. Several pieces of belting are shown constructed of narrow strips of leather laid side by side on edge and fastened together by steel bolts. Other styles of joints are exhibited, the old laced joint, the joint made by fine wire lacing and the cement. Over this case hangs a frame constructed of link belting in which are contained the various medals received by them for the superiority of their product.

At the left surrounded by a brass rail is the exhibit of link belts by the American Leather Link Belt Company. It contains the diploma and bronze medal received at the Paris Exposition of 1889, also various coils of the belt. One large roll contains 100 feet of 36-inch belt one-inch thick and others of 12, 14, 10 and 9-inch belt, all being eleven sixteenths of an inch thick and used for dynamo work are shown. Other coils fill out the geometric figures formed by these mentioned.

On the opposite side of the pavilion is the exhibit of perforated and electric belt the several smaller rolls being grouped tastefully about one large roll consisting of a "Zulu" electric belt, 72 inches wide. A roll of 14-inch perforated electric, one of 18-inch and another of 16-inch are also shown. A photograph of the New York factory at 41-51 Ferry street rests against this exhibit.

In section F, block 27, Machinery Hall, Schieren & Company have another very fine display. A large square pavilion of ebonized wood, touched out with gold leaf and gracefully hung with linen velour curtains of old rose color contains among others a roll so large that one doubts if the wheel has yet been built that will support it. This is a three-ply belt, 200 feet long and 96 inches wide and required the hides of 450 large steers for its construction. Grouped about it in graceful figures are rolls of the electric and perforated belt similar to those described in the exhibit in the Electricity Building. In a show case at the left are samples of the company's patent round and twisted belting, and in another case at the right are exhibited belt lacings both in rolls and in bunches tipped and plain. Belt dressing and coils of raw hide rope are also shown in the latter case. The names of the four principal offices of the company at New York, Chicago, Boston and Philadelphia are displayed on tasty signs on the front and back of this pavilion. Two diplomas received by the company at the Centennial and at Paris in 1889 adorn the walls. While the exhibit of belting at the Fair is large, Messrs. Chas. A. Schieren & Co. can be said to have held

their own in all ways and may well be proud of the display.

These pavilions do not contain all the belts of the firm at the Exposition. In Machinery Hall and the Electricity Building may be seen the Schieren belts in use on many of the various machines. Nearly all of the all-black belts in use at the Fair are of their make. The contracts from the Columbian Exposition Company amounting to nearly $8,000 are shown at the Chicago office of the company. Visitors at the Fair interested in belts and belting will find pleasure in examining these exhibits.

The Exhibit of the Belknap Motor Co.

One hardly looks to Maine for mechanical industries except shipbuilding, yet the exhibit of the Belknap Motor Company shows what Portland can do in the way of dynamo and motor construction. The Belknap Water Motor Company originally commenced the manufacture of water motors, eyclone coffee grinders and pulverizers.

The inability of the water motor to compete in many places with the electric motor was soon apparent, and the construction of electric motors was undertaken. The facilities have been greatly increased and the various sizes and styles of machines as shown in the exhibit are now manufactured.

THE EXHIBIT OF THE BELKNAP MOTOR CO.

Mr. George W. Brown is in charge of the exhibit, and takes especial pains to explain the different apparatus to visitors. The display is decorated by numerous signs and banners, fancy arrangement of colored lamps, etc. A large illuminated sign in the center of the exhibit has attracted a good deal of attention. A curtain constantly moved by a motor lets the lights concealed behind it flash out through openings, thus lighting up the glass sign, and as the opening in the curtain passes up to the top of the sign they give the appearance of waves of light passing over the sign.

The display includes various sizes of dynamos and motors, mills and pulverizers, water motors, fans, etc., in different combinations. A large exhaust fan and an electric drill that holds the piece of iron to be drilled by a magnet are shown. Among the exhibit of motors are motors of three and five horse power for a 220-volt circuits, a one-horse power for a 110-volt circuit, a one-quarter horse power, a one and one-half horse power running a large exhaust fan. This motor is of the four pole type. In dynamos the company has on exhibition 25, 60, 100 and 250-light capacity, at 110 volts. A combined dynamo and water motor of 10-lights capacity, and a similar one of 25-light capacity. These are very compact and efficient, and where water rates are not too high can be used to advantage.

The electric and water motors are shown in combination with fans, coffee mills and other small machines. A very convenient style of coffee mill and motor is one shown with the motor beneath the table. There is also shown six arc lamp on the constant potential 110-volt current. The various ammeters and voltmeters manufactured by the company are shown in use. The current for the different lights and motors in the exhibit is generated by 6.1 K. W. generator run by a 40 horse power 220 volt motor.

All the electrical devices shown are well designed, substantially built, and finely finished. Carbon brushes are used, and no sparking at the brushes is seen. The commutators are of tempered copper. The bearings are self-oiling. The dynamos are all compound wound. These machines have been designed by Mr. W. H. Chapman, the electrical engineer of the company.

The World's Congress of Electricians.

The World's Congress of Electricians that meets in Chicago in August, promises to be an interesting event. The response to the invitation for papers on appropriate subjects has been generous. It is desired that all papers be sent to Prof. T. C. Mendenhall, Washington, D. C., as early as the first of August to allow time for printing. Prod. H. S. Carhart, Ann Arbor, secretary of the Advisory Council, reports the following list of prominent electricians who have consented to read papers.

Mr. W. H. Preece, F.R.S., engineer in chief of the Post Office, London, "Signalling through space by means of the electromagnetic vibrations."

Prof. W. E. Ayrton, F.R.S., City and Guilds of London, Central Institute, London. "Variations of P.D. of the electric arc with current, thickness of carbons and distance apart."

Dr. Stephen Lindeck, Physikalisch-Technische Reichsanstalt, Berlin. "On materials for standards of electrical resistance and their construction."

Prof. S. P. Thompson, F.R.S. Finsbury Technical college. London. "Ocean Telephony."

Prof. Elihu Thomson has accepted the invitation of the committee to deliver one of the evening lectures which will be profusely illustrated.

An Exhibit of Incandescent Lamp Decorating.

One of the most ornamental features of the exhibit of the General Electric Company is the room in which are shown the meters, switches, instruments and other smaller products. The ceiling is ornamented and illuminated by pretty figures marked by miniature electric incandescent lamps of various colors. The frieze is also constructed of lamps set back in the wall making the entire effect very pleasing. In all 725 miniature lamps are used in the room, 388 of 5-candle power in the frieze; 60 of ten candle power in the cornice; and 277 of 4, 5, 8 and 32-candle power in the ceiling, making in all 3,892 candle power, equal to 243 gas jets of 16 candle power each. While this exhibit shows what can be done in the way of decorating and fancy work the switchboard in Machinery Hall for the 125-volt circuits of the General Company's plant shows what can be done in a mechanical way. The instruments are all of the latest pattern mounted on white marble. The connections are of massive copper bars and large cables.

A model of the tree system of wiring very accurately worked out and constructed is shown in one corner of the company's exhibit. The various sizes of wire and quantity of wire needed for a stated output are shown. Next to it is a model of the feeder system showing the same quantities as required by this system. Square blocks are shown that give the comparative quantities of copper needed for a given load and distance by each of the different systems used with the direct current. The old long magnet type of the Edison dynamo is shown and is quite a curiosity to some although there are a number of them still in use.

In looking over the exhibits of this company one sees nearly everything to which electricity could be applied; while it is doubtless a fact that some of the machinery designed has had little practical use it shows what can be done and in what direction inventive genius has worked. The applications are now becoming so numerous that it will not be long before the dynamo factories will be confined to the building of special apparatus and the smaller shops will take up electrical machine construction from special plans and specifications prepared by electrical engineers as is already done to some extent in Europe.

Coming Events at the Fair.

To day is Millers' Day, and the flour makers are given the privileges of the grounds.

For Saturday evening a special program is being prepared, including music and fireworks.

On Tuesday, the Fourth of July, preparations are being made for a grand old time celebration. The event will undoubtedly be a memorable one to all who attend, as the committee on ceremonies is arranging a suitable program to commemorate America's greatest day. The exercises will commence with the national salute at sunrise.

At 11 o'clock A.M.—The international procession of the peoples on Midway Plaisance.

At 3 o'clock P.M.—The reading of the Declaration of Independence. Speeches by distinguished men to be given in front of the Administration and Government Buildings, and in the Stock Pavilion.

At 8 o'clock P.M.—The illumination of the grounds, the Grand Basin and buildings.

At 8:30 o'clock P.M.—Fireworks in the Court of Honor, and at the Government Pier.

Theodore Thomas and Prof. Tomlins will give musical entertainments at Music and Festival Halls, and there will also be music at the band stands.

The Central Electric Company is completing its exhibit in the gallery of the Electricity Building as fast as possible. The main part of this exhibit consists of nine very handsome reels of Okonite submarine lead cased and aerial cables, for which the company is the western agent. Several photographs of the Okonite Company's factories are shown. The Central Company is also including in its exhibit two large boards containing samples of conduit in all the various forms made by the Interior Conduit and Insulation Company, for which it is the western agent. The boards contain all the necessary tools and fittings. Besides the exhibits mentioned the company is also making a handsome display of the Washington Carbon Company's product consisting of samples of pure carbon, battery carbons, light and brush carbons. When complete the Central company's display will certainly be very creditable and well worth seeing. Mr. H. H. Small is in charge.

ELECTRICAL INDUSTRIES.

Entered at Chicago Postoffice as second-class mail matter.

PUBLISHED EVERY THURSDAY BY THE

ELECTRICAL INDUSTRIES PUBLISHING COMPANY,
INCORPORATED 1893.
MONADNOCK BLOCK, CHICAGO.
Telephone Harrison 186.

R. L. POWERS, Pres. and Treas.   E. E. WOOD, Secretary.

E. L. POWERS . . . . Editor.

B. A. FOSTER, }
                                        } Associate Editors.
W. A. REMINGTON, }

R. E. WOOD, ... Eastern Manager.

FLOYD T. SHORT, ... Advertising Department.

EASTERN OFFICE, WORLD BUILDING, NEW YORK.
World's Fair Headquarters, Y 27 Electricity Building.

SUBSCRIPTION.

FIVE MONTHS, $1.00
SINGLE COPY, -   10
Advertising Rates Upon Application.

News items, notes or communications of interest to World's Fair
Visitors are earnestly desired for publication in these columns and will
be heartily appreciated. We especially invite all visitors to call upon us
or send address at once upon their arrival in city or at the grounds.
          ELECTRICAL INDUSTRIES PUBLISHING CO.,
                              Monadnock Block, Chicago.

The attendance at the Fair was not as large last week as during the previous week, being but 704,000. The special outside attractions undoubtedly limited the attendance, so that the large attendance on Thursday and Saturday of the previous week more than equalled the increased attendance of the other days. Next week the Fair will have the greatest crowds yet seen. The special attractions on the 4th of July and the inducements in the excursion rates offered by the railroads will bring large crowds to the city.

Although the buildings at the Exposition are large and well lighted, they seem to lack one very important feature, that of ventilation. The main floors are cool and comfortable, but it is a general complaint that the galleries are almost unbearable on warm days. As the days are all warm now and will continue so until toward the close of the Fair, steps cannot be taken too soon to remedy this defect. By a combined effort of the exhibitors in the Electricity Building and the Department some means can undoubtedly be found to effect a change. In some of the other buildings steps have already been taken.

The appointment of the jury of awards and the method to be employed in making the awards in the Department of Electricity is at present of the greatest interest to exhibitors. It is stated that the selection of the jury is now being made, the appointments to take place July 3rd, while the work of making awards will begin July 15th, and continue for two months. In the regulations governing the awards it is understood that sections 3 and 5 will be slightly modified, a matter which will be decided upon at once. From the great number of practical and theoretical electricians of the country we are sure that Chief Barrett and his department will succeed in selecting the best men available.

Since the initial number of our weekly World's Fair issue many compliments together with warm words of approval for our enterprise have poured in upon us from all parts of the country. It is, to say the least, very gratifying to us to know that our efforts to give our readers the freshest and most timely news connected with the great Exposition, are thus appreciated. We feel that our object has been fully attained in giving visitors to the Fair information not otherwise available. The illustrated articles and news items already published we have the satisfaction of knowing have been read far and wide and that the diagrams giving the names and locations of the exhibitors in Electricity Building are used universally as an official directory. We have also the satisfaction of knowing that strangers in the city use and appreciate the map showing the location of the electrical and allied business interests of Chicago. All we can say to our readers is that we thank them one and all for their kind words of approval, their subscriptions and good wishes, and hope they will be even better pleased with our subsequent issues.

Dynamos run by Gas and Oil Engines.

As yet very little has been done in the United States in the way of running dynamos by gas or oil engines. Many types, however, of both of these are shown in Machinery Hall. Among those deserving special mention is the safety oil engine, a three and a half horse power machine after the Hornsby-Akroyd patent, and made by R. Hornsby and Sons, Limited, of Grantham, England.

This engine using crude petroleum of 300 degrees flash test is said to consume but three-quarters of a pint of oil per horse per hour, and as run at this exhibition fuel for it is said to cost less than a cent per hour. The supply of oil is carried in a tank forming a part of the foundation box, and is automatically pumped into the combustion chambers as needed after the machine is started.

The engine is somewhat like the Otto gas engine, and works on the Otto-cycle principle. A hot combustion chamber at the rear end of the cylinder receives oil and air, vaporizes the former, and after mixing it with the air the gas thus formed is exploded by the compression of the piston forcing it into and against the hot combustion chamber. To start the machine, a lamp underneath this combustion chamber is lighted, and a blower provided for the purpose is turned by hand to heat the chamber up to a cherry red. When this is accomplished, which takes but three or four minutes, the light is extinguished and the chamber is kept hot by the compression and explosion of the oil gas, an explosion taking place once during each two revolutions of the fly-wheel.

This particular engine is belted to a Castle dynamo of 24 lamp capacity, and the regulation of speed is very perfect, no variation being noticeable when the load is all cut off or all thrown on. The dynamo was built by J. H. Holmes & Co. of Newcastle-on-Tyne, England, and is of much the same type as the Crocker-Wheeler dynamo of this country. It is very slow speed for the size, running at not over a thousand turns per minute. There is a large field here for engines of this type as the Americans have scarcely become aware of the fact that oil engines are very useful as well as inexpensive for producing electric currents. The foreign firms at the Fair have already noticed this fact and are preparing to supply the demand.

The Westinghouse Company is daily experimenting in alternating currents of very high potential in the room provided by it for the purpose. Some of the effects as now shown are grand and startling in the extreme. One needs his ears covered while sitting in the room, and the smell of ozone is strongly apparent. The advance made since the commencement of the exhibit has been very great, and new effects have been designed which will be ready for display very soon.

WORLD'S FAIR NOTES.

Numerous inquiries are made as to the whereabouts of the Edison Kinetograph. It is said to be advertised in the official catalogue, but has not been seen in the Electricity Building.

The Jenny Motor Co. has in operation two generators this week, current for which has been supplied by another exhibitor. The engine plant for supplying the regular power is not yet installed.

The large railway signaling exhibit of Siemens & Halske in the southeast end of the Transportation Building seems to be making progress. It is very nearly finished, but hardly attracts the attention it really deserves.

Mr. H. T. Paiste, Philadelphia, is just completing his exhibit in the gallery at the south end of the Electricity Building. Mr. Paiste is giving away a neat lead pencil to electricians, wiremen, and, in fact, to all who apply.

The numerous streams of water being pumped into the receiving basin in the annex to Machinery Hall from the various pump exhibits are continuous sources of amusement and wonder to the crowds of visitors in that part of the building.

Mr. Prentiss expects to have the large exhibit of the Brush Electric Company completed so as to have the formal opening on Saturday next. The pavilion is finished and all the heavy work done, and only a few details now need attention.

The current for charging the storage batteries in the electric launches is furnished by the two bipolar dynamos at the west end of the row of Edison dynamos in the Machinery Hall. For a time six boats were charged in series from the 500-volt power current.

The Electrical Engineering Department is at work placing 20-candle power series incandescent lamps under all the bridges crossing the lagoon. They are cut in on the regular arc lighting circuits and will make these spots decidedly more cheerful than at present.

The Ide and Ideal engines made by the Harrisburg Foundry and Machine Works of Harrisburg, Penn., attract much attention by their silent and smooth running. They are tandem compound, occupy little space and are in most most respects self oiling. The location is blocks 15 and 16, Sec. F.

The Briggs' Automatic Screw Machine, in the exhibit of the Western Electric Co., is one of the permanent attractions and bids fair to hold its favor with the public to the close of the Exposition, if one may judge by the number of people standing in line waiting the souvenirs it turns out with measured regularity.

A very handsome and smooth running engine is shown by the Siemens & Halske Company in their exhibit in Machinery Hall. It is vertical, triple compound condensing and drives their main lines of shafting by a peculiar arrangement of ropes. The engine was built by F. Schichau of Elbing Preussen, and is well designed.

The Walworth Manufacturing Company of Boston, Section O, No. 27, Machinery Hall, show samples of the several types of street railway poles, also a large assortment of steam valves in all sizes. A handsome ebonized wood glass paneled cabinet occupies the center of the space in which are shown the large number of smaller products of the factories, such as the Stillson wrenches, die plates and dies

and a great many other specialties. A very large proportion of the iron poles used by the electric street railways of the country has been supplied by this establishment.

The electric fountains are now in working order after various vexing delays. They are played three nights a week and have proved a drawing card. It must be admitted that they suffer somewhat from being placed so far below the level of the court and the best of the effect is thus lost to people who cannot get close to the railing.

The Vacuum Oil Company of Rochester and Olean, New York, exhibits in section K, No. 23, Machinery Hall, complete models of the manufacturing plants at both these places. They are complete in all detail and very instructive. This company manufactures the celebrated 600 w. cylinder oil, so long and favorably known to the trade.

The Exposition might well be called an electrical exposition since, with the exception of Machinery Hall where steam is used, all the power for every kind of apparatus is supplied by the electric current. The plan is carried out of driving each line of shafting by a separate electric motor; a more convenient method it would be difficult to find.

Taylor, Goodhue & Ames, have ready for their exhibit in Electricity Building a 10-horse power alternating current motor but are delayed in showing it in operation owing to their not being able to get alternating current in the space assigned them. This motor will, it is claimed, enjoy the distinction of being the only large alternating motor in operation on the grounds.

The Newark factory of the Westinghouse company has just contributed its quota of exhibits, and although long delayed, every one will be pleased to see the machines from this old established factory. All the new types and designs will be shown and when completed this exhibit will add one more to the numerous excellent displays of the Westinghouse Electric and Manufacturing Company.

The exhibit of the Phoenix Glass Company in the Electricity Building has usurped the place occupied by the Corliss engine at the Centennial Exposition as a meeting place. The tower has become one of the best known features of the Exposition, and Mr. Fox, the genial representative of the Phoenix Glass Company is daily the temporary guardian of lost children, packages, etc., and the custodian of various articles and messages.

Randolph & Clowes have an exceedingly appropriate booth in the Mining Building, made of different sizes and shapes of drawn brass and copper tubing. The general offices and factory of this company are in Waterbury, Conn., while the goods of Randolph & Clowes are not "quick winding" they quite evidently possess all the other points of merit that are indelibly impressed on the public mind as synonymous with the word "Waterbury!"

One can get some idea of where the copper comes from what is now so extensively used in the electric trades by a thorough examination of the fine large exhibit of the Calumet and Hecla Company in the Mines and Mining Building. Huge square cakes and ingots of solid copper are piled up on the floor and large cones of sheet copper and wire confront one on the west side of the exhibit. Models of some of its shafts and rock houses are shown and many fine pictures in colors adorn the walls of the section. The native ore is piled up about the space, and from the large maps and pictures one is able to get something of an idea of the size of the works.

Amusements.

Hooley's Theater—Mr. E. S. Willard, in "The Professor's Love Story." 149 Randolph street.

Columbia Theater—Miss Lillian Russell, in "Girofle-Girofla." Sixth week. 108 Monroe street.

Grand Opera House—Sol Smith Russell, in "April Weather." 87 Clark street.

Auditorium—Imre Kiralfy's Spectical "America." Congress street and Wabash avenue.

McVicker's Theater—"The Black Crook;" next week Denman Thomson, in "The Old Homestead." 82 Madison street.

Chicago Opera House—American Extravaganza Company, in "Ali Baba," or "Morgiana and the Forty Thieves." Fifth week. Washington and Clark streets.

Schiller Theater—Chas. Frohman's Stock Company, in "The Girl I Left Behind Me." Fifth week. Randolph, near Dearborn.

Haverly's Casino—Haverly's United Minstrels. Wabash avenue, near Jackson street.

Havlin's Theater—"The Cracker Jacks." Wabash avenue, near Eighteenth street.

Haymarket Theater—James J. Corbett, in "Gentleman Jack." Madison, near Halstead.

Windsor Theater—Rider Haggard's "She." 468 Clark street.

Tattersall's—Military Tournament. 16th and State streets.

Buffalo Bill's "Wild West." 63d street.

Pain's "Seige of Sebastopol," 60th street and Cottage Grove avenue. Opens Saturday, July 1.

PERSONAL.

Prof. W. E. Ayrton sails for New York early in July.

Prof. S. P. Thompson sails for New York on the Etruria July 15th.

Mr. Luke Lilly returned to Cincinnati Sunday night after a short visit to the Fair.

Mr. C. A. Coolige, of Centralia, Washington, is a visitor to the Fair, Chicago and vicinity.

Mr. Henry Villard, of the New York Financier, was a visitor at the Electricity Building last week.

Prof. George D. Shepbardson, of the University of Minnesota with his wife is spending a week at the Fair.

Mr. L. B. Stillwell, of the Westinghouse Electric & Manufacturing Company was in the city June 21st.

Prof. Dugald C. Jackson, of the University of Wisconsin, was at the Fair Saturday, accompanied by several of his pupils.

Jas. H. Mason of the Simplex Electrical Co., was in the city this week looking after the company's exhibit in George Cutter's space.

Prof. B. F. Thomas, of the Ohio State University, is paying a short visit to the Fair, expecting to return for the Electrical Congress.

Mr. Franklin Phillips, of the well-known firm of engine builders, the Hewes & Phillips iron works, is visiting the Fair with his wife.

Mr. G. E. Emmons, auditor of the Lynn factories of the General Electric Company, with his wife, has been visiting the Fair during the past week.

Mr. H. T. Paiste of Philadelphia is at the Fair looking after his exhibit that has been closed for a few days on account of the decorators working above it.

Mr. Ralph W. Pope, secretary of the American Institute of Electrical Engineers, arrived in Chicago last Friday and will make his home here until the close of the Fair.

Col. Geo. L. Beetle, the popular veteran in the electrical business, and in the service of the Western Electric Company, is confined to his house for a few days with a slight illness, due to over exertion during the recent warm weather.

Mr. John Kroosl, superintendent of the Schenectady works of the General Electric Company, was in the city for a few days early this week, making a short visit to the Fair.

Mr. Frank B. Rae of Detroit, was in Chicago last week. He has been retained as consulting engineer by a number of towns that contemplate putting in electric lighting plants.

Mr. W. J. Hammer, chairman of the committee on headquarters of the A.I.E.E. at the Fair, arrived last week, and is arranging the rooms of the Institute in the Electricity Building.

Mr. W. W. Primm, the electrical engineer of the Department of Electricity, and Miss Adliena M. Stark were recently married. The many friends of Mr. Primm at the Fair extend their congratulations and good wishes.

TRADE PUBLICATIONS.

The Charles E. Gregory Company, Chicago, has just issued a very neat and convenient pocket memorandum book containing a valuable telephone directory of the electrical interests in Chicago, and at the World's Fair grounds. Its value will be greatly appreciated by all who are so fortunate as to secure a copy.

The Ansonia Electric Company has just issued a very handy and neat edition of its house goods catalogue which will be known as B. 41. It is only issued temporarily for use until the larger and more complete catalogue comes out on which the company is now at work. A number of new goods are shown in this catalogue and also some changes in price. It may be had on application.

One of the many valuable and improved features in the catalogue just issued by Taylor, Goodhue & Ames is the manner of indexing. The index is placed in a convenient part of the book and includes a general index, an index of trade members and an index of all the numerous articles and supplies handled by this firm. The book is substantially bound with cloth covers, well printed, neatly and intelligently arranged. The house goods, central station supplies, line and general supply goods are all very finely illustrated on good paper and make the catalogue a valuable reference book for every user of electrical supplies.

BUSINESS NOTES.

McLean & Schmitt, Chicago, are keeping an increased force of men at work in their shop, doing general repair work, and a considerable amount of extra work for the World's Columbian Exposition.

Mr. Geo. W. Brown, general manager of the Belknap Motor Company, has a puzzle for his visitors when they attempt to tell which of his motors are running and which are not. The motors run so silently and even that it is almost a puzzle for an expert.

The Brush Electric Co., Cleveland, Ohio, are putting in a $100,000 plant at Montgomery, Ala. Mr. Ira J. Britton, for many years chief expert for the company, is in charge of the work. Over 300 men are engaged in the construction. The plant, when completed, will supply both light and power to the city of Montgomery.

The Interior Conduit and Insulation Company are distributing at the Fair very neat and useful souvenirs in the shape of pocket screw drivers. They are nickel-plated and close up in very convenient shape to carry in the pocket. The company's name is neatly etched on the handle.

The Electric Appliance Company's exhibit of Meston alternating current motor applied to the operation of a sewing machine continues to attract considerable attention and keep a crowd around their World's Fair space. The outfit is certainly a very complete and compact affair and the ease with which it can be controlled leaves nothing to be desired. The outfit is certainly one that will find a very large field.

The Telegraph & Telephone Construction Co., Detroit, Mich., has just removed to its new building containing the new exchange to which it has transferred all its old subscribers. This exchange is a model in every way, being equipped with a new multiple switch board made by the Western Electric Company, Chicago. All of the wires are run underground, and the building and its equipment constitute one of the most modern and best equipped telephone exchanges in the world.

STANDARD ELECTRIC CO.

General Offices, Suite 625 Home Ins. Bldg., } **CHICAGO.**
Works- 313-317 South Canal Street,

BUILDERS OF THE

STANDARD SYSTEM OF ARC LIGHTING.

FOR FURTHER INFORMATION

ADDRESS THE GENERAL OFFICES.

NOTICE: Central Station Managers, Municipal Officers, Owners of Isolated Lighting Plants and Prospective Purchasers everywhere are invited to inspect and investigate the **STANDARD SYSTEM** before contracting for apparatus.

For lighting stores, foundries, factories, or the streets of a city, this system has no equal.

OUR CONTRACT does not obligate the purchaser to buy apparatus or supplies exclusively of us for a term of years, but courts open competition and encourages fair tests for merit between any and all existing arc lighting systems and the **STANDARD**.

ELECTRICITY BUILDING—EXHIBITORS AND THEIR LOCATION.

GALLERY.

MAIN FLOOR.

Exhibitor.	Section.
Austria	Y
Atwoods Electric Co	Z
Nat. Inst. of Elec. Eng.	S
American Battery Co	T
Axtell, H. M	T
Allg. Elec. Gesellschaft	D
Baker Mfg. Co	S
Bryant Testing Co	R
Billings & Spencer	R
Brixey, W. B	T
Bishop Battery Co	S
Bell Telephone Co	Est.
Brush Electric Co	L
Caldwell El. Cloth Cut. Mch. Co	B
United Edi. Storage Co	R
Cutter, George	T
Canton Elec. Co	T
Chicago Elec. Ware Co	S
Copenhagen Fire Alarm Co	S
Central Electric Co	S
Commercial Cable Co	S
C. & C. Elec. Motor Co	S
Cleveland Elec. & Mfg. Co	S
Chicago Bridge Co	S
Delaney Clark Co	S
Department of Electricity	R
Eddy Elect. Instrument	S
Mott, Lowell & Nav. Co	S
Electric Signature	T
Edgerton, E. W	S
Elgin Telephone Co	S
Edison Elec. Mfg. Co	S
Enterprise Elec. Co	S
Eureka Temp. Copper Co	S
Electric Appliance Co	S
Elec. Sel. & Sig'l Co	S
Electric Heat Alarm Co	Y

Exhibitor.	Section.
Electrical Review	Y
Electricity	S
Electric Gas Co	Y
Electrical Engineer	S
Electrical World	V
Eddy Electric Motor Co	S
Excelsior Electric Co	D
Electrical Forging Co	S
Reynolds Byrnous Co	S
Elektron Mfg. Co	S
Electrical Conduit Co	S
England	O
Empire Chine Works	S
Franklin Elec. Appliance	S
French Piano Exhibit	S
Felten & Guilleaume	J
Fraser	K-P
Ft. Wayne Elec. Co	M
Gandt & Co. N	S
Gamewell Fire Alarm Co	V
General Electric Co	B-H-N-C & J
General Incand'sc't Arc Lt Co	S
Goerke, G. S. & Co	Y
German	S
Hubbard Wm. & Co	S
Robinson, C. J	S
Hart & Hegeman Mfg. Co	S
Hawes Elec. Appliance Co	S
Bell, Chas	S
Holmes, S. A	V
Johnson & Brass	S
Bonett & Von Winkle	S
Hotel, A. H	S
Brodmann, F. A. Co	S
Humric Alloy Co	S
Internat. Tel. Ltd & P't Co	S
India Rubber Comb Co	S

Exhibitor.	Section.
Jaeger, Chas. L	T
Jalus Mfg Co., H. W	T
Jewell Belting Co	S
Jenney Elec. Motor Co	S
Knapp Electrical Works	T
N. A. P. Elec. Novelty Co	S
Knapp & Buckley	S
Kennedy Electric Co	S
Lawson, H. A	S
LeVanche Battery Co	S
McNeil Tinder Elec. Co	S
Mascue, W. S	S
Meeker, Dr. G	S
McIntosh Bat. & Opt. Co	S
Mawson, F. L. Belting Co	D
Mather, A. C	S
Mather Electric Co	S
Newman Clock Co	S
New Magazine Watch Co	S
N. Y. Insulated Wire Co	S
National Carbon Co	S
Norwich Ins. Wire Co	S
North Am. Phonograph Co	S
N. Y. & E. R. A	S
Nat. Aut. Fire Alarm Co	S
Nat. Electric Machine Co	S
Owen, Dr. A	S
Phoenix Glass Co	S
Paiste, H. T	S
Pulsermacher Galv. Co	S
Pequolly, J. B	S
Pratt El. Med. Sup. Co	S
Powell, Wm. & Co	S
Phelps, A. H	S
Page Beling Co	S
Open & Co	S
Riegler, F. A	R

Exhibitor.	Section.
Robare Gauge Co	T
Roessler & Hasslecher Chem. Co	S
Street Railway Journal	A
Strecker Ant. Telph. Co	S
Standard Paint Co	S
Speakable, C. L	S
Star Iron Tower Co	W
Spon	Y
Schuere, Chas. A. & Co	S
Schachner & Achne	S
Siemens & Halske	E
Schuckert & Co	S
Short Electric Co	L
Sperry Elec. Railway Co	L
Standard Underg. Cable Co	L
Standard Electric Co	S
Sawyer Battery Co	S
Tate, Vol. El. Signal Co	Y
Todd, Applegate Co	S
Taylor, Goodhue & Ames	S
Thomson Elec. Welding Co	O
Telautograph, Elisha Gray	S
Union Electric Co	S
Votley, J. C & Co	W
Webb, G. F	S
Weston El. Instrument Co	S
Washburn & Moen Mfg. Co	S
Western Union Tel. Co	S
Waite & Bartlett Mfg. Co	S
White, S. S., Dental Mfg. Co	U
Western Electrician	U
Wilder Aut. Burglar Al. Co	S
Western Electric Co	A
Westinghouse El. & Mfg. Co	B-H-J
Wales & Neubold	S
Wing, L. J. & Co	P
Zucker & Levett Chem. Co	F

HAVE YOU EXAMINED

the new **6 C. P. PACKARD** lamps made to burn on regular voltages and to fit ordinary sockets.

SEE THEM ...in our World's Fair Exhibit.

ELECTRICITY "U 16" BUILDING

and place your orders with our representative there or send them to

242 Madison Street, CHICAGO.

ELECTRIC APPLIANCE COMPANY.

The "BUCKEYE" Lamp

A KNOWN Quantity
No EXPERIMENTING
Low Operating Cost
Improved Lighting Effects

Life-1000 to 4000 Hours
Western Deliveries can be made from Chicago Stock

"The **BUCKEYE** sets the pace"

THE BUCKEYE ELECTRIC CO., Cleveland, O.

Chicago
437 THE ROOKERY

New York
49 DEY STREET.

THE MATHER ELECTRIC CO.

MANCHESTER, CONN.

Dynamos, Motors, Generators,

Offices, 116 Bedford St., BOSTON.

—AND—

1002 Chamber of Commerce Bldg., CHICAGO.

THE "NOVAK" LAMP.

CLAFLIN & KIMBALL (Inc.)

General Selling Agents.

116 Bedford Street, BOSTON.

1002 Chamber of Commerce Bldg., CHICAGO.

Enterprise Electric Company

307 Dearborn Street, Chicago....

GENERAL WESTERN AGENTS

N. T. R.

Manufacturers' Agents and Mill Representatives for

Electric Railway, Telegraph, Telephone and Electric Light

SUPPLIES OF EVERY DESCRIPTION

Agents for Cedar Poles, Cypress Poles, Oak Pins, Locust Pins, Cross Arms, Glass Feeder Wire, Insulators, WIRES, CABLES, TAPE and TUBING

Map of Chicago.

Showing Location of its Electrical and Allied Business Interests, Principal Hotels, Theatres, Depots and Transportation Lines to the World's Fair Grounds. (Index numbers refer to the black squares.)

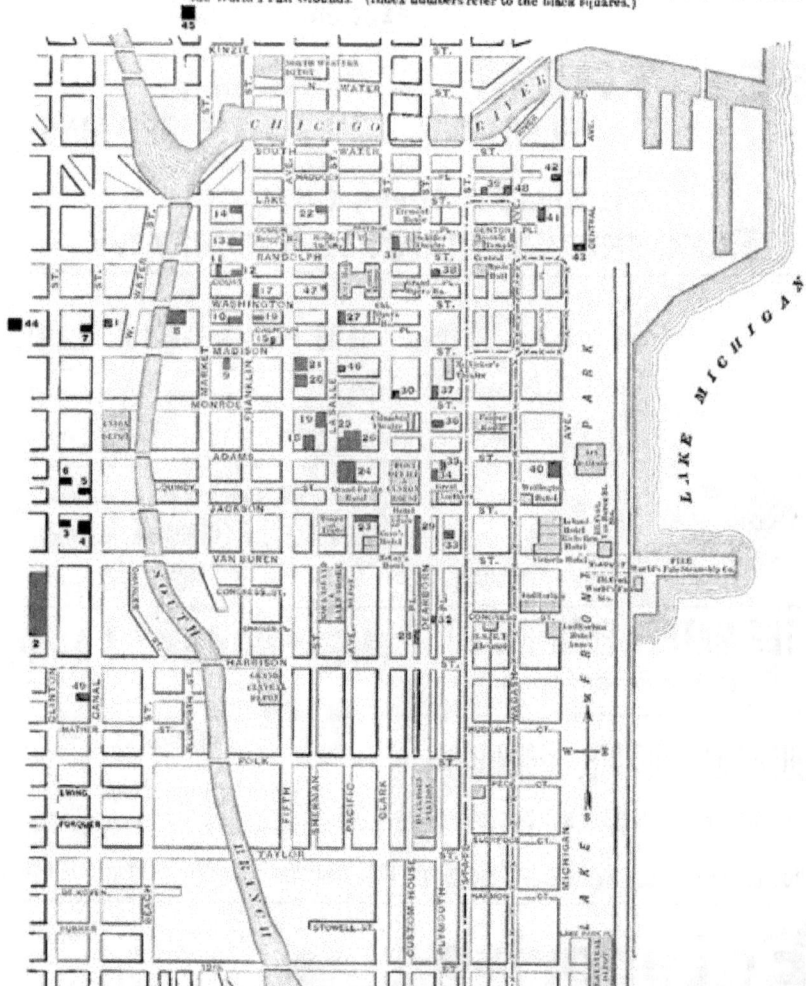

~Lundell~
Suspended Fan Outfit

Black

Japan

Finish

Self-Oiling

....and

Self-Aligning

Bearings

Electric Fans

Electric Fans

OKONITE WIRES
OKONITE === TAPES === MANSON
INTERIOR CONDUIT.

Batteries, Bells, Push Buttons, Annunciators, Volt Meters, Ammeters, Wheatstone
Bridges, Line Wire Cross Arms, Brackets, Pins, Insulators, Tools.

GENERAL SUPPLIES.

CENTRAL ELECTRIC CO.
116-118 Franklin Street,
CHICAGO, ILLS.

CLARK ELECTRIC COMPANY, NEW YORK.

192 Broadway and 11 John Street.

MANUFACTURERS OF ARC LIGHTING APPARATUS FOR EVERY PURPOSE A SPECIALTY.
The CLARK ARC LAMPS for use on EVERY CURRENT, have the reputation of being
the best and most durable of any ever made in the United States.

RAWHIDE PINIONS FOR ELECTRIC MOTORS
A SPECIALTY.
RAWHIDE DYNAMO BELTING

Greatest Adhesive Qualities.   A Non-Conductor of Electricity.
Causes Less Friction than any other Belt.

THE CHICAGO RAWHIDE MANUFACTURING CO.
THE ONLY MANUFACTURERS IN THE COUNTRY.

LACE LEATHER ROPE
AND OTHER RAWHIDE

GOODS
OF ALL KINDS
BY KRUEGER'S PATENT

This Belting and Lace Leather is
not affected by steam or dampness;
never becomes hard; is stronger,
more durable and the most economical Belting made. The Rawhide Rope for Round Belting
Transmission is superior to all
others . . . .

75 Ohio Street,   CHICAGO, ILL.

THE REGULAR MONTHLY EDITION

.... OF ....

Electrical Industries

Is read more widely and used more constantly for reference by actual buyers in the
electrical field than any other similar publication. WHY? Because it has in every issue handsomely illustrated special articles and descriptions of everything of interest in the electrical
world besides a complete

BUYERS DIRECTORY giving the names of all the manufacturers and dealers in the
trade; a complete directory of

ELECTRIC LIGHTING CENTRAL STATIONS and a complete directory of the
ELECTRIC RAILWAYS of North America corrected to the date of going to press,
features found in no other electrical journal in the world.

The Weekly World's Fair issue contains the most novel and unique features yet undertaken by an electrical Journal.

Sample copies free on application. Every electric lighting company should write for our
special terms. We are now making the most liberal offer to subscribers for both publications
during the next few months ever made by an electrical paper.

ELECTRICAL INDUSTRIES PUB. CO.

Monadnock Block, CHICAGO.

Western Electric Company,

CHICAGO.   NEW YORK.

Arc Lighting Apparatus
  High and Low Tension,
      Double and Single Service Lamps,
        All Night Single Lamps,
          Theater and Focusing Lamps.

ELECTRIC MOTORS

HIGH SPEED
    VARIABLE SPEED
        SLOW SPEED.

BUILT FOR SEVERE AND CONTINUOUS SERVICE.
SPECIAL TYPES FOR SPECIAL DUTY.

See Our
Exhibits:
East Gallery
Electricity Bldg.

And another:
851-853-855
The Rookery.

THE GEORGE CUTTER 1893 CO. THE ROOKERY CHICAGO

SIMPLEX WIRES

SIMPLEX

Ever Onward and Upward!

INSURE
HICH
INSULATION

Simplex Electrical Co.
620 Atlantic Ave.,

George Cutter, Chicago.   BOSTON, MASS.

XNTRIC

"That's the Switch"

And we control that movement.

H. T. PAISTE,
10 South 18th St.,
PHILADELPHIA,
PA.

Made   5 amp.   S. P.
10 amp.   S. P.
5 amp.   3 way.
10 amp.   3 way.

P: & S.
CHINA
SWITCHES

Simplest
Neatest

Made only by

PASS & SEYMOUR,
Syracuse, N. Y.
George Cutter, Chicago.

Consolidated Electric Co.

Manufacturers and Dealers in all kinds of

ELECTRICAL . SUPPLIES,

115 Franklin Street,

CHICAGO.

NEW YORK.   PITTSBURGH.   CHICAGO.
42 Murray Street.   43 Sixth Avenue.   19 & 21 Wabash Ave.

Phœnix Glass Company,

World's Fair Exhibit, Center Electricity Bldg.

Do not fail to call and leave your name for
our new catalogue of electric and gas globes
and shades.

...SEE AD...

Western Electric Co.,

PACE 15.

CHAS. A. SCHIEREN & CO.

MANUFACTURERS OF

Genuine Perforated Electric Leather Belting.

46 So. Canal Street,   -   CHICAGO.

Section 15, Dpt. F, Clm. 27.   Section D, Space 3.
MACHINERY HALL.   ELECTRICITY BUILDING.

WAGNER ELECTRIC FAN MOTORS

For Direct or Alternating Currents.

These motors give a stronger breeze with less consumption of current than
any other fan motor on the market. They are ½ to 1-½ horse power. Six bladed
12-inch fan. Self-oiling. Furnished with or without guards.

IT WILL PAY YOU TO SEE THE WAGNER BEFORE BUYING ELSEWHERE.

TAYLOR, COODHUE & AMES,
348 Dearborn Street, CHICAGO.

See Our Exhibit of **ELECTRICAL FIXTURES**

IN SECTION "N", BETWEEN COLUMNS 62 AND 64. MANUFACTURES BUILDING.

CLOBE LICHT & HEAT CO.. 52 & 54 Lake St., CHICAGO.

Weekly World's Fair Supplement.

ELECTRICAL INDUSTRIES

DEVOTED TO THE ELECTRICAL AND ALLIED INTERESTS OF THE WORLD'S FAIR, ITS VISITORS AND EXHIBITORS.

Vol. I, No. 4.   CHICAGO, JULY 6, 1893.   FIVE MONTHS $1.00 TEN CENTS A COPY

The Exhibit of the Weston Electrical Instrument Company.

The exhibit of the Weston Electrical Instrument Company is located in the gallery of the Electricity Building, just east of the main stairway leading to the offices of the Department of Electricity. A square showcase around a square upright glass faced cabinet occupies the center of the space in front of a comfortably equipped office. Arranged around the outside are showcases in which are shown the various instruments and apparatus made by the company. The office itself is provided with current and various forms of instruments are shown in use.

The front side of the office is formed by a large enameled slate switch board furnished by J. T. Murphy, of New York; the enameling on this board is an imitation of the grain of oak, and the effect is quite pleasing. The instru-

ments on the switchboard consist of one ammeter reading to 1,500 amperes, and four reading to 300 amperes, four volt meters and one potential indicator. All are of the regular switchboard illuminated dial type now so extensively made by the Weston company. Ajax switches are used on the switchboards and include six of 300 ampere and one of 1,000 ampere capacity, the latter being double-throw and changing from a three-wire to a two-wire system.

THE EXHIBIT OF THE WESTON ELECTRICAL INSTRUMENT COMPANY.

At the bottom of the board on a slab of onyx enameled slate is a voltmeter arranged as a ground detector. It has two keys for closing on either side of the circuit, the line potential being read on the indicator above, the resistance of the ground is calculated by the following formula. R being the resistance of the voltmeter; V, the total line voltage, and v^1 and v^2 being the readings on the positive and negative sides respectively; thus X, the resistance of the ground

$$\left(\frac{v^2(v^1+v^2)}{v^1}\right)R \text{ for the positive side and } X^2 \left(\frac{v^1(v^1+v^2)}{v^2}\right)R$$

for the negative side. Mr. R. O. Heinrich, who is in charge of the exhibit has made a very thorough study of the detection and measurement of grounds with the Weston instruments and has simplified many of the usual formulae. The company manufactures high grade, portable, and laboratory standard resistance boxes in sizes from 100,000 ohms up. The construction of these boxes is such that the hard rubber top may be removed without disturbing the coils. The alloys used in these boxes which do not require a temperature co-efficient are the invention of Mr. Weston and the fundamental patents were issued to him in 1888, on nickel-copper and manganese nickel-copper alloys.

The central square cabinet contains samples of all the instruments made by the company separated into all the parts used in their construction, even to the smallest screw or jewel. Various samples of standard voltmeters, alternating current voltmeters, ammeters both for laboratory use, testing and switchboard work are shown. A variety of shunts for use with the switchboard ammeters are exhibited. It

Weston instruments may seem expensive to the ordinary central station superintendent, the saving in lamp breakages and other faults by knowing the actual condition of machines and circuits, more than compensates for this cost. For switchboard use in isolated plants and smaller stations they make instruments which, while not having the fine exterior finish found in the standard instruments, are made of exactly the same working parts and are in every way just as reliable as the standard instruments and at about one-half the price. Mr. Richard O. Heinrich, in charge, will be found willing to explain in a very concise and clear manner all the various exhibits, and a chat with him on the use of these instruments is sure to be of benefit to anyone interested.

The Western Electric Company's Scenic Theater in the Electricity Building.

While travelling abroad one of the officials of the Western Electric Company saw a small scenic theater in operation, all the different effects being produced by electric

FIG. 1.—THE WESTERN ELECTRIC COMPANY'S SCENIC THEATER IN THE ELECTRICITY BUILDING.

will be remembered that the Weston ammeter is simply a milli-voltmeter, reading the fall of potential around a shunt. Several samples of laboratory voltmeters and other instruments are shown. A new form of bridge for laboratory work is here exhibited for the first time. All contact blocks are placed underneath the hard rubber top, and the plugs are long and extend through holes in the same. This arrangement prevents dust settling between the blocks and facilitates cleaning to a great extent. All coils in this and the other instruments are wound with the patent composition wire made by this company and which has practically no temperature co-efficient. Various sizes of multiplying coils for use with the ordinary voltmeters for reading very high voltages are shown.

Of other instruments deserving special mention are various samples of the astatic ammeters for use in places where there is a strong magnetic field; an inspector's set including voltmeter, and four adapted plugs for use with the Edison, T. H., Westinghouse, and the old U. S. sockets, together with flexible cord and plugs. While the higher grades of

lighting. The result was so pleasing that when the company was searching for new features for showing the applications of electricity this gentleman expressed a desire that such a theater might be constructed for the company's exhibit at the World's Columbian Exposition, and the matter was placed in the hands of Mr. A. L. Tucker for execution. Dr. Hornsby, the Assistant Chief of the Department of Electricity, had also seen the same thing and was able to add some information to that already obtained. Messrs. Sosman & Landis of Chicago were given the order to make the scenery under the direction of Mr. Tucker, a small model eight or nine inches wide having been completed about February 14th. The scene is supposed to be Swiss, although no special location is given.

As shown in the engraving, a snow capped mountain is in the rear of the scene representing foot hills, in front of which is a lake of real water overflowing into a brook that tumbles over several little falls and emerges under an old Roman bridge in the foreground. At the right is an abrupt crag or precipice surmounted by a castle, with a road winding away from the bridge up the hillside. At the left on the road leading from the bridge and overshadowed by a rugged mountain is a church and several small houses and outbuildings nestling against the hillside. The object of the arrangement is to show by means of electric lighting the varied light of the complete cycle of a day.

Commencing in the evening with the lamps lighted in the houses, the church is illuminated, and the whole scene is enveloped in moonlight. As the moonlight dies out and dawn approaches, the soft grey light first appears and gradually turns to red, orange and yellow as the sun gets farther above the horizon. The lights first strike the mountain peaks and creep gradually down the sides into the valley until the full light of the gorgeous sun is spread over all, being tempered by a bluish light to give the proper atmospheric effect to the scene. After the sun has passed the meridian a storm appears, the first indication being a dimming of the sunlight and a gradual darkening of the sky. As the storm arrives sheet and chain lightening fill the sky, after which

To the electrical engineer the method of accomplishing these wonderful lighting effects is full of interest, and we are pleased to be able to give a description of the apparatus used.

The amphitheatre is about 20 feet broad by 15 deep, the floor sloping toward the stage. Between the curve of the front of the amphitheatre and the proscenium arch is a space of 10 feet in width in which are placed two miniature electric fountains which play constantly and by means of clockwork change colors from red to green, red, yellow, blue and white. A 10 horse power electric exhaust fan is placed under this floor and discharges upward into the room to cool the interior.

The stage is 20 feet wide by 15 deep, the proscenium arch being 12 feet wide by nine feet in height. The sky is represented by a painted canvas hung in a curve on the back wall and being perforated with holes, the 30 16 candle power lamps behind it gives all the appearance of stars in the sky. In front of this scene and in rear of the mountain

FIG. 2.—THE WESTERN ELECTRIC COMPANY'S SCENIC THEATER IN THE ELECTRICITY BUILDING.—THE RHEOSTATS.

the sunlight begins to emerge and a rainbow appears. The ends of the bow are first seen at the sides of the mountain creeping up and finally arching over, becoming as vivid as in nature. The sun gradually increases in brilliancy and the bow disappears in the same mysterious manner in which it came, breaking first in the centre and gradually disappearing as if into the ground. By this time evening has arrived and the sun begins to set, the first evidence of which is a purple glow and the light slowly creeps back up the mountains and changes through the yellows, oranges and reds glowing in the sky and over the mountains, lake and foot hills until it entirely disappears. As darkness approaches, the stars peep out in the sky, twinkling as in nature, the houses are lighted, shooting stars and a comet appear and the Aurora Borealis spreads its glow over all the scene. The cycle is thus completed, taking about 20 minutes. Visitors wait for a long time to get inside the amphitheatre, and this feature is in every way as well patronized as the others of this company's most attractive exhibits.

scene are, first a tin reflector on each side sloping from a higher point in the center downward toward the sides, each containing 15, 16 candle power lamps under a red glass screen. These lights are used to illuminate the sky during the rising and setting of the sun. Just forward of the red lights are two more tin reflectors slanting in the same way and containin 20, 32-candle power blue lamps used for making blue sky.

In front of the mountain and immediately back of the foot hills scene is one long tin reflector containing 24, 32-candle power blue lamps used for tinting the mountains and for producing atmospheric effects.

Placed in reflectors one on each side back of the foreground scene are 10, 32-candle power blue lamps for lighting the sides and foot hills and for the general atmospheric effects. On each side of the proscenium arch is a vertical row of 10, 16-candle power lamps behind red glass for rising and setting sun effects, with a row of 10, 32 candle power blue lamps on either side for producing atmospheric effects on the front scenes. The moon consists of a group of three 16 candle power green lamps placed directly over the arch, over which is a drum about three feet long by about 12 inches in diameter with part of the surface covered with red and orange gelatine for running the sun light down or up the mountains at the proper time.

The sun is made up of a group of 15, 32-candle power lamps on a carriage which travels across the stage from right to left, the color effects being obtained by graduated glass screens on either side running from a deep red down through the lighter reds, oranges and yellows until the clear light appears on the scene, the same being repeated on the opposite side but in reverse order. The border lights are in the usual long inverted reflectors, the rear one having 26, 32-candle power blue lamps; the middle, 20, 16-candle power lamps behind a red glass screen; while the front one for grey effects contains 20, 16-candle power lamps behind ground glass. An extra front border is being put in place for lighting the front of the scenery more effectually. Of course the lamps would be of no effect

The mechanism is simple, very effective, and provided with cut outs which act automatically to stop the screw at any point desired. Drums with cords attached and revolved from the machine shaft work the sun and color drum by watching the progress of the scene the prompter is able to start up the lights at the proper time. The theatre runs very smoothly and requires but one man for its operation.

The company intends to move the theatre into the southeast corner of the Electricity Building early in July, and the present pavilion will then be used as an office by the representatives of the company located at the Fair.

Exhibit of the General Incandescent Arc Light Company.

In contemplating the decorative lighting of interiors one cannot help wondering why the use of the arc lamp for that purpose has been so long neglected. The amount of illumination obtainable, the economy of production and the

THE EXHIBIT OF THE GENERAL INCANDESCENT ARC LIGHT COMPANY.

unless means were provided for turning on the light gradually and at the proper time. For this purpose all circuits from the lamps are led to a frame work placed just north of the theatre on the outside. On this are 11 rheostats, made of cylinders of sheet iron, about 20 inches long by 10 inches in diameter. These cylinders are wrapped with asbestos and wound with german silver wire of the proper size, the lower end being connected to one terminal of the bus bars the other being free. A slide which is made to travel up and down the side of the cylinder is attached to the wire leading to the lamps and cuts in more or less resistance according to its position on the cylinder. Switches and safety fuses are provided for cutting out any circuit.

The slide screws are revolved by means of round belts running over grooved pulleys on a shaft along the frame of the machine and as the belt doubles back to the screw, another grooved pulley above the first can be thrown in by a guide so as to turn the slide screw either up or down thus cutting the resistance in or out.

large field for decorative lighting and effects which needed but a good lamp and experience, would seem to commend its use. In spite of this the same old time shapes of lamps are still in use and it is exceedingly difficult to convince central station managers that a lamp handsome in appearance and ornamental in design is desirable, and can be made financially successful.

The General Incandescent Arc Light Co. seems to have recognized this as shown by the interesting and attractive exhibit, located in section E, Electricity Building, a space 68 feet long by 13 feet wide. It is surrounded by a substantial railing with iron posts supporting a canopy of red and yellow, the Spanish colors, and tastefully draped from post to post. The rear wall is partially covered by a large American flag. A table and comfortable chairs are provided.

The specialty of the company is ornamental arc lamps for every service and any current for which the arc lamp is to-day adapted. The regular series arc lamp with auto-

matic cut-out, lamps for low tension multiply circuits either
continuous or alternating current; railway lamps used 10
in series on 500 volt circuits; long lamps with rack feed
and short lamps with ribbon or chain feed are all shown in
beautiful designs and suspended in many different ways.

The company's standard arc lamp is of the rack feed
type, plain in finish and when for use out of doors is provided
with a very neat protecting hood inside of which the extra
resistance is placed when used in multiple. The resistance is
separate when the lamp is used for interior lighting and is
placed in some convenient place. This resistance is of
German silver wire wound on an iron frame having porcelain
covering where wires touch and a cover of sheet iron over
all. These lamps are ornamented in many different ways,
by an extra finish of brass; in dead black or in a steel
finish. Handsome designs in fancy brass and iron work
are shown, and also of designs suitable for any place. For
places where it is not desirable or convenient to use the
standard lamp, the bijou or short lamp is provided. This
lamp has a so-called ribbon feed, the ribbon being formed
of fine brass or composition wires braided together thus
forming a very strong and flexible suspension as well as a
good conductor to the positive carbon. The works are
otherwise precisely the same as in the rack feed lamp with

THE BRYANT COMPANY'S EXHIBIT.

the exception of the drum which holds the cord. These
lamps are only 30 inches long over all, and are made in all
the standard sizes of four, six, eight, ten and 20 ampere
current capacity, and are subject to the same degree of
ornamentation as the standard lamps. Another new feature
is the use of side brackets and electroliers for the hanging
of small lamps, samples of the latter for two lamps and six
lamps being shown; by this means the small arc lamp
should be not only very effective for illuminating purposes
but can be made highly decorative.

The lamp mechanism is very simple in construction,
having but one electro magnet for feeding, the lifting of
the upper carbon to draw the arc being accomplished by
means of a spring, which may be tightened or loosened to
increase or decrease the length of arc. The side rods are of
small brass tubing, the lower carbon holder being made of
a small split tube with a conical screw on the end which
enters the lower base framed from the bottom. It can thus
be removed with the lower carbon attached without disturbing the globe, a small feature but one that will be much
appreciated by the practical central station man. All parts
of the lamp are thoroughly insulated from the circuit excepting those holding the carbon and electrical contacts.

This company is a special advocate of arc lamps on the
low tension continuous currents now so largely supplied in
our large cities at from 110 to 220 volts. They state as
reasons for it, the simplicity of installation, safety and reliability of service, and great range of candle power as 500,
700, 1,000 and 1,500 or 2,500 candle power lamps can be
taken from the same circuit, thus supplying any kind or
quality of illumination that may be desired. It is said that
many of the local Edison companies have some hundreds of
the lamps already in use and are increasing their orders.
The company's factory is at 53rd street and First ave.,
New York city and its western office is 169 Adams street,
Chicago. Mr. S. Bergmann being the President of the
company is in itself an assurance that the ornamental
features will receive the best of attention. Mr. Phillip Klein
is the Secretary.

The Bryant Company's Exhibit.

Every electrical man who attends the fair becomes familiar with the sight of the Bryant D. P. switch, as all the
incandescent lights throughout the fair grounds are controlled by the Bryant switch. The company's exhibit,
which is in the gallery of the Electricity Building in the
southeast corner, however, gives an idea of the great variety of specialties which are manufactured by this company.

Plug cutouts, horseshoe cutouts, of white porcelain, key
and keyless sockets, polished yellow brass Paiste S. P. fire
and ten ampere and three way switches, also the double
pole Bryant switch with shining nickel caps, together
with branch and main line cutouts of various kinds and a
beautiful assortment of finely decorated china switches, all
grouped in a most artistic manner on the black back ground
forms a picture interesting and pleasing to the eye. The
arrangement of these appliances in this manner is so
striking that visitors pause to examine them. Even the
railing that surrounds the exhibit is constructed of the
Bryant specialties.

The harmony of color and the beautiful arrangement of
these useful devices show that artistic talent exists in a
marked degree among some of our electrical fraternity.
Mr. Edward R. Grier, who is associated with Mr. Thos. G.
Grier in the management of the western office, personally
arranged and erected the entire exhibit. The reputation
of the Bryant company and the merits of its goods are both
well established and so well known here to the electrical
trade they need no comment. Their western office is located in the Monadnock Block.

A very pleasant event happened on June 24 in the
department offices of the Electricity Building. It was
the birthday of the genial chief of the department,
Prof. Barrett, and his friends, the press representatives
from newspaper row presented him with a handsome floral
emblem in the conventional shape of a magnet. Mr. M. J.
Sullivan made the presentation speech in a very short and
appropriate manner, but the Professor was too surprised to
frame a reply.

ELECTRICAL INDUSTRIES has endeavored always to keep in
the foremost rank in giving its readers the latest and best
information on current subjects, but never has it received a
more graceful compliment than it did the other day when
a representative discovered one of its contemporaries using
the map of Electricity building, published in the Weekly
World's Fair Supplement, as a guide in locating the different exhibitors. The compliment is appreciated.

ELECTRICAL INDUSTRIES.

Entered at Chicago Postoffice as second-class mail matter.

PUBLISHED EVERY THURSDAY BY THE

ELECTRICAL INDUSTRIES PUBLISHING COMPANY,
INCORPORATED 1892.

MONADNOCK BLOCK, CHICAGO.
TELEPHONE HARRISON 185.

E. L. POWERS, PRES. AND TREAS.   E. E. WOOD, SECRETARY.

E. A. POWERS   -   -   -   -   -   EDITOR.

B. A. FOSTER,   }
                                                   } ASSOCIATE EDITORS.
W. A. REMINGTON, }

E. E. WOOD,   -   -   -   -   EASTERN MANAGER.

FLOYD T. SHORT,   -   -   ADVERTISING DEPARTMENT.

EASTERN OFFICE, WORLD BUILDING, NEW YORK.
World's Fair Headquarters, Y 27 Electricity Building.

SUBSCRIPTION:

FIVE MONTHS $1.00
SINGLE COPY   10

Advertising Rates Upon Application.

News items, notes or communications of interest to World's Fair
Visitors are earnestly desired for publication in these columns and will
be heartily appreciated. We especially invite all visitors to call upon us
or send address at once upon their arrival in city or at the grounds.
ELECTRICAL INDUSTRIES PUBLISHING CO.,
Monadnock Block, Chicago.

THE Fourth of July was most fittingly celebrated at the
Fair. Special efforts were made, especially in the Electricity building to make the exhibits look as attractive as
possible. The grounds and buildings were most elaborately decorated, and everything had a holiday appearance.
Although a program was prepared including speeches,
parades and music, the electric fountains and fireworks
were the great attractions. The attendance exceeded
300,000.

THE presentation to Prof. J. P. Barrett, Chief of the
Department of Electricity, a few days ago, of a handsome
floral design by the electrical press, on the occasion of his
birthday was a most pleasant occurrence. This compliment,
however, only reflects the sentiment and high esteem in
which he is held by the entire electrical fraternity. Prof.
Barrett is without doubt one of the most popular chiefs of
any department of the Fair, and his efforts in behalf of the
electrical interests to act impartially and bring about a display of credit to these important interests are certainly appreciated. ELECTRICAL INDUSTRIES most heartily unites
with all his friends in wishing him many more and pleasant
birthdays.

THE various outdoor attractions, the fountains, the bands,
the boats and the attractions of Midway Plaisance seem to
draw the evening crowds. With the exception of the Electricity building, the buildings have very few visitors during
the evening, so few that the exhibitors do not consider it to
their advantage to be with their exhibits nor to place
attendants in charge. A program is being discussed by
the chiefs of the departments, and the management for
keeping open during the evening the Electricity building
and one other building, taking the buildings in order.
If this program is adopted, as it now has the appearance
of being, the attendance at the Electricity building will be
greatly increased, much to the advantage of the exhibitors.

The Jury of Awards in the Department of Electricity
has not as yet been announced. It will include from 21 to
31 members, as the occasion and necessity of the situation
demand. Of this number from 6 to 8 will be selected from
the eminent foreign electricians. The jury will be divided
into 9 committees, one for each section in the department.
A meeting of the committee will be held to determine the
method of conducting examinations and tests. The actual
work of each section will be done by one or more members
of the committee forming a sub-committee. The report
will be made by the sub-committee to the committee of the
section by whom it will be referred to the Department Jury
of Awards for consideration. From the Department Jury
of Awards a report will be referred to the Executive Jury
of Awards, from whom it will pass to the National Commission, who will order the awards to be made by the secretary
of the treasury. A bronze medal, accompanied by a diploma,
stating the points of excellence and advancement, will be
awarded to those selected by the jury.

At noon, on July 1, and entirely without ceremony of
any nature, the decorations by the Westinghouse company
on the south end of the Electricity building were unveiled.
Over the top, following the curve of the arches of the
building, is the name of the company. Immediately underneath is a large painting of the head of Columbus, large
handsome scrolls being painted on either side. Lower
down are the dates, 1492 on the left, and 1892 on the right,
with the name, Columbus, directly underneath the picture.
All these letters and figures, which are painted on the wall
with a back-ground of terra cotta, are filled in with incandescent lamps for evening illumination. There are in all
1988 16-candle power lamps used in this decoration, 580
being for the head of Columbus, 588 in the scrolls, 260 in
the name and dates, and 560 for the sign of the company.
These lamps are of different voltages, varying from 85 to
110 in order to get the proper shadings. Owing to the
hurry to get ready for the celebration on the 4th, white
lamps will be used at present, but eventually colored lamps
will take their place in order to produce the best effects.

A fine exhibit of leather belting is shown by A. Domage
of 74 Boulevard Voltaire, Paris, in the French section of
Machinery Hall. Specimens of single and double belting,
round twisted and link belting, which by the way differs from
ours in having double links, that is, two of the ordinary links
side by side. A 32-inch belt is shown made of strips of
leather cut about three-quarters of an inch wide sewed
together so that the edge of the leather is the bearing surface. This appears like a very stiff and strong belt,
but must be very expensive. A frame is shown containing
samples of leather of different quality and shape and
samples of belting are arranged at the sides in ornamental
forms.

The electric light buoys from Van Buren street wharf to
the World's Fair Pier were started Saturday evening, July
1st. The light is said to be very brilliant and highly satisfactory to the lake pilots.

The Lamp Case at Milwaukee.

Hearing in the suit for injunction brought by the General Electric Company against the Electric Manufacturing
Co. of Oconto, Wisconsin, was begun in the U. S. District
Court at Milwaukee, Monday the third; the court adjourned
over the fourth to Wednesday morning. The Oconto Company is represented by its local counsel, Mr. Webster and
Messrs. Witter & Kenyon, who so ably defended the St.
Louis case. The General Company has its full force of
legal talent, including Mr. Fish.

WORLD'S FAIR NOTES.

The Western Electric Co.'s scenic theater is attracting considerable attention, and it is especially amusing to listen to the comment of children who eagerly watch the changing effects. One little tot, as the shadows followed the fading light, inquired anxiously: "Is it going to rain, mamma?"

Saturday, July 1st, was Dominion Day at the Fair. The programme included meeting at the Canadian pavilion at at 2 P. M.; a procession to Festival Hall at 3 P. M. and speeches after that hour by the Canadian Commissioner and others. Music was furnished by Tattersall's military and other bands.

The Electric Heat Alarm Co.'s exhibit is worthy of inspection. If it be visited on a sunshiny day, Mr. E. Nasbold, the manager of the Chicago office and who is in charge of the exhibit, will show the visitor how sensitive are the thermostats used by the company by simply exposing one of them to the sun's rays for a few seconds.

Albert & J. M. Anderson, Boston, have an interesting exhibit of electric railway, light and power specialties in the Transportation building, showing, among other goods, the Ajax switchers and lighting arresters manufactured for C. S. Van Nuis of New York. Mr. T. A. Matthews is in charge of the exhibit for a New York exhibitors' association.

The Western Electric Company added still another feature to its already large exhibit, last Saturday, in the shape of a large focusing lamp with the front lens concealed by a disc formed of richly colored pieces of cut glass, with the name of the company inlaid in glass of a different color around the edge. It is located just north of the scenic theatre and is destined to be one of the evening attractions.

A great deal of interest is manifested in the electric welding of small samples of metal, shown by the attendant at the exhibit of the Thomson Electric Welding Co. Short pieces of iron or steel wire are almost instantaneously joined by the automatic welder. On the opposite side of the building the electrical heating of metals for working, shown by the Electrical Forging Company attracts a large crowd whenever any of the machines are in operation.

One elderly lady approached Mr. Fox under the Tower of Light and inquired if the surrounding exhibits were those of the General Electric Co. Upon being assured they were, she watched the pumps, gazed on the drilling machine with a puzzled expression, and finally her eyes wandered to the illuminated monogram in colored light in Section B. Her face lit up and she exclaimed in pleased surprise: "How kind of them! C. E., Christian Endeavor"!

The Miyoshi Electric Works, of Tokyo, Japan, have just commenced the installation of an exhibit in the northeast corner of the Electricity Building, near that of Queen & Co. They already show some very handsome electroliers made entirely of bamboo in various forms; also a number of enlarged photos of views of the effects of earthquakes in Japan. The company was established in 1883 and manufactures dynamos, motors, telephones, telegraph and various other electrical material.

A very finely finished alternating current dynamo of the Zipernowsky pattern is exhibited in Machinery Hall, along with other products of the firm of Schneider et Cie of Creusot, France. At 500 revolutions the capacity is stated as 25 amperes at 2,000 volts. The shape and design is very much different from those in this country and the machine

would have deservedly received much more notice in the Electricity Building than it does now, but away behind a lot of high power cannon and revolving turrets.

One of the finest exhibits of electrical fixtures ever placed on view is that of the Archer & Pancoast Manufacturing Co., New York, located in the New York State Building. The two $15,000 electroliers in the Ball Room contain 72 lights each of 16 candle power. Four 12 light gold plated standards valued at from $2,500 to $3,000 each are also to be seen in the Ball Room. Scattered throughout the building are 11 six light standards of artistic design. A 56 light electrolier illuminates the main stairway, while over the main entrance is a 40 light gold lantern that has already been sold to Chicago parties.

Among new steam engines at the Fair is one that has not heretofore been seen in the East. In section M, No. 13, of Machinery Hall is an engine exhibited by the Golden State & Miners Iron Works, of San Francisco, Cal. This engine, the invention of I. F. Thompson, is known as the slide-valve Corliss, and the leading features are said to be one eccentric; all flat slide valves; exhaust valves fixed; steam valves quick opening and closing; and the fewest joints to accomplish the above results of any engine yet produced. The arrangement of lugs, dashpots, governor and other small detail is peculiar and deserve the great deal of attention from engineers the engine is now receiving.

The Javanese Village, Midway Plaisance, was the scene the other day of a most interesting series of experiments with the electrical current. The attendant was replacing burned out lamps and had one native assistant, to whom he tried to explain the nature of the current. First wetting his finger the instructor placed it in the socket, being careful not to make a connection. The little man from Java put in his finger, minus the water. No effect. Then he wet it and tried again. The standing jump he made would have turned a college athlete green with envy! Not a word of complaint from the bright-eyed little foreigner, but the attendant had no sooner left than another native was initiated into the mysteries of electricity by the original investigator! And so it went; each in turn bringing another, till now it would be hard to find a more "highly charged" community.

The Wm. Powell Company, of 225 Spring Grove Ave., Cincinnati, has a fine exhibit of engine and sight feed lubricators in the southwest gallery of the Electricity Building. There are grease cups for all purposes, for locomotive and for street car motors, special sight feed oilers for triple expansion engines, for large dynamo bearings, in fact oilers for most places were lubricating oils are used. A fine exhibit is also made of Powell's patent regrinding seat globe valve. Not wishing to be inconvenienced by using oil for showing the working of these cups, the Powell Company has installed a small induction coil plant for driving sparks between two points inside the various sight glasses. This plant consists of a half horse power Jenney motor, belted to a small generator of the same capacity, this in turn runs a large induction coil, the spark from which produces the required effect in the oilers. As none of the regular spring make and break vibrators would do for the heavy currents used in this device Mr. Powell has constructed a very ingenious make and break, by the use of a mercury cup. The three cabinets of ebonized wood, trimmed in gilt, with curved glass fronts, show off to good advantage against a background of heavy dark curtains draped against the back of the space. It is intended to show some further electrical effects as soon as the apparatus can be got ready.

PERSONAL.

R. M. Bayles, of New York, formerly of the C. & C. Motor Co., is a visitor at the Fair.

Mr. W. G. Brixey and family, of Ansonia, Connecticut, is spending a couple of weeks at the Fair.

Mr. G. T. Williams, formerly Supt. of the Nickel Plate Railroad telegraphs attended the Fair a few days last week.

Mr. C. F. Scott, electrician of the Westinghouse Electric & Mfg. Co., was in Chicago for a few days at the Fair last week.

Mr. J. H. Rhotehamel, of the Columbian Incandescent Lamp Co., accompanied by Mr. Henry Gobel, visited the Exposition this week.

Mr. D. H. Dorsett of Jamestown, New York, and a pioneer in underground conduit construction was at the Fair during the last of June.

Prof. and Mrs. Thos. French, of the University of Cincinnati, are in town for a few days, and will return later for the electrical congress.

Mr. G. R. Schallenberger, consulting electrician of the Westinghouse Electric & Manufacturing Co., with his wife visited the Fair last week.

C. M. Lungren, of the Siemens-Lungren Lamp Co., is doing some work in the Department of Electricity for Appleton's Popular Scientific Monthly.

Mr. F. H. H. Paine of St. Louis, and the representative of Siemens & Halske Co., in that city was registered at the Institute headquarters last week.

Mr. Edwin F. Morse, of Morse, Williams & Co., Philadelphia, who have an exhibit of electric elevators in the Transportation Building, was in town last week.

Prof. W. E. Ayrton and daughter arrived in New York on the Germanic a few days ago. After a brief stay in New York they will come to Chicago for six weeks.

Prof. A. Macfarlane, of the University of Texas, stopped in the city for a day or two last week. He will return again later to attend the Congress of electricians.

Mr. Henry A. Reed, general manager of the Electrical World, arrived in Chicago on Thursday of last week and is spending a few days at the Fair.

Mr. Nelson W. Perry, editor of the Electrical World, was married to Miss Marie Eugenia Bedell, at Brooklyn, N. Y., June 28. ELECTRICAL INDUSTRIES extends congratulations.

Mr. James G. Kaelber, the Rochester agent of the Western Electric Company, is visiting the Fair this week, in company with J. S. Hays, electrician of the Carnagie Steel Company.

Mr. S. K. Bullard, of Sedalia, Mo., Supt. of telegraphs for the M. K. & T. Railroad is in the city with his family on his way home from the meeting at Milwaukee. He will spend a few days visiting the Fair.

OBITUARY.

Mr. Wm. Stanley Sr., the father of Wm. Stanley Jr., died at his summer house in Great Barrington, Mass., June 28. He was a well known lawyer, and at the time of his death was 65 years of age.

NEW PUBLICATIONS.

Price list Number 6 has just been issued by the Okonite Company, 13 Park Row, New York, containing complete price lists of its insulated wires and cables. The wires and cables manufactured by this company include those adapted for high and low tension currents, aerial and sub-marine telegraph and telephone cables, electric light and railway wires insulated by Okonite and Candee insulation.

The new catalogue of the American Institute of Electrical Engineers has been received. It is corrected to June 1st, 1893, and contains alphabetically arranged lists of the members and associates together with a geographical distribution of the same. Names of past offices, rules of the Institute and a calendar of the dates of meetings for the ensuing year are also included. The book is marked with the cut of the new badge which will hereafter be found on all literature of the association. The membership has increased greatly in the past two years and doubtless many new names will be added during the Fair.

BUSINESS NOTES.

The Phoenix Glass Co. furnished the globes used by the Westinghouse Company in lighting its Kiosks.

THE WESTERN ELECTRIC COMPANY, Chicago, has been awarded the contract for wiring the Ferris wheel for incandescent lights.

Notice was given Saturday, July 1st, that a semi-annual dividend of four per cent has been declared by the Great Western Electric Co., and is payable on and after July 1st at the office of the Company, 207 South Canal street, Chicago.

The W. S. Edwards Mfg. Co., Chicago, has recently completed contracts for gas and electrical fixtures for the following buildings: Hotel Edinburgh, the Kedzie Building, the Teutonic Building and a new Government building at Fort Sheridan.

The American Electrical Works, Providence, R. I., is presenting its friends with a harmless piece of fire works which calls attention, not only to the national holiday, but also to the insulated wires, electric light and Faraday cables and other goods manufactured by this company.

THE ELECTRIC APPLIANCE COMPANY has been working for some time in the telephone line with the idea of getting up a first class non-infringing electric telephone. They have at last succeeded in securing an instrument that is satisfactory to themselves, which is a sure guarantee that it is a first class instrument. They promise to proceed at once to make a few ripples in the telephone puddle.

Amusements.

HOOLEY'S THEATER—Mr. E. S. Willard, in "The Professor's Love Story." 149 Randolph street.

COLUMBIA THEATER—Miss Lillian Russell, in "La Cigale." Seventh week. 108 Monroe street.

GRAND OPERA HOUSE—Sol Smith Russell, in "April Weather." 87 Clark street.

AUDITORIUM—Imre Kiralfy's Spectacle "America." Congress street and Wabash avenue.

McVICKER'S THEATER—Denman Thompson, in "The Old Homestead." 82 Madison street.

CHICAGO OPERA HOUSE—American Extravaganza Company, in "Ali Baba," or "Morgiana and the Forty Thieves." Sixth week. Washington and Clark streets.

SCHILLER THEATER—Chas. Frohman's Stock Company, in "The Girl I Left Behind Me." Sixth week. Randolph, near Dearborn.

HAVERLY'S CASINO—Haverly's United Minstrels. Wabash avenue, near Jackson street.

HAVLIN'S THEATER—"The Cracker Jacks." Wabash avenue, near Eighteenth street.

TROCADERO—Concert. Michigan avenue near Monroe street.

THE GROTTO—Vaudeville. Michigan avenue near Monroe street.

Buffalo Bill's "Wild West." 63d street.

Pain's "Siege of Sebastopol." 60th street and Cottage Grove avenue.

One of the best known advertisers in the trade writes us as follows: "We were well impressed with the idea of your weekly issue and more so with the way you carried it out. As a bright and timely enterprise it certainly deserves a generous patronage."

For a new sensation try a camel ride.

The donkey drivers in the Cairo street do not need electric lights to see a dime.

Central Station and Isolated Plant Astatic Ammeter (One-Half Size).

The Weston Standard Portable Direct Current Voltmeters and Ammeters are used in all parts of the Civilized World.

They are recognized by all authorities as the very best instruments ever produced.

During the past year Mr. Weston has given much time to the further study and improvement of these instruments, and has been led to make many changes of great accuracy, durability, accuracy and excellence than were ever thought possible. The New Model is to raise its electrical superiority to the such best is vastly superior to the latter as all that constitutes this first-class instrument of them the same of perfection, and is absolute to become and to last.

THE STANDARD PORTABLE INSTRUMENT OF THE WORLD.

We have perfected a larger model of these instruments especially adapted for use as absolute standards in stations and laboratories.

WESTON PORTABLE ALTERNATING CURRENT VOLTMETERS.

These instruments are absolutely permanent and very accurate, and are completely superior to any form of hot wire instrument known. They will be found invaluable in recording and measurement uniformly and economy of operation of alternating current points.

We make multipliers for use with these instruments to extend their range so as to measure the highest voltage used with any primary circuit.

We also make voltmeters of different ranges of readable period, both read.

DEAD-BEAT SWITCH-BOARD AMMETERS. (See illustration.)

The capacity from 5 to 2500 amperes to 750 amperes. These instruments are a new type, and are very accurate and well made. The scales are very regular, the instruments very sensitive, thoroughly durable and absolutely permanent, and the making of one sir and sensitive affecting the most powerful electro-motive fields. In addition, we are designing a full line of Dead-beat ammeters.

STATION AND ISOLATED PLANT VOLTMETERS.

in three ranges and many ranges. These instruments are very dead-beat, and can have a great new and standard testers/ballasters.

They are especially adapted for railway and power plants and are well adapted for all sorts of instrument, light circuits, and all other work requiring good and thoroughly trustworthy Voltmeters. They are especially designed for work of kinds.

We are also making a fine line of high grade

STATION AMMETERS.

in ranges from 5 to 2500 amperes with absolutely proportional scales through out the entire range. The sensibility and accuracy of these instruments have never been approached. A 15 ampere instrument at 5 inches dia. with full load can be read. The same range of a a single range and can be read in any part of the scale to the scale. They are very dead beat, handsome in design and finish, permanent, durable and will undershine under use in any.

SEE OUR ELABORATE EXHIBIT AT WORLD'S FAIR S. E. CORNER SECOND FLOOR, ELECTRICITY BUILDING, WHERE OUR REPRESENTATIVE WILL BE PLEASED TO RECEIVE VISITORS.

THE WESTON ELECTRICAL INSTRUMENT CO.
114-120 William St., Newark, N. J., U. S. A.
CORRESPONDENCE SOLICITED.

ELECTRICITY BUILDING—EXHIBITORS AND THEIR LOCATION.

GALLERY.

MAIN FLOOR.

Exhibitor.	Section.	Exhibitor.	Section.	Exhibitor.	Section.	Exhibitor.	Section.
Austria	Y	Electrical Review	Y	Jaeger, Chas. I.	T	Reliance Gauge Co.	T
Ansonia Electric Co.	Z	Electricity	Y	Johns Mfg. Co., H. W.		Rossiter & Hasslacher Chem. Co.	S
Am. Inst. of Elec. Eng.		Electric Gas Co.	R	Jewell Belting Co.		Direct Railway Journal.	
American Battery Co.	T	Electrical Engineer	Y	Jenney Elec. Motor Co.	x	Strowger Aut. Telph. Co.	
Ashton, R. M.	Y	Electrical World	Y	Knapp Electrical Works	T	Standard Paint Co.	T
Alg. Elec. Gesellschaft	D	Edgly Electric Motor Co.		K. & P. Elec. Novelty Co.		Sprehulis, C. F.	
Bates Mfg. Co.		Excelsior Electric Co.	B	Knapp & Berkley		Star Iron Tower Co.	W
Bryant Electric Co.	R	Electrical Forging Co.	D	Kennedy Electric Co.		Spahr.	S
Billings & Spencer		Equitable Protector Co.		Lawson, B. A.		Schuren, Chas. A. & Co.	P
Brixey, W. R.		Alektros Mfg. Co.		LeClanche Battery Co.	Y	Schomberg & Sohle	
Bishop Motor Co.	K	Electrical Conduit Co.		McNeil Tinder Elec. Co.		Siemens & Halske	
Bell Telphone Co.		England	P	Marcus, W. N.		Schuckert & Co.	
Brush Electric Co.	L	Empire China Works	S	Mocker, Jr. G.	N	Short Electric Co.	
Caldwell El. Cloth Cut. Mch. Co.	Y	Franklin Elec. Appliances		McIntosh Bat. & Opt. Co.		Sperry Elec. Railway Co.	
Conrad. Elec. Storage Co.		French Piano Exhibit		Munson, C. Belting Co		Standard Underg. Cable Co.	
Cutter, George	T	Felton & Guilhaume		Mather, A. C.	K	Standard Electric Co.	
Canton Elec. Co.		France	R-P-J	Mailer Electric Co.	N	Simpson Battery Co.	S
Chicago Elec. Mnte Co.		Ft. Wayne Elec. Co.	M	Newman Clock Co.		Tate Aut. El. Signal Co.	Y
Copenhagen Fire Alarm Co.		Gash & Co., N. C.		Non-Magnetic Watch Co.		Todd, Applegate Co	Z
Central Electric Co.		Gamewell Fire Alarm Co		N. Y. Insulated Wire Co.		Taylor, Goodhue & Ames	A
Commercial Cable Co.		General Electric Co.	B-H-N-C-J	National Carbon Co.		Thomson Elec. Welding Co	O
C & C. Elec. Motor Co.	x	General Incand't. Arc Li. Co.		Norwich Ins. Wire Co.		Tobin Spring, Kiehn Gray	W
Cleveland Elec. & Mfg. Co.	A	Greely, E. S. & Co		North Am. Phonograph Co.	Y	Union Electric Co.	
Chicago Belting Co.		Siemsald		N. Y. & I. E. A.		Vetter J. C. & Co.	Y
Delaney Clock Co.		Babbit, Wm. J. & Co.		Nat. Aut. Fire Alarm Co.		Webb, G. F.	Y
Department of Electricity	H	Barleman, C. J.		Nat. Engraving Machine Co.	A	Weston El. Instrument Co.	
ELECTRICAL INDUSTRIES		Hart & Bremaus Mfg. Co.		Owen, Dr. A.		Washburn & Moen Mfg Co.	
Ele. Launch & Nav. Co.		Hope Elec. Appliance Co.		Phoenix Glass Co.		Western Union Tel. Co.	V
Electric Separator	T	Hall, Chas.		Paiste, M. T.		Ware & Bartlett Mf'g. Co.	
Edgerton, E. M.	T	Hoffner, N. S.	K	Polytechnische Gdis. Co.		Witte, S. S., Dental Mfg. Co.	T
Elgin Telephone Co.		Hartmann & Braun		Pasquelly, J. K.		Western Electrician	
Edison Elec. Mfg. Co.		Hansen & Van Winkle		Prott El. M-d. Sup. Co		Wilder Aut. Hespeler Al. Co.	
Enterprise Elec. Co.		Heys, J. M.		Powell, Wm. A. Co.		Western Electric Co.	
Eureka Temp. Copper Co.	I	Houtlmuth, F. & Co.		Phelps, A. H.		Westinghouse El. & Mfg. Co.	B-H-J
Electric Appliance Co.		Hilmate Alloy Co.		Page Belting Co.		Wheel Bradfeld	
Elec. Sel. & Sig'l Co.		Internat. Okl. El. & P't Co.	S	Queen & Co		Wing, L. J. & Co.	
Electric Heat Alarm Co.	Y	India Rubber Comb Co.	S	Ritzler, F. A.	R	Zucker & Levett Chem. Co.	P

STILL FORGING AHEAD

Every one takes a certain pleasure in watching a meritorious article grow in popularity and fill an ever increasing field of usefulness. It is undoubtedly this appreciation of a good thing that pushed the sales of the

PACKARD LAMP

up to the present high water mark. And the good work still goes on.

The reputation of the lamp is its constant advertisement. **USE THEM**

Electric Appliance Company, 242 Madison St.
ELECTRICAL SUPPLIES. CHICAGO.

The "BUCKEYE" Lamp

A KNOWN Quantity
No EXPERIMENTING
Low Operating Cost
Improved Lighting Effects

Life 1000 to 4000 Hours
Western Deliveries can be made from Chicago Stock

"The **BUCKEYE** sets the pace"

THE BUCKEYE ELECTRIC CO., Cleveland, O.

Chicago
437 THE ROOKERY

New York
49 DEY STREET.

THE MATHER ELECTRIC CO,
MANCHESTER, CONN.

Dynamos, Motors, Generators,

Offices, 116 Bedford St., BOSTON.

—AND—

1002 Chamber of Commerce Bldg., CHICAGO.

THE "NOVAK" LAMP.

CLAFLIN & KIMBALL (Inc.)

General Selling Agents.

116 Bedford Street, BOSTON.

1002 Chamber of Commerce Bldg., CHICAGO.

Enterprise Electric Company

307 Dearborn Street
Chicago . . . .

N. I. R.
GENERAL WESTERN AGENTS

Manufacturers' Agents and Mill Representatives for

Electric Railway,
Telegraph, Telephone and
Electric Light

SUPPLIES OF EVERY DESCRIPTION

Agents for Cedar Poles,
Cypress Poles, Oak Pins,
Locust Pins, Cross Arms, Glass
Feeder Wire, Insulators,
WIRES, CABLES, TAPE and TUBING

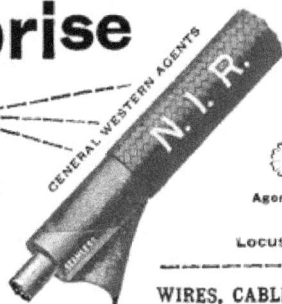

Map of Chicago.

Showing Location of its Electrical and Allied Business Interests, Principal Hotels, Theatres, Depots and Transportation Lines to the World's Fair Grounds. (Index numbers refer to the black squares.)

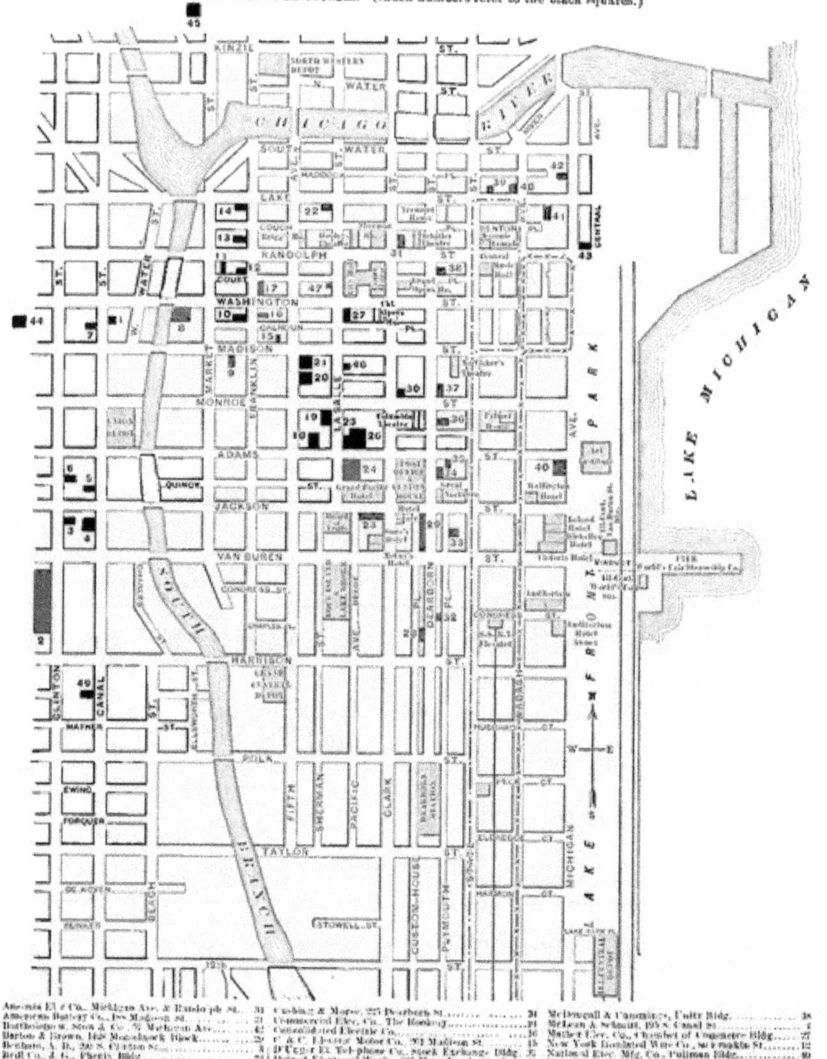

~Lundell~
Suspended Fan Outfit

Black
Japan
Finish

Self-Oiling
....and
Self-Aligning
Bearings

Electric Fans   **Electric Fans**

OKONITE WIRES
OKONITE === TAPES === MANSON
INTERIOR CONDUIT.

Batteries, Bells, Push Buttons, Annunciators, Volt Meters, Ammeters, Wheatstone
Bridges, Line Wire Cross Arms, Brackets, Pins, Insulators, Tools.

GENERAL SUPPLIES.

CENTRAL ELECTRIC CO.
116-118 Franklin Street,
CHICAGO, ILLS.

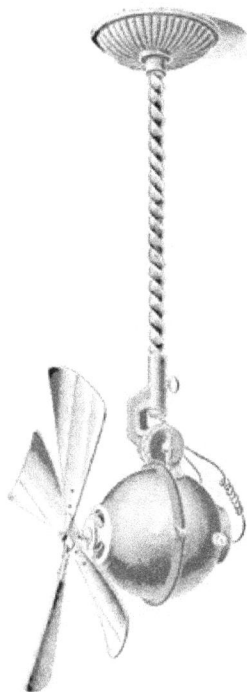

CLARK ELECTRIC COMPANY, NEW YORK.

192 Broadway and 11 John Street.

MANUFACTURERS OF ARC LIGHTING APPARATUS FOR EVERY PURPOSE A SPECIALTY.
The CLARK ARC LAMPS for use on EVERY CURRENT, have the reputation of being the best and most durable of any ever made in the United States.

RAWHIDE PINIONS FOR ELECTRIC MOTORS

A SPECIALTY.

RAWHIDE DYNAMO BELTING

Greatest Adhesive Qualities.   A Non-Conductor of Electricity.
Causes Less Friction than any other Belt.

THE CHICAGO RAWHIDE MANUFACTURING CO.

THE ONLY MANUFACTURERS IN THE COUNTRY.

LACE LEATHER ROPE
AND OTHER RAWHIDE

GOODS
OF ALL KINDS
BY KRUEGER'S PATENT

This Belting and Lace Leather is not affected by steam or dampness; never becomes hard; is stronger, more durable and the most economical Belting made. The Rawhide Rope for Round Belting Transmission is superior to all others

75 Ohio Street,   CHICAGO, ILL.

The REGULAR MONTHLY EDITION of ELECTRICAL INDUSTRIES is the most complete Electrical Journal published— every issue containing descriptions of all the new applications of electricity, complete directories of the Manufacturers and Dealers, the Electric Lighting and Railway Companies in North America, revised and corrected to the date of going to press. These special features are found in no other Electrical Journal in the world, and consequently it is read by more actual buyers than any other publication, which fact makes it without a superior as an advertising medium.

ELECTRICAL INDUSTRIES PUBLISHING CO., Monadnock Block, CHICAGO.

STANDARD ELECTRIC COMPANY.

GENERAL OFFICES: 625 Home Insurance Building.
WORKS: So. Canal Street,

CHICAGO.

STANDARD SYSTEM

AT THE

WORLD'S FAIR.

MACHINERY HALL, Sec. Q, 2 Standard Arc Dynamos.
Sec. S, 20 " " "
ELECTRICITY BUILDING, Sec. P, Space 2, Arc Lighting Exhibit.

The Standard Lamps Light the Power Plant, Machinery Hall, Agricultural Hall, Shoe and Leather Building, and Other Buildings and Portions of the Grounds.

See our Double Service All Night Lamp Before Buying an Old Style Two Rod Lamp.

Western Electric Company,

CHICAGO. NEW YORK.

Arc Lighting Apparatus
　　High and Low Tension,
　　　　Double and Single Service Lamps,
　　　　　All Night Single Lamps,
　　　　　　Theater and Focusing Lamps.

ELECTRIC MOTORS

HIGH SPEED
　　VARIABLE SPEED
　　　　SLOW SPEED.

BUILT FOR SEVERE AND CONTINUOUS SERVICE.
SPECIAL TYPES FOR SPECIAL DUTY.

GEORGE CUTTER 1893 THE ROOKERY CHICAGO

See Our Exhibits:
East Gallery
Electricity Bldg.

And another:
851-853-855
The Rookery.

George Cutter, Chicago.

SIMPLEX WIRES

SIMPLEX — Ever Onward and Upward

INSURE
HIGH
INSULATION

Simplex Electrical Co.
620 Atlantic Ave.,
BOSTON, MASS.

XNTRIC

"That's the Switch"

And we control that movement.

H. T. PAISTE,
10 South 18th St.,
PHILADELPHIA,
PA.

Made 5 amp. S. P.
10 amp. S. P.
5 amp. 3 way.
10 amp. 3 way.

P. & S.
CHINA
SWITCHES

Simplest
Neatest

Made only by
PASS & SEYMOUR,
Syracuse, N. Y.
George Cutter, Chicago.

NEW YORK, 42 Murray St.   PITTSBURCH, 43 Sixth Ave.   CHICAGO. 19-21 Wabash Ave.

PHŒNIX GLASS COMPANY,

World's Fair Exhibit, Center Electricity Building.

Do not fail to call and leave your name for our new catalogue of electric and gas globes and shades.

Consolidated Electric Co.

Manufacturers and Dealers in all kinds of

ELECTRICAL . SUPPLIES,

115 Franklin Street,

CHICAGO.

CHAS. A. SCHIEREN & CO.

MANUFACTURERS OF

Genuine Perforated Electric Leather Belting.

46 So. Canal Street,   -   CHICAGO.

Section 15, Dpt. F, Clm. 27.
MACHINERY HALL.

Section D, Space 3.
ELECTRICITY BUILDING.

...SEE AD...

Weston Electrical Instrument Co.,

PACE 9.

WAGNER ELECTRIC FAN MOTORS

For Direct or Alternating Currents.

These motors give a stronger breeze with less consumption of current than any other fan motor on the market. They are full 1-8 horse-power. Six bladed 12-inch fan. Self-oiling. Furnished with or without guards.

IT WILL PAY YOU TO SEE THE WAGNER BEFORE BUYING ELSEWHERE.

TAYLOR, COODHUE & AMES,

348 Dearborn Street, CHICAGO.

See Our Exhibit of ELECTRICAL FIXTURES

IN SECTION "N", BETWEEN COLUMNS 62 AND 64, MANUFACTURES BUILDING.

CLOBE LIGHT & HEAT CO., 52 & 54 Lake St., CHICAGO.

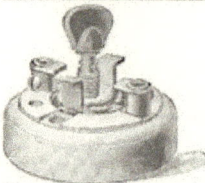

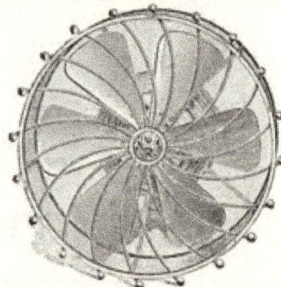

Weekly World's Fair Supplement.

Electrical Industries

DEVOTED TO THE ELECTRICAL AND ALLIED INTERESTS OF THE WORLD'S FAIR.
ITS VISITORS AND EXHIBITORS.

Vol. 1, No. 5. CHICAGO, JULY 13, 1893. FIVE MONTHS $1.00
TEN CENTS A COPY

The Siemens & Halske Central Station Exhibit.

During the past two or three years there has been a strong tendency towards the direct combination of dynamos and engine on one shaft and foundation. This has been connected dynamos and compound engines and this station has grown until now there are 20 of these direct connected machines of the same size and type as the one shown in this exhibition. On the live wire systems, as shown here, stations have been constructed or are under contract in the follow

FIG. 1.—THE SIEMENS & HALSKE CENTRAL STATION EXHIBIT.

noticeable especially in the heaviest machines of the railway power house and it has grown in favor with central stations and isolated plants.

As early as 1887 the Siemens & Halske Co. commenced the equipment of the Berlin central station with direct

ing cities abroad. Two in Vienna, one in Sector Clichy, Paris, one in Trient, one in Rotterdam, and another now being built at Cape Town, South Africa. It will be seen that up to the present date European engineers have given more attention to this class of apparatus than those of our own country. The plant here shown is part of the exhibit of Siemens & Halske Electric Co., being located almost in the center of Machinery Hall and consists of a large vertical engine, a dynamo connected directly to the end of the crank shaft, and the equalizer for distributing the current to the five wires used in the system of conductors.

The engine, built by F. Schichau, Elbing, Prussia, is of the vertical triple expansion type running 100 revolutions per minute and rated at 1,000 to 1,200-horse power. The high pressure cylinder is 22.86 inches diameter, the intermediate 37.4 inches, and the low pressure 57.03 inches; all having a stroke of 27.56 inches. The valves are balanced pistons with the cut-off valve surrounding the inner piston which is called the expansion valve and being controlled by the governor regulates the speed of the engine.

The control of speed is very accurate and it is said to vary but one to two per cent from no load to full load. The governor is one of the ordinary centrifugal type which through a simple system of levers turns the inner piston valve on its axis, the stem of the valve extending through the upper end of the steam chest for that purpose.

The ports in both valves are diagonal slots somewhat after the style of the old Ryder cut-off, and the turning of the inner valve changes the relative position of one opening to the other very rapidly and this cuts off the steam earlier or later in the stroke as the case may require, in order to lower or raise the rate of revolutions. This valve is said to admit steam for seven eighths of the full stroke when necessary.

The framing of the engine is light and graceful, but is so braced as to make the most of every pound of metal contained in it. For the capacity of the engine it takes but little space, about 9 by 18 feet and runs under full load without the slightest noise or jar. The lubrication, as in all machines of this type, is accomplished by means of small tubes leading to each and every joint from oil tanks on the rear of the cylinders. All stuffing boxes, bolts, nuts, and in fact every part that could in any way work loose, has been secured in some special manner, advantage having been taken of the firm's very extensive marine practice and experience on these points.

The dynamo known as type 1 is shunt wound and is of the regular external armature type made by Siemens and Halske, having ten inside pole pieces, ten brush holders, about 1,500 windings on the armature which also serves as a commutator, the external surface of the drum having been turned true for that purpose. The method of taking the current directly from the armature in this manner is said to increase the output. At 100 revolutions the total output is 1,000 amperes at 500 volts, and as but 22 amperes are used in the shunt at full load and the resistance of the armature is extremely low, the electrical efficiency of the machine is rated at 98½ per cent and the mechanical efficiency approaches 96 per cent.

The construction of the dynamo is especially interesting. A heavy ring of soft iron forms a continuous yoke and support for the ten field magnets which radiate from the outer surface. The ends of the pole pieces spread out so that they form almost a continuous circle, thus presenting a large field surface to the armature revolving about them. This frame with the radiating fields is secured to the engine foundation. The engine shaft passes through the center of it and supports the arms to which the outside armature is attached. The armature core is constructed of thin soft iron plates that overlap each other, thus breaking the joints in the construction of the ring secured on the transverse

pins which extend out from one side of the radial arms. This frame is supported on a large bearing on which it may be moved away from the fields, thus giving plenty of room for inspection or repair. The windings of the armature are thin copper bars shaped to fit the core and securely soldered in place. By having no lead wires or connections many of the small attending evils are avoided and great simplicity of construction secured, allowing the armature resistance to be so low as not to require compounding. The same bearing that supports the armature spider also supports two other frames, one having radial arms, supporting from its ends the brush holder rods which are of steel and extend across the face of the armature.

This frame rotates a short distance for adjusting the brushes to the non-sparking point. The other frame supports a gear and short levers attached to the brush holder rods in such a way that when the frame and gear are rotated a short distance all the brushes are lifted from the armature surface at once. The insulation for all exposed parts is thick hard rubber, and the connections from the brushes to the connecting rings and from thence to the terminals are layers of thin sheet copper. The connecting of all the brushes of like polarity is accomplished through a very substantial ring supported on the armature bearing and to which the flexible leads from the brushes are connected, heavy flexible connections leading from these rings to the regular machine terminals.

Where it is necessary to make connections for very heavy currents, the surfaces of the contact blocks are grooved to give better contact and permit of greater pressure. Three brushes about 1½ inches wide and of small straight copper wire are used on each brush rod. It is said that the commutator surface of the armature of the machine first installed in the Berlin station in 1887 is as good today as when installed and shows but the slightest signs of wear.

From the dynamo heavy conductors are led to the small marbleized switch board on which are placed two large single pole switches, one for each side, an ammeter showing the total out-put of the machine, a volt meter recording to 600 volts, and a rheostat for the shunt the contact wheel being on the front of the board and the resistance back of it. There is also a special carbon contact switch for breaking the shunt field circuit gradually to avoid trouble from extra currents when stopping the machine. When the machine is running the switch is closed but when it is desired to stop, the switch is opened throwing the current through the carbon pencils which are then gradually separated. The current is carried from this board by two feed wires to the centers of distribution where are placed the equalizers. These equalizers are specially constructed shunt motors with very low resistance armatures, all connected on one shaft, the four separate machines being exactly alike. The fields are connected in series from one terminal of the 500 volt circuit to the other, the armature circuits being connected in series between the same terminals. A resistance is placed in series with the armature circuit and used only for starting, in exactly the same manner as that in use with a shunt wound motor, and is cut out after the machines are up to speed. By this method the action of the equalizer is briefly as follows; the fields, being in series across the terminals, are practically independent of the machines themselves and exert the same influence all the time; in case one of the four circuits is loaded heavier than the others the particular machine on that circuit acts as a generator and being driven by the others adds its current and makes up the balance; if the circuit is underloaded as compared with the others the machine is carried as a motor and thus balances for the overflow. Of course if the circuits are all equally balanced no action takes place in the machines,the counter electromotive force simply balancing the consumption of current excepting that needed to overcome the friction and machine losses, the machines need only be of sufficient capacity to take care of the largest difference in load between any two of the circuits, care being taken in mapping out the system to as nearly as possible balance the different circuits one with the other. In this special case the equalizers are wound for 125 volts pressure, 50 amperes of current and run 1650 revolutions per minute. At full load they are said to have 90 per cent efficiency, the armatures being of very low resistance and the brushes taking current directly from the end face of the windings without special commutator, the machines are practically automatic, the variation in pressure being not over one and one-half per cent from nothing to full load. It may be said that where

terminals, the second and fourth the intermediates and the third the central wire which is used as one of the outsides in cases where only three wires are run.

The system permits of a great degree of flexibility and the fact of needing but two feed wires tends to reduce the apparent complexity due to the great number of wires. On the continent these circuits are laid directly in the ground about two and a half feet below the street surface and without other protection than the iron or steel armor of the cables themselves. The insulation consists of a layer of jute saturated with some ozokerite compound next to the wire, then lead, this being covered with another layer of of jute and wound over all with two layers of iron or steel ribbon both in the same direction in order to cover the entire surface and give the proper amount of flexibility.

All connections for customers are made through junction boxes buried in the street and are quite similar to those made by some manufacturers in this country. Large feeder

FIG. 2.—THE SIEMENS & HALSKE CENTRAL STATION EXHIBIT.

the load is very uneven storage batteries are advocated by this company in place of the equalizers, as having a somewhat higher efficiency. They are placed across the terminals and connected up in precisely the same manner as the motor equalizers and have the same effect.

At the center of distribution the equalizer is placed between the terminals and five wires are run from it as distributing mains, one from each of the outside terminals and one from each junction of the armature circuit. As the potential difference at the dynamo is generally 500 volts and the lamps used are 110 volts a drop of 60 volts may be allowed between the primary dynamo and the lamp socket. It is not necessary to run live wires in every case, as side streets and buildings requiring but few lights need but two or three wires, the system being balanced at some other point.

The comparative sizes of the five wires are about as follows: 1, ⅓, ⅓, ⅓, 1; the first and last being the outside

boxes are placed at street junctions for connecting up and testing the system.

The plant exhibited here is used for lighting the Terminal Building, Festival Hall, Wooded Island, German sections in Machinery Hall and in the Electricity Building. Equalizers are placed in Festival Hall, Terminal Building, Electricity Building, the one for Machinery Hall forming part of the exhibit. On this system are carried 450 arc lamps, 3,000 incandescent and 250 horse power in stationary motors.

It was first intended to erect a complete generating plant for tri-phase work and show a street car truck equipped with a tri-phase motor running on a piece of track, but the proper space and concessions not being obtainable, the generator, transformers and motor truck are at present located in the German section of the Electricity Building.

Outside the plant just described is also shown a 600 ampere 110-volt bi-polar direct current dynamo. The fields of this machine are very heavy and are flattened on the insides where they face each other, otherwise the machine is of much the same type and design as bi-polar machines of American make. The speed is 650 revolutions. This exhibit although not as extensive in its variety as some others, shows the great care and skill of the company in the design, construction and erection of its machinery.

A comfortable office, as shown in the engraving, is provided and Mr. W. Fricke, who is in charge, is always ready to explain any apparent intricacies of the five wire system used.

The steam engine part of the exhibit is in charge of Mr. Ludwig Gelbrecht for F. Schichau, his office and quarters being located on the west end of the engine platform.

The Electric Launches at the World's Fair.

One of the marked features of the Exposition grounds and one which renders them much more attractive than those of any previous Exposition, is the system of lagoons and water ways within the grounds. Much credit is due to

feet 10 inches long over all by six feet three inches beam, draw 28 inches of water and will seat comfortably 30 persons. They were built and are operated by the General Electric Launch & Navigation Company of New York.

The motive power is furnished by 72 cells of the S type of storage battery cells made by the Consolidated Storage Battery Co. of 120 Broadway, New York. These cells have seven positive and eight negative plates each, the containing jar being of hard rubber four and a quarter inches by seven and a quarter by nine inches high, with a hard rubber tight fitting cover, the connecting lugs coming out through a small hole in the center of the top. The plates are separated by the regular corrugated perforated hard rubber sheet now used by this firm and a new feature is the covering of the positive plate with a very thin perforated sheet of lead, which is folded around the bottom and prevents the active material falling out while in use. This sheet soon becomes formed and is said to add much to the economy and durability of the cell. The motor is five-horse power capacity, of the two field and four pole type,

THE ELECTRIC LAUNCHES AT THE WORLD'S FAIR.

the landscape architect, Mr. Frederick Law Olmsted, for the improvement and development of this part of the grounds. The excellent taste displayed by those who selected the forms of craft which should be used on these water ways is also a credit to those who had the affair in charge. The gondolas with the Venetian boatmen not only afford a pleasant ride and excellent means of going around the grounds but are also an ornamental feature of the Exposition. These boats make regular trips for which a fee of 50 cents is charged, or they may be rented by the hour.

The electric launches, forming a prominent and attractive feature of the lagoons, are greatly appreciated by the visitors. They are a matter of interest to all as this means of propulsion is new and it has never before been tested on so large a scale. To the many visitors, who are unfamiliar with the rapid development and numerous applications of electricity, these boats gliding through the water swiftly with no visible means of propulsion are a matter of wonder and astonishment. The competition for this concession was very sharp at the time it was awarded, for the manufacturers of steam, naptha, electric and other launches entered into competition for the concession. The boats selected are 35

designed and made for this special purpose by the General Electric Company. The armature is cross connected, thus requiring but two brushes, is made thoroughly moisture proof as is attested by the fact that motors have at times been run while half submerged.

There are four speeds at which the boats can be run forward, but two only have been provided for backing. The speed is controlled by a special electromagnetic switch, having mercury cup contacts, and an electromagnetic arc breaker, all being controlled by a handle in the bow of the boat. A special auxiliary switch is also provided, by which the circuit to the motor may be opened in case of trouble on the regular switches. Backing is accomplished by reversing the motor armature terminals in the usual manner. The four speeds are, three and a half, five, six and eight miles per hour, the second being the one most generally used. Following is the arrangement of cells for the different rates of speed:

First.—Three series of 24 cells, each in parallel with a rheostat in series to the motor.

Second. Same as above with the rheostat short circuited.

Third.— Two series of 36 cells, each in parallel with the rheostat in series to the motor.

Fourth.— Same as third with rheostat short circuited.

The first two are the only ones used for backing. Some idea of the smoothness and freedom from trouble with which these boats work, may be gained from the fact that on an average 48 boats are out all the time, each averaging 13 trips per day, a trip being three and one half miles, and consuming from 15 to 30 minutes.

As soon as the boats come in at night the wires are connected, and charging commenced. A current of 52 amperes is used, the cells being arranged in three series of 24 each in parallel, thus taking about 17½ amperes per series; the time taken for a full charge after using all day is seven hours. Current is supplied by the large bi-polar generators of the General Electric Company in Machinery Hall, through two heavy sets of feeders run in the Edison three-wire tubing to the launch house at the southeast corner of the Agricultural building wharf.

A stall and charging leads are provided for each boat, the latter being taken from one side of the three wire system through a double pole fuse and switch, one side running through an ampere meter, and rheostat of two ohms

EXHIBIT OF THE ELECTRICAL FORGING COMPANY.

hence on to the boat, the other lead going directly from the switch to the boat.

The course of most of these launches is from the starting point on the north front of the Agricultural building, to and along the south side of the Hall of Liberal Arts, thence up the north canal around by the Fisheries building and Clam Bake to the Art building, returning by the way of the Lagoon, in front of the Woman's building, Horticultural Hall, Transportation building, Mines and Mining and Electricity buildings, to the starting point. One or more stops are made at all prominent points, the fare for the round trip being 50 cents, or 25 for a half trip. The boats are provided with carpets of perforated rubber mats, the seats with leather cushions stuffed with hair, and awnings of canvas in red and yellow stripes. The pattern and colors were selected by F. D. Millet, director of colors for the Exposition. The officers in immediate charge of the concession on the grounds are, General Manager, C. D. Wyman; Superintendent, C. H. Barney; Electrical Engineer, R. N. Chamberlain. A pilot and guard is provided for each boat, and a guard is also stationed at each wharf, where passengers enter through a turnstile, after buying tickets at the small booth provided. Handsome blue uniforms with brass buttons are worn by all employees, the different classes being designated by appropriate letters on the cap.

There is now no doubt of the success of the electric launch as a means of transportation, and after riding in one of these boats, gliding over the water so smoothly, one can

have no doubt of the future growth of the business. The record of one of these boats is deserving of special notice; after running over 1,200 miles without repairs, it made a continuous trip of 60 miles on one charge.

Exhibit of the Electrical Forging Company.

A very interesting exhibit of the Electrical Forging Company is located in section D, Electricity Building, just east of the German section. It contains three large C. & C. motors used to drive the two welding dynamos and the shafting over head to which are belted the special tools used in forging. There are also a number of other tools, rollers, etc., used in the practical exhibition of the process.

The company manufactures electrical welding and forging tools and machines, a large number of which are shown and used in this exhibit. The dynamo used for generating the current is an alternating intermittent dynamo having ten field coils and five armature coils which have each a special winding by which it is claimed that not only more pulsations are generated per minute but the peculiar windings of the armature adds another feature in that the current is not evenly alternated but has an intermittent strength which stirs up the molecules more violently. The controlling rheostat is placed in the shunt field circuit of the exciter, the field circuit of the dynamo itself being in connection with the armature only. While this arrangement is not new as applied to lighting it is said to be new in this application. The two welding dynamos shown are of 75 horse power capacity at 1,600 volts pressure the exciter having a capacity of 220 volts and six amperes.

The transformer has a core formed of pieces of bare iron telegraph wire in the form of a large ring, the wires being placed so as to break joints. Wound around this ring at short distances apart are the primary coils of fine insulated copper wire, two coils being connected in series and the pairs in multiple. The secondary coils are sheets of copper cast in the shape of a square spiral and slipped on to the iron core between the primary coils. The terminals are extensions from the connecting rings. Working terminals can be taken from any part of these rings thus permitting a number of kinds of work being done at once. The terminals shown with these machines are of different kinds and adapted to several kinds of work. A system of heating and welding by arc currents is also shown. Various goods manufactured by this process are exhibited among which is an ornamental fence in three sections one being made of copper, one of brass and the other of iron. The president of the company is Mr. George Burton, who is at present at the exhibit ready to explain the process to interested parties.

An Injunction Granted.

Judge Lacombe, of the United States circuit court of New York, has granted injunctions to the Edison Illuminating Company, of New York, against the Holland House and Imperial Hotel prohibiting them from using incandescent lamps infringing the Edison patent. The Edison Illuminating Company is sole licensee for the city of New York for the Edison lamps.

Willans & Robinson, Ltd., are the English makers of the Willans engines erroneously call the Willans engines in a description of the engines at the World's Fair which recently appeared in this paper. The M. C. Bullock Mfg. Company, 1170 W. Lake St., Chicago, Ill., is the American manufacturer of these engines.

ELECTRICAL INDUSTRIES.

Entered at Chicago Postoffice as second-class mail matter.

PUBLISHED EVERY THURSDAY BY THE

ELECTRICAL INDUSTRIES PUBLISHING COMPANY,
INCORPORATED 1891.

MONADNOCK BLOCK, CHICAGO.
TELEPHONE HARRISON 393.

R. L. POWERS, Pres. and Treas.   E. E. WOOD, Secretary.

R. L. POWERS   Editor.

R. A. FOSTER   }
W. A. REMINGTON }   Associate Editors.

E. E. WOOD   Eastern Manager.

FLOYD T. SHORT   Advertising Department.

EASTERN OFFICE, WORLD BUILDING, NEW YORK.
World's Fair Headquarters, Y 27 Electricity Building.

SUBSCRIPTION.
FIVE MONTHS, ... $1.00
SINGLE COPY, ... 10
Advertising Rates Upon Application.

News items, notes or communications of interest to World's Fair
Visitors are earnestly desired for publication in these columns and will
be heartily appreciated. We especially invite all visitors to call upon us
or send address at once upon their arrival in city or on the grounds.
ELECTRICAL INDUSTRIES PUBLISHING CO.,
Monadnock Block, Chicago

Two notable and suggestive events have happened during
the past week. The arrival of the caravels, models of the
ships of Columbus in which he made his first voyage, and
the arrival of Capt. Anderson and his Viking ship, a supposed model of Lief Ericson's craft. Thus is brought to
mind the discovery of the western world by two distinct
races. One in about the year 1000, but whose discovery
made no stir in the world, and the voyage of Columbus in
1492 which brought civilization to this continent.

The small attendance on Sundays at the Fair has been a
surprise to all. On last Sunday, July 9th, the attendance
was but 14,000. The reduction of the price of admission
to 25 cents is strongly advocated by some of the daily
papers. As many of the exhibits are closed, and but few
attendants are with them that day, it would not seem to
be bad policy to reduce the price of admission for the one
day each week. The American public does not like the
idea of paying the full admission price for seeing half a
show.

It was confidently expected that the railroads would
establish excursion rates to the Fair from all points; but in
this there has been so far only disappointment. What has
been granted applies only to distant points and has been
made in such a way as to deprive the traveler of all the
comforts of travel. The half of the Fair is nearly over, and
but a small portion of the people of this country has visited
it. Some inducement in the railroad fare and some comfort and accommodation while on the way is the thing
that will bring the people.

The disastrous fire of last Monday afternoon was a never
to be forgotten spectacle to the thousands of visitors present. The death of nearly a score of firemen throws a
shadow over the festivities of the Fair. The building, which
was of staff construction, was used for cold storage and ice
making machinery. It had also a skating rink on the upper floor, for which the first coating of ice had just been
frozen. The great tower in the center of the building
formed a trap from which there was no means of escape.
It is probable that no other building on the grounds could

furnish the same conditions, but this fire has its lesson
and will undoubtedly lead to greater precautions. As the
first step, however, every building on the grounds should
be provided with fire escapes. The fire department is vig-
ilant and well organized but nothing should be left undone
to make the buildings safe. In the Electricity Building
the matter of insulation should be looked after with the
greatest care so that no fire may be attributed to the electric current. We are confident the department will take
this matter in hand at once.

One of the finest models shown at the Exposition is that
of the four cylinder triple expansion engine made by Har-
lan & Hollingsworth, in the south end of Transportation
building. Engines of this style were built by the firm for
the steamers "Maine" and "New Hampshire" of the Ston-
ington Line, and are said to have been the first four cylinder triple expansion engines built in the United States.
The dimensions of the cylinders of these engines are respectively as follows: 28, 45, 51 and 51 inches with 42-inch
stroke.

The Chapman Valve Manufacturing Co., of Indian
Orchard, Mass., have an exhibit in Machinery Hall, Main
28, K. 28, where are shown the various sizes and styles of
high pressure steam valves, such as are now in use by
nearly all the steam and electric power stations. This
valve is extra heavy in all its parts; has a gate giving full
opening of pipe, outside screw and packing, and in the
very large sizes, such as 24 inches, has a small by-pass
valve admitting steam around the regular gate in order to
even up the pressure on both sides before opening it. These
valves all have moveable bronze seats, thus allowing quick
and easy repairs. Samples of high class extra heavy flanges
for steam pipe work are also shown. This company furnished
the throttle for the large Allis Corliss engine, all the valves
for the Intramural Power Station, and all valves used on the
Heine boilers in the power house. Samples of hydrants
and hose bibs for water works are shown, also samples of
the large water gates made by the company. The exhibit
is nicely located, well-displayed, and Mr. Edward L. Ross,
the well-known agent of this company, is in charge and
will take pleasure in making it comfortable for any visitors
stopping at his gate.

The French section, in the Manufactures and Liberal
Arts building, contains several exhibits of high grade electrical fixtures which are worthy of more than passing
notice. A. Rollet, Paris, exhibits several beautiful electroliers, nearly all of which are designed after water fowls and
aquatic plants. In price these goods range from $100 to
$200. Compier & Dromart, Paris, show among other goods
many artistic designs in bronze of one, two, three and four
light fixtures. Female figures predominate, though one of
the most attractive of the lot is a group of three children
running holding incandescent lamps in their hands. In
price the goods of this firm range from $130 to $350 for
articles mentioned; Mr. Ch. Jacobson is in charge. The
works of Compier & Dromart were Hors Concours at Paris
Exposition, 1891. H. Baudidine, Pere et Fils, show one
pair of vases, with thirteen light clusters, valued at $2,000;
also two female figures, life size, with raised arms supporting basket filled with lights; the pair are held at $5,000.
Maison F. Barbedienne, Paris, makes an exceedingly elaborate display of electroliers, the principal feature of which is
two life-size female figures supporting a cornucopia filled
with flowers, in which are sockets for lamps; the pair is
valued at $10,000.

WORLD'S FAIR NOTES.

The managers of the Turkish Cafe on Midway Plaisance gave a reception to the press on Friday evening last.

A unique feature in the exhibit of the New York Insulated Wire Co. is a log cabin built entirely of vulca duct. It is high enough to stand in, and makes a valuable addition to the display.

The Worthington Pump Co. have a very large display of its celebrated steam pumps at the extreme southeast corner of Machinery Hall. Some of the pumps are in constant use for the service plant, others are on exhibition only.

The arc lamp poles in the Javanese village have been given native decorations. The hoods have been thatched with straw, and the poles covered with the long hairlike bark that grows on the sugar palm. The decorations match the native houses, but the arc lights look rather strange under their Javanese covering.

Siemens & Halske are fast completing their excellent exhibit in the Electricity Building. Among the articles shown is a resistance or rheostat in which the coil is made of copper wire cloth. It has been little used in this country although possessing several advantages, among others the rapid dissipation of the heat. An electric railway truck with tri-phase motor, together with generator, exciter and transformers are shown.

The Sloss-Stein Electric Co. of 218 La Salle street, Chicago, is making an exhibition of electric gas lighters, in which a battery and spark coil are concealed under the canopy of each fixture, thus doing away with lead wires. The pavilion is located in the southeast corner of Electricity building, and is decorated in a very unique manner with spider webs formed of cord with exaggerated flies and spiders fastened on the surface.

The headquarters of the American Institute of Electrical Engineers are now completely furnished. The rooms are comfortable, cheerful and offer an excellent resting place. The files of electrical and daily papers, together with the library of electrical books presented by Van Nostrand, furnish reading material for visitors. A stenographer is located in the offices, and long distance telephone and telegraph lines are being put in so that every convenience will be afforded members. Various photographs and relics are distributed about the rooms, making a visit to the headquarters interesting and instructive.

An exhibit of the application of electric motors to heavy machine work is made by Heinrich Ehrhardt, of Dusseldorf, in the German section, Machinery Hall. A special motor made by Siemens and Halske, of Berlin, is used, and it is of much the same form and shape as the late Edison arc dynamo, the size being modified to meet the special requirement. Most of the machinery is for sawing or cutting iron and steel, and is rather heavy in its nature. There is one upright band sawing machine for sawing scroll work in heavy iron plates; another for cutting large bars of iron or steel in pieces and others for doing various kinds of work. One special feature is a method of keeping a belt tight between the machine and counter shafts.

A very complete exhibit in its line is that of the James Morrison Brass Mf'g. Co., Limited, of Toronto, Canada, in Section F., No. 2, Machinery Hall. It consists of oil cups of all kinds; gauges, gauge glasses, cocks, clocks for steam uses, steam gauges, revolution counters, steam ship controllers, gongs, bells, steam whistles, inspirators of various sizes, cylinder oilers of different forms and sizes, in fact almost everything in the way of a steam engine fitting is shown in this space. The sizes of the different pieces of apparatus are such as are in general use and many of extra large sizes are also shown. The exhibit is well arranged.

Messrs. Schaeffer & Budenberg of 66 John street, New York and 22 West Lake street, Chicago, have a very fine exhibit of their steam specialties in Machinery Hall, in the main hall and aisle near the entrance to the annex. Arranged around the edge of their space are a number of different sizes of the exhaust injector, while inside a cabinet of ebonized wood are samples of all the smaller material used or sold by this firm. Among others were noticed samples of heavy gauge glass, which are made in sections so as not to destroy but a small section in case of blow out. The exhaust injector sold by this company is a feature which deserves mention. It is placed in the exhaust pipe of an engine and produces a partial vacuum, which forces water into the boiler and at the same time taking most of the remaining heat out of the exhaust steam. In a frame hung over the desk are some forty medals that have been received at different expositions by the firm. Mr. A. L. Portong is the western manager and is in charge of the exhibit. He is pleased to meet his old friends at his exhibit and hopes to make a long list of new ones before the exposition is over.

The American Battery Company of 188 Madison St., Chicago, is showing a storage battery carriage in the northwest corner of the Electricity Building. The carriage has three seats and is somewhat heavier in construction than carriages of the same type built for horses. There are twenty-four cells, eight under each seat. The capacity is 150 ampere hours and when fully charged are 2.2 volts each and are never run below 1.9 volts. The motor is a small bi-polar type placed on the running gear underneath the body and geared through three reductions to the rear axle. The speeds are 4, 8 and 12 miles per hour and the carriage is said to be capable of running 60 miles on one charge. As high as 75 miles on one charge has been obtained. This carriage has been in use for the past 18 months and is said to show no deterioration whatever in the batteries. About three times a week the carriage is run around the grounds, accommodating three or four people and excites very much interest from the visitors on the grounds.

Foreign Visitors to the World's Congress.

The following letter has been received from Professor H. S. Carhart.

Dear Sir, Dr. Elisha Gray has just received a letter from Professor von Helmholtz announcing that he will take part in the conference of the World's Congress of Electricians in Chicago, Aug. 21st, as the official delegate of the German Government. He will be accompanied by Drs. Foerster, Kurlbaum, Lenau, Lindeck, Lummer and Pringsheim, his assistants in the Reichsanstalt in Berlin. These gentlemen have made important contributions to our knowledge of the values of the modern practical electrical units, and their work will doubtless form a basis of Dr. von Helmholtz's recommendations. The presence of von Helmholtz will render the Electrical Congress a notable occasion.

Yours truly,

(Signed) Henry S. Carhart, Secretary.

PERSONAL.

Mr. W. J. Hammer returned to New York on Saturday last.

Mr. W. R. Brixey, of Day's Kerite, returned to New York last Saturday.

Prof. N. M. Terry of the U. S. Naval Academy, Annapolis, is visiting the Fair.

Mr. E. Hornell, managing director for the de Laval Steam Turbine Co., is visiting the Fair.

Mr. C. A. Benton, of the Sprague Electric Elevator Company, New York, paid the Fair a visit last week.

Mr. James H. McGraw, President of the Street Railway Journal, was a visitor at the Fair last week.

Mr. J. C. Chamberlain is stopping at the Auditorium for a few days, and is spending some time at the Exposition.

Mr. Julian A. Moses of the Electrical Review, New York, registered at the Institute headquarters on the 16th.

Mr. Charles Seldon, of Baltimore, recently registered at the headquarters of the American Institute of Electrical Engineers.

Mr. J. H Rhotehamel, president of the Columbia Incandescent Lamp Company, St. Louis, visited the Fair and Chicago last week.

Mr. Brainard Borison, secretary of the Fort Wayne Electric Company, is in the city with his family and will spend considerable time visiting the Fair.

Mr. R. O. Heinrich, of the Weston Electrical Instrument Co., has returned to his home in Newark, N. J., after supervising the installation of the company's exhibits.

Mr. J. Hammar, of Guttenburg, Sweden, member of the Institute of Electrical Engineers of Great Britain, is the engineer in charge of the de Laval Steam Turbine exhibit in Machinery Hall.

Mr. F. O. Blackwell, chief engineer of the mining department of the General Electric Co., at Lynn, Mass., is spending a few weeks at the Fair, with headquarters at the company's office in the Electricity building.

Mr. Carl K. MacFadden, formerly with Taylor, Goodhue & Ames, and for two years or more engaged as expert on the Nutting arc lamp, has recently been engaged by Bartholomew, Stow & Co., of Chicago, the general selling agent for the Nutting system of arc lighting.

BUSINESS NOTES.

The Enterprise Electric Company, Chicago, is still finding a good demand for fan motors.

Mr. L. E. Frolary, western manager of the General Incandescent Arc Light Co., must be doing a lively business from the appearance of his headquarters at 169 Adams St., and his cheerful countenance.

The Electric Appliance Company with its usual enterprise has come to the front with a Weston fan motor wound particularly for the low frequency alternating current at the World's Fair. It has succeeded in securing results that have been a surprise to all. The efficiency of the motors is very high and the company is making arrangements to have the motors rewound after the Fair at a nominal expense to operate on alternating circuits of ordinary frequency where customers so desire.

DEPARTMENT OF ELECTRICITY.

OFFICES SECTION R, ELECTRICITY BUILDING.

Chief, JOHN P. BARRETT.
Assistant Chief, J. ALLEN HORNSBY.
General Superintendent, J. W. BLAISDELL.
Electrical Engineer, W. W. PRINE.

DEPARTMENT OF MECHANICAL AND ELECTRICAL ENGINEERING.

OFFICES SOUTH OF MACHINERY HALL.

Mechanical Engineer, C. F. FOSTER.
Electrical Engineer, R. H. PIERCE.
First Asst. Mechanical Engineer, JOHN MEADEN.
First Asst. Electrical Engineer, S. G. NEILER.

AMERICAN INSTITUTE OF ELECTRICAL ENGINEERS.

World's Fair Headquarters,
SECTION S, ELECTRICITY BUILDING.

RALPH W. POPE, Secretary.

Open from 9 a.m. to 5 p.m.

CHICAGO WORLD'S CONGRESS OF ELECTRICIANS.
ADVISORY COUNCIL.

President, DR. ELISHA GRAY, Highland Park, Ill.
Secretary, PROF. H. S. CARHART, Ann Arbor, Mich.

EXECUTIVE COMMITTEE.

Chairman, PROF. ELIHU THOMSON, Lynn, Mass.

COMMITTEE ON INVITATIONS.

Chairman, T. COMMERFORD MARTIN, 253 Broadway, New York.

COMMITTEE ON PROGRAM.

Chairman, PROF. T. C. MENDENHALL, Washington, D. C.

COMMITTEE ON FINANCE.

Chairman, B. E. SUNNY, 175 Adams Street, Chicago.

Amusements.

HOOLEY'S THEATER—Mr. E. S. Willard, in "The Professor's Love Story." 149 Randolph street.

COLUMBIA THEATER Miss Lillian Russell, in "La Cigale." 108 Monroe street.

GRAND OPERA HOUSE—Sol Smith Russell, in "April Weather." 87 Clark street.

AUDITORIUM—Imre Kiralfy's spectacle "America." Congress street and Wabash avenue.

McVICKER'S THEATER—Denman Thompson, in "The Old Homestead." Second week. 82 Madison street.

CHICAGO OPERA HOUSE—American Extravaganza Company, in "Ali Baba, or Morgiana and the Forty Thieves." Seventh week. Washington and Clark streets.

SCHILLER THEATER—Chas. Frohman's Stock Company, in "The Girl I Left Behind Me." Seventh week. Randolph, near Dearborn.

HAVERLY'S CASINO Haverly's United Minstrels. Wabash avenue, near Jackson street.

TROCADERO—Concert. Michigan avenue near Monroe street.

THE GIOTTO Vaudeville. Michigan avenue near Monroe street.

Buffalo Bill's "Wild West," 63d street. Daily at 3 and 8.30 p.m.

Pain's "Siege of Sebastopol," 60th street and Cottage Grove avenue. Tuesday and Thursday nights.

Next week Mr. E. S Willard will appear in "The Middleman" and "Judah" at Hooley's.

Sol Smith Russell began the eleventh week of his successful engagement at the Grand Opera House last Sunday evening. This will be the last week of "April Weather" as Mr. Russell will appear next week in his famous play "A Poor Relation."

Ali Baba, the 500th performance of which occared last Wednesday evening, goes on record as having been played the greatest number of times of any play in the west. It was intended to have replaced it with "Sinbad" before this, but its continued popularity will keep it on for some time.

Kiralfy's spectacle "America" is having an extraordinary success at the Auditorium. The Shaffers are popular favorites and a number of wonderful feats have recently been introduced by them which are surprisingly skillful and intensely interesting.

Siemens & Halske Electric Company

of America.

Chicago, Illinois.

Electrical Machinery.

Siemens & Halske,

Berlin:
Charlottenburg:
Vienna:
St. Petersburg:

1000 H. P. DIRECT CURRENT GENERATOR.

AGENCIES.

GENERAL OFFICES:
Monadnock Bldg., Chicago.

NEW YORK OFFICE:
136 Liberty Street.

CINCINNATI OFFICE:
Perin Building.

ST. LOUIS OFFICE:
Bank of Commerce Bldg.

MILWAUKEE OFFICE:
413 Broadway.

1000 H. P. ALTERNATING GENERATOR

.....MANUFACTURERS OF.....

Direct Current Multipolar Dynamos and Generators

These machines are constructed with outside, revolving armatures, without and with special commutator, as desired. They have proven remarkably efficient and economical. Used largely in European Central Stations. They are slow speed machines, made for direct connection to engine without belting, and in sizes from 20 H. P. to 1,500 H. P.

HIGH SPEED BELTED SIEMENS'S DYNAMOS.

These machines are copied extensively in this country as "Drum Type." We are building these machines in sizes from 1 H. P. to 150 H. P.

ALTERNATING AND MULTIPHASE CURRENT DYNAMOS.

Dynamos with laminated field and armature, in sizes from 1 H. P. to 4,000 H. P. for belted or direct coupling.

MOTORS.

Motors of every speed for direct, alternating or multiphase current, in sizes from one tenth H. P. up to 4,000 H. P. Street car motors and kindred appliances; these motors have been made double reduction, single reduction and gearless. Durable and economical in operation.

SIEMENS' BAND LAMPS.

Lamps for direct and alternating current, for constant, potential and series machines.

CARBONS.

Siemens' arc light carbons, with either solid or soft cores, which are the most economical in the world, and give a steady light.

INSTRUMENTS.

All instruments requisite for the regulation of electrical apparatus; also the Siemens voltmeters and ammeters and Automatic Devices made in Berlin.

POWER TRANSMISSION.

Profiting by our European experience and exhaustive tests, we are prepared to estimate on power transmission intelligently, and guarantee successful operation.

ELECTRICITY BUILDING—EXHIBITORS AND THEIR LOCATION.

GALLERY.

MAIN FLOOR.

Exhibitor.	Section.	Exhibitor.	Section.	Exhibitor.	Section.	Exhibitor.	Section.
Austral	Y	Electrical Review	Y	Jaeger, Chas. L.	T	Reliance Gauge Co.	
Ansonia Electric Co.	X	Electricity	Y	Johns Mfg. Co., H. W.		Roessler & Hasslacher Chem. Co.	S
Am. Inst. of Elec. Eng.		Electric Gas Co.		Jewell Belting Co.		Street Railway Journal.	
American Battery Co.	T	Electrical Engineer	Y	Juney Elec. Motor Co.	T	Sprague Elec. Railway Co.	T
Ashton, R. M.		Electrical World	Y	Knapp Electrical Works	T	Standard Paint Co.	
Alleg. Elec. Gen. Elec. Co.	D	Edd's Electric Motor Co.		K. A. P. Elec. Novelty Co.		Sponholz, F. L.	
Ball & Mfg. Co.		Interior Electric Co.		Knapp & Buckley		Star Iron Tower Co.	Y
Bryant Electric Co.		Electrical Forging Co.	D	Kennedy Electric Co.	L	Spain	
Belmar & Spencer	H	Equitable Dynamo Co.		Lawton, B.	Y	Schierer, Chas. A. & Co.	X
Brixey, W. H.		Allison Mfg. Co.		LeChocho Battery Co.		Schonburg & Sohne	
B-Knapp Motor Co.	K	Electrical Conduit Co.		McNeil Tudor Elec. Co.	L	Siemens & Halske	X
Bell Telephone Co.	E-A	Enclad		Mercer, W. M.	S	Schubert & Co.	E
Bush Electric Co.		Empire China Works		Meeker, Dr. G.		Short Electric Co.	
Caldwell El. Cloth Pol. Mch. Co.	Y	Franklin Elec. Appliances		McIntosh B. L. & Opt. Co.	W	Sperry Elec. Railway Co.	
Consol. Elec. Storage Co.	R	French Piano Exhibit		Munson, C. Belting Co.		Standard Underg. Cable Co.	
Cutter, George	T	Felsen & Guilleaume		Mather, A. C.		Standard Electric Co.	
Cotton Elec. Co.		France	K P-S	Mather Electric Co.	M	Samson Battery Co.	
Chicago Elec. Works	T	Ft. Wayne Elec. Co.	M	Newman Clock Co.		Tate Ast. El. Signal Co.	T
Copenhagen Elo Manu. Co.	N	Gould & Co., N. C.		Non-Magnetic Watch Co.	T	Todd, Applegate Co.	
Central Electric Co.		Gamewell Fire Alarm Co.	Y	N. Y. Insulated Wire Co.	P	Taylor, Goodhue & Ames	A
Commercial Cable Co.		General Electric Co.	B-H-S-C-J	National Carbon Co.	T	Thomson Elec. Welding Co.	O
C & C Elec. Motor Co.		Goforth and Co.'s the Lt. Co.		Norwich Iron Wire Co.		Telautograph, Elisha Gray	F
Cleveland Elec. & Mfg. Co.	A-J	Greeley, E. S. & Co.		South Am. Phonograph Co.	N-P	Union Electric Co.	
Chicago Belting Co.		Germania		N. Y. & R. E. A.		Vettel F. C. & Co.	
Dabney Clock Co.		Highland Mtg. & Co.		Nat. Ant Fire Alarm Co.		Welch, C.	
Department of Electricity	H	Helmann, C. J.		Nat. Engraving Machine Co.		Weston El. Instrument Co.	H
Electrical Department		Hart & Hegeman Mfg. Co.	S	Owen, Dr. A.		Washburn & Moen Mfg. Co.	Y
Elec. Lamb & Ins. Co.	H	Hope Elec. Appliance Co.		Booth Glass Co.	I	Western Union Tel. Co.	
Electric Separator	Y	Holt, Chas.		Poize, F.		Ware & Bartlett Mfg. Co.	
Edgerton, K. W.	T	Holtzer, N. S		Palmemother Galv. Co.		White, S. S. Dental Mfg. Co.	
Fisk Telephone Co.	T	Hartman & Braun	R	Ponnvelly, J. S.		Western Electrician	
Edison Elec. Mfg. Co.	T	Hanson Art Works		Pratt El. Mcd. Sup. Co.		Wilder Ast. Burglar Al. Co.	
Enterprise Elec. Co.		Hirsch, J. M.		Powell, Wm. & Co.		Western Electric Co.	
Eureka Trap Copper Co.		Breinesch, F. & Co.		Phelps, A. H.		Westinghouse El. & Mfg. Co.	B-H
Edison Appliance Co.	I	Hanson Alloy Co.		Pals Belting Co.	D	Wiley & Penfield	
Elec. Acc. Steel Co.		Internat. Ind. Lt. & P'r Co.	T	Queen & Co.		Wing, L. J. & Co.	
Electric Heat Alarm	X	India Rubber Comb Co.		Rupler, F. S.		Zucker & Levett Chem. Co.	

ELECTRICAL INDUSTRIES.   11

ARE YOU KEEPING COOL
COOLING MACHINE
THE MESTON MOTOR

We have the means to that end.
We have the only

guaranteed to work in a satisfactory manner on
an alternating current.

It will operate with less than half the
current of any other equally powerful
alternating current fan motor.

This means that you will save the price of the motor in a few months.

Electric Appliance Company, 242 Madison St.

ELECTRICAL SUPPLIES.   CHICAGO.

THE MATHER ELECTRIC CO,	THE "NOVAK" LAMP.
MANCHESTER, CONN.	
Dynamos, Motors, Generators,	CLAFLIN & KIMBALL (Inc.)
	General Selling Agents.
Offices, 116 Bedford St., BOSTON.	
—AND—	116 Bedford Street, BOSTON.
1002 Chamber of Commerce Bldg., CHICAGO.	1002 Chamber of Commerce Bldg., CHICAGO.

THE REGULAR MONTHLY EDITION OF
ELECTRICAL INDUSTRIES

Is the most complete Electrical Journal published,

Every issue containing descriptions of all the new applications of electricity, complete directories
of the Manufacturers and Dealers, the Electric Lighting and Railway Companies in North America,
revised and corrected to the date of going to press. These special features are found in no other Electrical
Journal in the world, and consequently it is read by more actual buyers than any other publication, which
fact makes it without a superior as an advertising medium.

ELECTRICAL INDUSTRIES PUBLISHING CO., Monadnock Block, CHICAGO.

Enterprise
Electric
Company

307 Dearborn Street,
Chicago . . . .

GENERAL WESTERN AGENTS
N. I. R.

Manufacturers' Agents and Mill Representatives for

Electric Railway,
Telegraph, Telephone and
Electric Light

SUPPLIES OF EVERY DESCRIPTION

Agents for   Cedar Poles,
Cypress Poles, Oak Pins,
Locust Pins, Cross Arms, Glass
Feeder Wire,   Insulators,

WIRES, CABLES, TAPE and TUBING

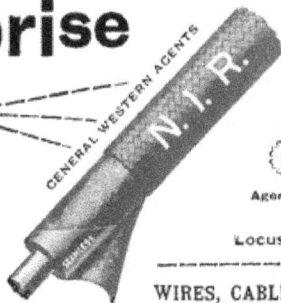

Map of Chicago.

Showing Location of Its Electrical and Allied Business Interests, Principal Hotels, Theatres, Depots and Transportation Lines to the World's Fair Grounds. (Index numbers refer to the black squares.)

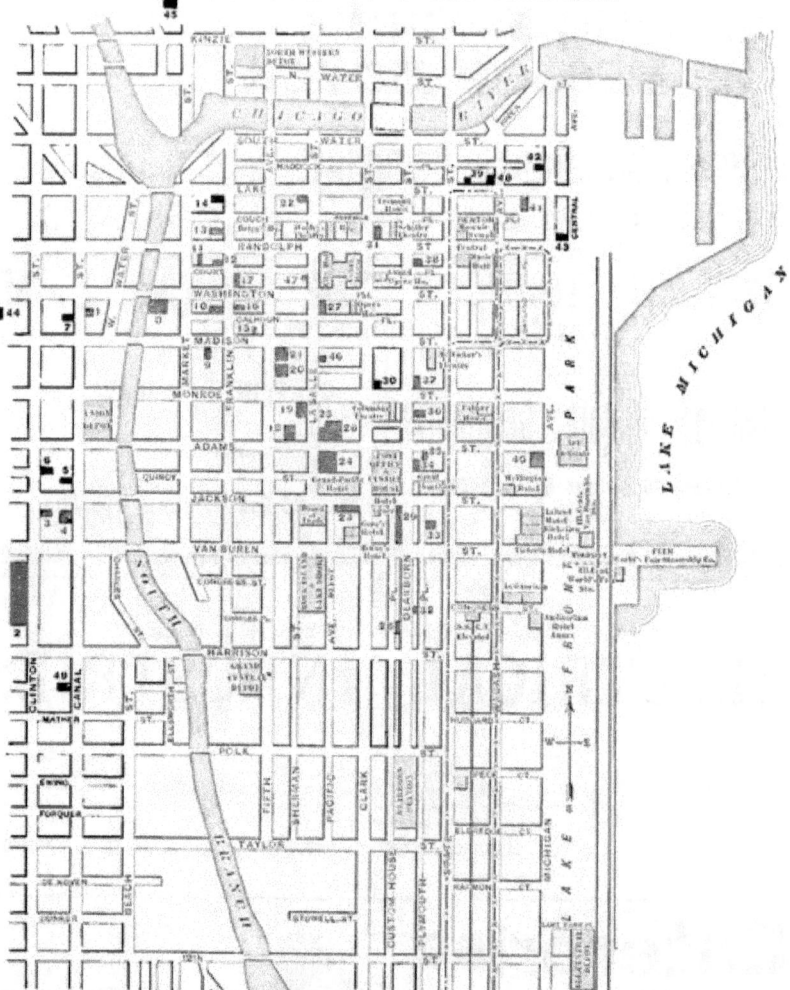

~Lundell~
Suspended Fan Outfit

OKONITE

Black

Japan

Finish

Self-Oiling
....and
Self-Aligning
Bearings

Electric Fans   **Electric Fans**

OKONITE WIRES
OKONITE === TAPES === MANSON
INTERIOR CONDUIT.

Batteries, Bells, Push Buttons, Annunciators, Volt Meters, Ammeters, Wheatstone
Bridges, Line Wire Cross Arms, Brackets, Pins, Insulators, Tools.

GENERAL SUPPLIES.

CENTRAL ELECTRIC CO.
116-118 Franklin Street,
CHICAGO, ILLS.

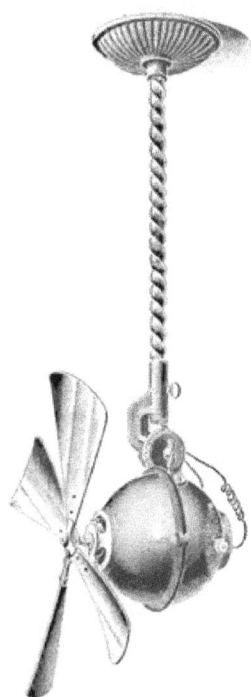

CLARK ELECTRIC COMPANY, NEW YORK.

192 Broadway and 11 John Street.

MANUFACTURERS OF ARC LIGHTING APPARATUS FOR EVERY PURPOSE A SPECIALTY.
The CLARK ARC LAMPS for use on EVERY CURRENT, have the reputation of being the best and most durable of any ever made in the United States.

RAWHIDE PINIONS FOR ELECTRIC MOTORS
A SPECIALTY.
RAWHIDE DYNAMO BELTING
Greatest Adhesive Qualities. A Non-Conductor of Electricity.
Causes Less Friction than any other Belt.

THE CHICAGO RAWHIDE MANUFACTURING CO.
THE ONLY MANUFACTURERS IN THE COUNTRY.

LACE LEATHER ROPE
AND OTHER RAWHIDE
GOODS
OF ALL KINDS
BY KRUEGER'S PATENT

This Belting and Lace Leather is not affected by steam or dampness; never becomes hard; is stronger, more durable and the most economical Belting made. The Rawhide Rope for Round Belting Transmission is superior to all others

75 Ohio Street, CHICAGO, ILL

STANDARD ELECTRIC COMPANY.

GENERAL OFFICES: 625 Home Insurance Building.

WORKS: So. Canal Street,

CHICAGO.

STANDARD SYSTEM

AT THE

WORLD'S FAIR.

MACHINERY HALL, Sec. Q, 2 Standard Arc Dynamos.
Sec. S. 20 " " "
ELECTRICITY BUILDING. Sec. P, Space 2, Arc Lighting Exhibit.

The Standard Lamps Light the Power Plant, Machinery Hall, Agricultural Hall, Shoe and Leather Building, and Other Buildings and Portions of the Grounds.

See our Double Service All Night Lamp Before Buying an Old Style Two Rod Lamp.

Western Electric Company,

CHICAGO.　　NEW YORK.

Arc Lighting Apparatus
　High and Low Tension,
　　Double and Single Service Lamps,
　　　All Night Single Lamps,
　　　　Theater and Focusing Lamps.

ELECTRIC MOTORS

HIGH SPEED
　　VARIABLE SPEED
　　　　SLOW SPEED.

BUILT FOR SEVERE AND CONTINUOUS SERVICE,
SPECIAL TYPES FOR SPECIAL DUTY,

Cutter's "Boulevard" Streethood

THE LATEST, NEAREST AND BEST.

GEORGE CUTTER.

851 853 855 The Rookery...CHICAGO.

SIMPLEX WIRES

**INSURE
HICH
INSULATION**

SIMPLEX

Ever Onward and Upward!

Simplex Electrical Co.

620 Atlantic Ave.,

George Cutter, Chicago.   BOSTON, MASS.

XNTRIC

"That's the Switch"

And we control that movement.

H. T. PAISTE,

10 South 18th St.,

**PHILADELPHIA,
PA.**

Made   5 amp. S. P.
      10 amp. S. P.
       5 amp. 3 way.
      10 amp. 3 way.

P. & S.

WIRING INSULATOR,

Saves TIME
TROUBLE
and TIE WIRE

Made only by

Pass & Seymour,

SYRACUSE, N. Y.

George Cutter,

CHICAGO.

Consolidated Electric Co.

Manufacturers and Dealers in all kinds of

ELECTRICAL . SUPPLIES,

115 Franklin Street,

CHICACO.

...SEE AD...

Siemens & Halske Electric Company

PACE 9.

CHAS. A. SCHIEREN & CO.

MANUFACTURERS OF

Genuine Perforated Electric Leather Belting.

46 So. Canal Street,   -   **CHICACO.**

Section 15, Dpt. F. Clm. 27.   Section D, Space 3.
MACHINERY HALL.   ELECTRICITY BUILDING.

J. HOLT CATES,

Manager Western Dept.

THE WADDELL ENTZ CO.

**MULTIPOLAR
DIRECT CONNECTED DYNAMOS.
SLOW SPEED MOTORS.**

1122 Monadnock Block,   -   **CHICAGO.**

WAGNER ELECTRIC FAN MOTORS

For Direct or Alternating Currents.

These motors give a stronger breeze with less consumption of current than any other fan made on the market. They are full 1-8 horse power. Six bladed 12-inch fan. Self-oiling. Furnished with or without guards.

IT WILL PAY YOU TO SEE THE WAGNER BEFORE BUYING ELSEWHERE

TAYLOR, COODHUE & AMES,

348 Dearborn Street, CHICAGO.

See Our Exhibit of ELECTRICAL FIXTURES

IN SECTION "N", BETWEEN COLUMNS 62 AND 64, MANUFACTURES BUILDING,

CLOBE LICHT & HEAT CO.. 52 & 54 Lake St., CHICAGO.

Weekly World's Fair Supplement.

ELECTRICAL INDUSTRIES

DEVOTED TO THE ELECTRICAL AND ALLIED INTERESTS OF THE WORLD'S FAIR,
ITS VISITORS AND EXHIBITORS.

Vol. I, No. 6.　　　　　CHICAGO, JULY 20, 1893.　　　　FIVE MONTHS $1.00
　　　　　　　　　　　　　　　　　　　　　　　　　　　　TEN CENTS A COPY

Exhibit of Allgemeine Elektricitats Gesellschaft.

In the very extensive display made by Germany in the Electricity Building the exhibit of the Allgemeine Elektricitats Gesellschaft, Berlin, attracts universal attention. It trimmings. The whole partition is surmounted by a very ornate bronzed wrought iron sign giving the name of the company and carrying 220 16 candle power incandescent lamps. The trusses on the side next to the main aisle are concealed from view by decorated panels. The column in

FIG. 1. EXHIBIT OF THE ALLGEMEINE ELEKTRICITATS GESELLSCHAFT.

is situated in the German section just east of the long distance department of the Westinghouse Company. The partition wall between this exhibit and that of Felten & Guilleaume at the south end is handsomely paneled in brown cloth, the columns being of wood painted white with tinted the center of the exhibit is draped in brown, the ruling color of the decorations, and has two very handsomely ornamented arc lamps suspended from it. Just under these is a crown-shaped decoration of gilded wrought iron, supporting 80 or more incandescent lamps. White posts, sup perting heavy brown ropes with brass ornaments and tassels surround the exhibit and divide it into parts. At the two entrances to the exhibit are placed cones formed of coils of bare copper wires and cables.

The principal feature and the one which attracts much attention from the electrical engineer, is the power transmission plant, consisting of a direct current shunt wound motor running 620 revolutions a minute and with a capacity of 450 amperes at 500 volts. This motor is supplied with current from Machinery Hall and drives by a belt a three phase generator made especially for the exhibit. The generator has a capacity of 450 volts and 500 amperes, separately excited by a continuous current of 110 volts. It has 44 poles and the speed is such as to give 50 periods per second; the efficiency is said to be 92 per cent. The three phase generator drives through proper wire connections a 50 horse power three phase motor running at 725

revolutions and 120 volts pressure. It is said to have an efficiency of 91 per cent. This motor is directly connected through a brush coupling so called, to a direct current generator of 300 amperes capacity at 120 volts, the current from which is used to supply the arc and incandescent lamps for lighting the exhibit. A peculiarity in the construction of the direct current motor and generator used in this display is that a complete cast iron shell surrounds the armature between it and the faces of the pole pieces. The shell is about one inch thick over the pole pieces but thinner between them. It is pierced by a row of half inch holes half way between the pole pieces. The effect of this shell is said to prevent the changing of the position of the neutral point when the load is varied to any great extent.

The switch board, on which are located the instruments and switches for the control of the current used in the exhibit, is mounted on an iron frame and is arranged in three

panels. The front of the board is divided into two panels; the upper one is placed in a vertical position and contains the ammeters, volt meters, etc. The second panel, placed at an angle of 15 degrees from the horizontal and just below the first, contains the main switches and fuse blocks. The third panel, placed on the rear of the board, in a vertical position, contains the distributing bus bars, switches and fuses. The body of this board is made of the new insulating material called "stabilit", which is said to have all of the insulating properties of hard rubber and to be very much like it in every way excepting that it does not soften from heat. The particular material used on this board is of a reddish color, having the appearance of our American vulcanized fibre.

This switch board is supplied with all the necessary switches, cut-outs, fuses, bus bars and resistances, ampere meters, and volt meters necessary to control the current

for the various machines and lamps in the exhibit. The switches, and in fact all of the apparatus, show a very high degree of finish and have some points of excellence that might well be copied by American makers. The starting resistances for the direct current machines are situated underneath the frame of the board with the wheels for the same on the front. Another important feature in the exhibit is the theatrical outfit, consisting of an iron frame with a simple but comprehensive arrangement of levers and controlling stands. Tables are provided all about the space for displaying the minor products of the company. They are all handsomely draped in brown cloth and tassels. On a table near the main aisle is a sample of automatic resistance for controlling the shunts of dynamos and several samples of small three phase motors.

A handsomely arranged circular table displays a great number of samples of incandescent lamps with different

FIG. 2.—EXHIBIT OF THE ALLGEMEINE ELEKTRICITÄTS GESELLSCHAFT.

bases and bulbs variously decorated. A cone under a glass cover, on top of the table, shows the filaments in all stages of preparation. Another table, near the front on the main aisle, exhibits various sizes of direct current motors from one-sixteenth-horse power to six-horse power. Other tables are loaded down with material for outside construction such as are used for line construction and for railroad work. The various grades and sizes of volt meters and ampere-meters, switches and cut-outs in great number, instruments for testing and registering pressures, one of the latter deserving special mention in that it is a registering volt meter which makes a curve, showing the state of the voltage for the entire 24 hours. The fuse-cut-outs made by this company are similar to the Edison screw base cut-outs. The plugs for the different capacities of fuse are of different lengths, graduated from that of the smallest capacity in the longest plug to that of the greatest capacity used in this form of cut-out which is the shortest plug. The screws are the same for all sizes but it will be seen that the plug with the heaviest fuse if screwed into the base requiring but a small fuse will not reach the bottom and therefore will not close

the circuit. Nearly all metallic contact surfaces are tinned over to prevent oxidizing.

This company also exhibits a complete outfit of electrical heating devices such as teapots, flat irons, pipe and cigar lighters, and one instrument which has not been noticed among American exhibits of like material, being a long rod with a flat circular disc on the end containing the electrical heating device used for heating water in a pitcher or other receptacle. Numerous forms of switches for use with accumulator installations are shown. The special feature the company is exhibiting is a small shunt arc lamp which it is said will burn on constant potential circuits, constant current circuits and alternating current circuits, and with the same mechanism will burn on one ampere of current or any amount above that to 30 amperes at 45 volts. A small opal globe, about three inches in diameter, is so suspended from the upper carbon holder as to conceal the arc at all times, the regular globe being put on over the lamp in the usual manner.

Another feature is a system of self winding and self setting clocks that is attached to the regular low tension light

ing circuit of an electrical system of distribution instead of having special electrical circuits. An automatic device attached to the shunt of a dynamo and actuated by a master clock placed in the central station, at a certain time of day, reduces the regular potential of the circuit to 100 volts and the clocks by this action are set and wound. The whole arrangement takes but an instant of time and the shunt apparatus being immediately short circuited, the circuit potential returns to its former standard. A device for indicating the time of departure of trains and the stations at which they stop, for use in railway stations, is also shown. The exhibit is one of the handsomest shown in the Electricity Building, is very comprehensive and includes practically all the details required in either central station or isolated work either for lighting or power.

Exhibit of Queen & Company.

This company, located at the end of the German section in the northeastern part of the Electricity Building, is only a part of the company's entire exhibit at the Fair, as

THE EXHIBIT OF QUEEN & COMPANY.

they have many branches of business other than electrical. Outside of the few instruments necessary for use in the general study of physics which are located in the hall of Manufactures and Liberal Arts all the electrical apparatus made or sold by the company is shown here.

The location being at the end of a section allows of a very open arrangement of the show cases. A handsome designed office occupies the center, the cabinets and show cases being arranged around the outer edge of the space. The office besides being fitted out very comfortably with desks and other necessary furniture is supplied with a well stocked electrical library. On one wall arranged in tasteful designs are samples of the Carden hot wire volt meters and several sizes of magnetic vane volt meters and ammeters made by this company; these latter instruments while being very reliable and consuming little energy while in circuit are very reasonable in price. On another side of the office are arranged samples of Lord Kelvin ampere gauges reading in one instance as high as 6,000 amperes and of the type designed for switchboard use. There are also samples of switchboard form of multicellular volt meter of the same make. These voltmeters being of the electrostatic type consume no energy whatever when kept in circuit.

Back of the office and entered by a door leading from it, is a dark room for showing working samples of photometers, and in that connection the several forms of sight boxes, including the Lummer Brodhun type. There is also a very complete collection of Geissler crooks and spectrum tubes, which will be shown in use as soon as current can be furnished. In the show cases and cabinets outside of the office, while some attempt at classification has been made, no fixed rule has been followed, so that the instruments have been arranged with a view to decorative effect. In galvanometers there are shown several new forms of the reflecting type, all of them having the finely drawn quartz fibre suspension. Corrugated hard rubber support posts are used in nearly all cases for increase of insulation, and all parts of the instrument have been simplified so as to be easily and conveniently reached for adjustment.

In the high grade type of Thomson instruments the coils are all hinged in pairs on the frame of the instrument and contacts are all closed when the coils are shut against the frame. In some cases the terminals of the coils are brought outside the frame to a small plug switchboard by which all combinations of coils are easily and quickly made. The ballistic galvanometer shown is one of the finest and most complete of that type made. The suspension is quartz fibre and is got at by removing one screw from the outer case, which turns back and allows the front coil to be opened out; the needle is then entirely free for inspection. The peculiar construction of the needle and its magnets allow its sensibility to be increased or decreased without the introduction of any outside force. Other types are an Anthony tangent galvanometer of the Helmholtz Gaugain pattern; a modification of the Wiedemann dead beat reflecting galvanometer; a sample of the Moler swinging arm tangent instrument which has an exceedingly wide range of reading, and samples of the new Queen pattern of D'Arsonval galvanometer. This last instrument has a very delicate suspension, and is very convenient for use in intense and varying magnetic fields, its own field being so strong as to be entirely unaffected by outside influences.

Several types of scales, lamps and reading telescopes, as well as shunt coils for use with the galvanometers are also shown. One form of shunt coil shown is provided with a compensating coil which is introduced in addition to the regular multiplying shunt in order to make the resistance of the circuit the same as at first. There are also several new forms of discharge and contact keys. A very large assortment of resistance boxes, rheostats and bridge sets are exhibited, one very small testing set known as the Acme deserving especial mention, as it is but eight inches long by six inches wide and high, and yet includes a battery of 10 cells, a D'Arsonval galvanometer, a rheostat of 11,112 ohms, that will measure by means of the reversing arrangement, as high as 11 meg-ohms, a carbon rheostat useful in calibrating with large currents. Two large Anthony bridges which it will be remembered are provided with temperature coil and so arranged that the coils of the rheostat can be checked against each other, several forms of standard ohm coils which include one of a new type having besides the standard resistance a temperature coil and heating coil for determining temperature co-efficients.

In batteries the Carhart-Clark cell is shown. This is said to have an exceedingly small temperature co-efficient and is provided with a coil of 100,000 ohms in

series with a battery to prevent short circuiting. Samples of the Queen aluminium iodine testing battery are also on exhibition. This cell is said to have one and-a-half volts, and to be a great improvement over old forms. An extensive exhibit of Siemens dynamometers is made, the form shown by this company having but one support, a pivot suspension in place of fibre and various other detail improvements, and arranged in sizes from a five ampere instrument to 500 amperes. There are samples of the Mascart electrometer and also of the Ryan modification of the same. The latter instrument has taken very well and several of them have already been sold.

For school purposes a large case of statical electrical apparatus for lectures and a fine set of apparatus, consisting of a hand dynamo capable of furnishing either direct or multiphase currents, two motors for use with the same, and various other pieces of special apparatus are shown.

One unique exhibit is a large **Hertz** mirror for illustrating the oscillatory theory of electricity according to **Hertz**. Among the newer and special forms of apparatus shown is a new form of slide wire bridge embodying Willyoung's improvements for very accurate work, a set of apparatus for determining the difference in resistance according to the Cary-Foster method, and another which is the reverse of this for determining the conductivity of specimens. These last instruments are entirely new, and have not yet been placed freely on the market. The exhibit is in its line very comprehensive, and includes about every instrument required in the study and practice of electrical science. It is thought that a careful study of the instruments included will convince the most prejudiced mind that it is no longer necessary to go abroad for this class of material. The fact that the entire collection is valued at something like $15,000 will indicate the amount and class of material shown. A large number of these instruments have already been purchased by the Armour Institute of Chicago for use in their course of electrical engineering and their name will be found on many of the exhibits.

About half the material has never before been shown to the public, and has nearly all been made by Queen & Company. Mr. C. W. Pike is in charge of the exhibits, and is ably assisted by Mr. E. E. Keller, M. E. Both gentlemen take pleasure in showing and explaining the use of the instruments.

Exhibit of the Washburn & Moen Manufacturing Company in Electricity Building.

This well known firm has a very extensive exhibit of such of its products as are used in electrical work, in Section V in the east gallery of the Electricity building. A platform raised about 6 inches above the floor, surrounded by a handsome brass rail gracefully hung with brown cord and tassels covers the entire space. The name "Salamander," made up of incandescent lamps, is supported over the center of the front of the space, and American flags are gracefully draped on each side. A desk and chairs occupy one corner of the platform; a cabinet containing a battery of secondary cells used for showing the fire proof qualities of the new Salamander wire, occupies the center of the front side, and directly back of this is a large blackwalnut cabinet, or show case. Carpeted walks, railed off with brass, lead up to the large cabinet and around the testing platform.

On the left of the space next to the office, are arranged a

number of coils of bare copper rod. Next to them are six large reels of weather proof wire running from No. 8 to No. 14 B. & S. gauge. In the same space are also four large reels of weather proof insulated feeder wire of sizes between number 0000 and 0 B. & S gauge. The last mentioned reels are surmounted by coils of magnet and annunciator wire of various sizes and styles. One large reel of stranded cable 300,000 circular mills cross section; a reel each of number 0000 and 00, 37 strand feeder wire, and a coil each of No. 2 and No. 4 B. & S. gauge cotton covered magnet wire are shown, on top of which is a coil of bare twisted copper wire such as is used for lamp cord. By a peculiar process used by this company, it is made to retain its bright surface and not oxidize.

At the left of the testing cabinet and fronting on the aisle are a number of solid copper ingots just ready for rolling; a number of reels of magnet wire, coils of bare copper and iron wire, spools of fine annunciator wire and a ten-foot sample of an insulated copper cable about an inch

merous short samples of black and white covered wire are lined up under the scrolls and around the center cone.

The battery cabinet contains 15 cells of accumulator and through suitable terminals and switches on top the current is used to produce a great degree of heat in short samples of different makes of insulated wire. Samples up to six inches in length of No. 14 of different makes are joined in series with a similar sample of the Salamander brand. On turning about 120 amperes of current through this series the heat immediately destroys the insulation of all but the Salamander wire. At the right of the battery cabinet are coils of galvanized iron wire, bare copper wire and coils of various sizes of Salamander and cotton covered lamp cord.

On the corner at the extreme right is a glass show case containing materials used for the rubber insulation of wires, consisting of four grades of rubber, pure, red, white and black. The space also contains large reels of stranded railroad feeder wire of 500,000 C. M. cross section, a reel each of Nos. 0000, 00, and 0, rubber covered, taped and

EXHIBIT OF THE WASHBURN & MOEN MANUFACTURING COMPANY IN ELECTRICITY BUILDING.

and a half in diameter, cut from a piece, supplied a plant used for the electrolytic deposition of metals. In this same space are also sample coils of red, blue and green cotton and silk covered Salamander lamp cord, which lend color to the exhibit.

The glass paneled cabinet back against the wall contains as the most prominent object a half section of a large cone covered with alternate layers of red, blue and green Salamander lamp cord, surmounted by two copper whips made from copper rods about one half inch diameter, the size being reduced every three inches by drawing down, until at the end of the lash the copper is very fine, probably not larger than No. 40 B. and S. gauge, showing the quality of the material and the high degree of efficiency to which the workmanship and machinery has been brought.

Arranged on the bottom of this case are coils of all kinds of lamp cord, rubber, silk and cotton covered. The back of the case being lined with black velvet, the fancy scrolls and geometrical figures in copper and brass wire arranged on this dark background forms an attractive display. Nu-

braided, all being surmounted by a cone formed of coils of rubber covered wire in sizes running from No. 6 to No. 18 B. and S. gauge. Cables of iron wire for guys and cross over wires are exhibited in coils. The exhibit is very comprehensive, including as it does about all the numerous styles of electrical conductors made by the company, but especial prominence is given to the new brand called Salamander. This wire is insulated as follows: First, a thin layer of rubber to render the insulation water proof, then a layer of "Salamander," that is, a composition somewhat like a fine clay or kaolin which is fire-proof, this being covered with an outer braiding of soft white cotton over which is applied a heavy coat of black insulating paint. The covering, as made up, is said to be fire and water proof, and from the tests shown here there can be little doubt of the former. The company refines its copper from the ore electrolytically, having erected during the past year or two a large plant for this purpose. Mr. F. A. Warren is in charge of the exhibit and will show the numerous samples and explain the insulation to those interested.

ELECTRICAL INDUSTRIES.

Entered at Chicago Postoffice as second-class mail matter.

PUBLISHED EVERY THURSDAY BY THE

ELECTRICAL INDUSTRIES PUBLISHING COMPANY,
INCORPORATED 1892.
MONADNOCK BLOCK, CHICAGO.
Telephone Harrison 181.

E. L. POWERS, Pres. and Treas.   E. E. WOOD, Secretary.

E. L. POWERS,   Editor.
H. A. FOSTER,   }
W. A. REMINGTON,   }   - Associate Editors.
E. E. WOOD,   - Eastern Manager.
FLOYD T. SHORT,   Advertising Department.

EASTERN OFFICE, WORLD BUILDING, NEW YORK.
World's Fair Headquarters, Y 27 Electricity Building.

SUBSCRIPTION.
FIVE MONTHS $1.00
SINGLE COPY   10
Advertising Rates Upon Application.

News items, notes or communications of interest to World's Fair
Visitors are earnestly desired for publication in these columns and will
be heartily appreciated. We especially invite all visitors to call upon us
or send address at once upon their arrival in city or at the grounds.
ELECTRICAL INDUSTRIES PUBLISHING CO.,
Monadnock Block, Chicago

The Western Passenger Association meets tomorrow to again consider the question of World's Fair rates. It is hoped, for the good of the Fair and the people, the Association will decide on a general one way rate. The proposition to be submitted is an advance in the right direction, but why should people living within 200 miles of Chicago be excluded from this rate?

The gate receipts of last Sunday, which was the last open Sunday of the Fair, were appropriated to the benefit of the sufferers by the recent fire at the Exposition grounds. The fund raised by contributions large and small together with what has been raised from benefits given by concessioners and others has become large enough so that the sufferers will be relieved from want. The action of the directors in thus disposing of the receipts on the last open Sunday was most appropriate.

In another column will be found a list of the jurors so far appointed in the Department of Electricity. The selection of the Jury of Awards has been a most difficult task but has thus far been most creditably done. In looking over the list of names the number of men connected with educational institutions is noticeable. It was advisable to select men not connected with electrical companies and for that reason the choice was confined almost entirely to prominent electricians connected with educational institutions. At a meeting of the jury held on Tuesday afternoon Prof. H. S. Carhart was elected president.

Last week the Board of Directors by a vote of 24 to 1 decided to close the Fair on Sunday. This decision of the directors seems to give, with a few exceptions, universal satisfaction. It will undoubtedly be a relief to the public to know that the matter is settled and the army of employes at the Fair will appreciate a day of rest each week. The wish of the people in this matter was shown by their staying away from the Fair on Sunday; after a sufficient time for making a test the above decision has been reached. Chicago has many beautiful parks and places of interest, so that visitors if they desire will find abundant places to spend the Sabbath, and the increased attendance during the week will more than make up for the slight profit there would have been secured on Sundays.

The German Commission is arranging some very handsome decorations in the center of the east gallery. The plan is copied from the memorial to Dr. Von Siemens erected by the Emperor at Berlin. The panels will be of finest crimson silk plush with gold trimmings. A statue of Siemens is shown in the back ground while arranged in the foreground on tables are numerous cases of instruments and apparatus of historical interest.

The Hope Electric Appliance Co. of Providence is showing a number of interesting pieces of apparatus in the Gallery of the Electricity Building Section S. A wooden arch over the center of the space supports a number of incandescent lamps. A converter on a frame back of the arch supplies current for the lamps and serves to show the special double-pole dead cut-out switch for opening the primary circuit of an alternating current. The switch is in an iron box, has a very quick motion, is insulated with porcelain and can be made with or without fuses. Samples of series are circuit cut-out switches are also shown with the mechanical action and construction very similar to that of the primary alternating switch mentioned above. Other special switches are also shown in operation. A very clever device, Wright's automatic mast arm, is secured to one of the posts running up to the roof through the space. This arm consists of one long piece of pipe extending out from the post and strongly braced in all directions, secured to a casting at the post end and parallel to it underneath is a swinging arm, to the outer end of which the arc lamp is fastened. A thin tape of strong bronze metal about $\frac{3}{8}$ inch wide is attached to the lamp, runs over a pulley on the outer end of the stationary arm, through that pipe to the post where it runs over another pulley and down the pole to a windlass. By paying out the tape at the windlass the lower arm swings down with the lamp attached. The wires from the circuit are first carried to a series cut-out on the post, from thence through the movable arm to the lamp. In dropping the lamp for trimming the circuit is entirely cut off, making it absolute safe to handle. The windlass is boxed in to prevent trouble from storms, and the bronze tape runs in a cleat down the pole. The company has also a mast arm for use on streets where there are trolleys; the lamp is in this case drawn in to the pole on the same level by means of an endless cable the circuit wires being looped for that purpose. The company owns numerous patents on and is making a specialty of these devices.

The Electricity Building Gallery Exhibitors' Club.

During the past week an organization has been perfected of the exhibitors in the galleries of the Electricity Building. It is called the Electricity Building Gallery Exhibitors' Club, and the object of it as stated is to advance the interests of the gallery exhibitors.

The officers are: President, Thomas R. Lombard, of the North American Phonograph Co.; Vice President, C. E. Lee, Chicago; Secretary and Treasurer, M. J. Sullivan, of the Electrical World. At a meeting on Monday afternoon a resolution was passed that members should contribute five dollars each to the treasury for a preliminary fund for immediate expenses, and committees were appointed as follows: General Committee: C. E. Lee, George Clark, Herr Lobach, M. J. Sullivan, Dr. A. Owen. Committee on entertainment: Dr. Waite, E. Nashold, Prof. J. P. Barrett, Mr. Eckert, Dr. A. Owen. Committee to confer with American Exhibitors' Association: Lieut. E. J. Spencer, M. J. Sullivan, C. C. Breckner.

WORLD'S FAIR NOTES.

The exhibit of the Newark factory of the Westinghouse Company is fast approaching completion and by the middle of next week will be in very good shape.

During the severe storm of Thursday last lightning struck the top of the northeast tower of Machinery Hall tearing the staff off for a considerable distance.

The Equitable Manufacturing and Electric Company is showing an exhibit of printing telegraphs in the gallery of the Electricity Building under the big Westinghouse sign.

Much interest is still manifested in the British Post office exhibit and Mr. Chapman, who is in charge, is kept constantly busy explaining the different pieces of apparatus to interested listeners.

In order to have some repairs made the Wellington restaurant in the west gallery of the Electricity Building has been closed for the present. This restaurant has been one of the most popular on the grounds.

The Excelsior Electric Company has nearly completed its exhibit. The decorative arrangement of fancy colored lamps and moving illuminated signs will probably be entirely complete by the end of this week.

Mr. E. Nashold, western manager of the Electric Heat Alarm Company, has added new decorations to his exhibit in the west gallery. A canopy over the top with neatly draped curtains at the sides and an electric fan has made the booth attractive and comfortable.

Jury of Awards Department of Electricity.

About one half the total number of judges for the Department of Electricity have now been confirmed. The names are as follows:

Henry S. Carhart, University of Michigan.
Harris J. Ryan, Cornell University, Ithaca, N. Y.
Benj. F. Thomas, Ohio State University, Columbus.
Geo. F. Barker, University of Pennsylvania, Philadelphia.
T. C. Mendenhall, United States Coast and Geodetic Survey, Washington, D. C.
Robt. B. Owens, University of Nebraska, Lincoln.
M. O'Dea, Notre Dame, Ind.
W. M. Stine, Chicago.
Sam'l Reber, U. S. Army.
Henry A. Rowland, Johns-Hopkins University, Baltimore.
E. P. Warner, Chicago.
Dr. Chas. E. Emery, New York City.
A. E. Dolbear, Tuft's College, Mass.
W. E. Ayrton, F. R. S., London.
Geo. Forbes, M. A., F. R. S., London.
Director Rathenau, Berlin.
Danrath Ulbricht, Berlin.
Pierre Dehausse, Belgium.
Wm. Shrader, University of Missouri.
S. Brown Ayres, Tulane University, Louisiana.
D. C. Jackson, University of Wisconsin, Madison.
S. Thompkins, Clemson College, S. C.
R. W. Pope, Sec. Am. Inst. E. E.

An informal meeting of the jury was held at 2 o'clock Monday afternoon in the office of Prof. Barrett, nine members being present. The meeting was called to order by Dr. Barrett, who acted as temporary chairman, and a general discussion of methods of procedure took place. It was decided that the president of the jury should be an American and that two vice-presidents shall be chosen, one of whom should be English and the other German. Other officers were not decided on, and the candidates were not discussed.

How the Writing is Done.

In the crowd in front of the "W. E. Co." writing machine there is always some one ready to explain its operation. The following explanations were overheard in the course of an hour the other evening:

"O! Mr. Smith! isn't that perfectly lovely; I wonder how that funny little pointer lights those lamps."

"Why, Miss Jones, that pointer doesn't light the lamps. It doesn't touch them. I saw how it was done the first time."

"O! please tell me; I'm just dying to know."

"Well, the stick doesn't light the lamps but the current that lights the lamps makes the stick move. You know that a current of electricity will attract a magnet; just as in the telephone. When you talk into a telephone you turn the crank a while first; that charges the wire with electricity. When you talk, the vibrations of the air cause the electricity charged into the wire to vibrate too. The diaphragm at the other end is a magnet, and as the electricity in the wire vibrates the magnetic diaphragm is attracted and repelled; it follows the movements of the charged electricity exactly and so produces the sound. Now, in this case, there is a magnet in the end of the stick. As the current is turned on to light the lamps it passes along from one lamp to another and the magnet is attracted and follows the current, and that is what makes the pointer follow the letters."

"O, I see now; but I thought electricity went faster than that."

"So it does, normally. But when we pass it through an apparatus called a rheostat it absorbs so many lines of force that it is retarded. You saw the rheostat over by the column that retarded the lightning there."

"Perhaps they do that by a brush discharge from the end of the pointer. It could be done that way easily enough; but I think more likely it is a chemical process, because you see the wires on the back of the frame are painted with something. The stick may contain another chemical, or more likely is hollow and the chemical in the form of gas is blown through it against the chemical on the wires and so current enough is generated to light the lamps."

"Yes, I see; it could be done that way; but how does the stick put the lights out when it goes back?"

"There is probably an automatic device to reverse the current when it goes back."

"O, aunt Mary! you must see this! see! it looks as if that pointer lighted the lamps but it don't. George read all about it to me out of a paper the other day; that box is full of cans and things and the electricity goes through the cans before it gets to the lights and the pointer doesn't have anything to do with it at all."

"Say, Mickey, tell us what makes de stick go?"

"O, I'm onto it; dere's a kid in de box w'at wiggles de stick, see?"

"Aw, come off; de kid couldn't do it all day; he'd be tired."

"Say, you makes me tired; dere's a hole in de floor, and w'en de kid is tired he goes down and anudder one comes up, see?"

PERSONAL.

Prof. George Forbes, of London, is registered at the Victoria Hotel.

Prof. and Mrs. Harris J. Ryan, of Cornell University, are visiting the Fair.

Mr. Walter L. Flower, of the Acme Filter Company, St. Louis, is visiting the Fair this week.

Mr. C. F. Scott, of the Westinghouse Company, Pittsburgh, is visiting the Fair with his family.

Prof. W. E. Ayrton, of London, arrived Monday, in time for the first meeting of the Jury of Awards.

Mr. Wellington W. Cuomer, of the Cadillac (Mich.) Electric Light Company, and family were at the Fair last week.

Mr. Thos. W. Stewart, of the Thackara Manufacturing Company, Philadelphia, was a visitor at the Exposition last week.

Prof. Benj F. Thomas, of the Ohio State University, is in the city to visit the Fair and attend the meetings of the Jury of Awards.

Mr. S. A. Drake, superintendent of the People's Gas & Electric Light Company, Canton, Ill., is spending a few days at the Fair with his family.

Mr. A. A. Dion, superintendent and electrician of the Chaudiere Electric Light and Power Company, of Ottawa, Ont., is visiting the Exposition.

Mr. Clarence J. Reddig, treasurer of the Shippensburg Electric Light Company, of Shippensburg, Pa., while doing the Fair last week, called on ELECTRICAL INDUSTRIES.

Mr. Fred. A. Gilbert, of the Boston Electric Light Company, is spending a couple of weeks at the Fair and making a special study of the exhibits in the Electricity building.

Col. Henry S. Kearney, engineer of the New York Board of Electrical Control, is visiting the Fair and devoting a considerable time to the Electrical features of the Exposition.

Dr. Charles E. Emery, of New York, is visiting the Exposition and also attending the meetings of the Jury of Awards. The Department of Electricity should be congratulated on securing the services of Dr. Emery on the jury.

DEPARTMENT OF ELECTRICITY.

OFFICES: SECTION E, ELECTRICITY BUILDING.

Chief, JOHN P. BARRETT.
Assistant Chief, J. ALLEN HORNSBY.
General Superintendent, J. W. BEARDSELL.
Electrical Engineer, W. W. PRIOR.

DEPARTMENT OF MECHANICAL AND ELECTRICAL ENGINEERING.

OFFICES SOUTH OF MACHINERY HALL.

Mechanical Engineer, C. F. FOSTER
Electrical Engineer, R. H. PIERCE.
First Asst. Mechanical Engineer, JOHN MEADEN.
First Asst. Electrical Engineer, S. G. NEILER.

AMERICAN INSTITUTE OF ELECTRICAL ENGINEERS.

World's Fair Headquarters.
SECTION S. ELECTRICITY BUILDING.

RALPH W. POPE, Secretary.

Open from 9 a. m. to 5 p. m.

CHICAGO WORLD'S CONGRESS OF ELECTRICIANS.

OPENING SESSION, MONDAY, AUGUST 21st, 3 P. M.

ADVISORY COUNCIL.

President, DR. ELISHA GRAY, Highland Park, Ill.
Secretary, PROF. R. S. CARHART, Ann Arbor, Mich.

EXECUTIVE COMMITTEE.

Chairman, PROF. ELIHU THOMSON, Lynn, Mass.

COMMITTEE ON INVITATIONS.

Chairman, T. COMMERFORD MARTIN, 203 Broadway, New York.

COMMITTEE ON PROGRAM.

Chairman, PROF. T. C. MENDENHALL, Washington, D. C.

COMMITTEE ON FINANCE.

Chairman, R. E. SUNNY, 175 Adams Street, Chicago.

BUSINESS NOTES.

THE CHICAGO ELECTRIC WIRE COMPANY has secured from the government a contract for the wire for the submarine cables to be attached to the automobile torpedoes which are being tested at Willetts Point, N. Y. The contract calls for the expenditure of $75,000. This contract will take from one to two months to complete, and will require the uninterrupted work of the factory.

THE ELECTRICAL APPLIANCE COMPANY report that it is very interesting to note the way in which the orders commence to pour in for Meston fan motors after a few consecutive hot days. It has already sold several hundred of these machines and is filling large orders every day for country as well as city trade. A very large number is now in use at the Fair and the Electric Appliance Company expects to have several hundred motors installed there before the fan season is over.

THE ELECTRICAL ENGINEERING COMPANY, St. Louis, Mo., through its manager, Mr. E. G. Bruckman, has been awarded the contract for the installation of the electric light work of the new Union Depot at St. Louis. This depot is one of the largest of its kind and the electrical work is of correspondingly large size. The contract for some 6,000 lights required in the Union Trust Company's new building has also been awarded to the same company.

Amusements.

HOOLEY'S THEATER—Mr. E. S. Willard, in "The Middleman." Saturday evening, "The Professor's Love Story." 149 Randolph street.

COLUMBIA THEATER—Miss Lillian Russell, in "La Cigale." 108 Monroe street.

GRAND OPERA HOUSE — Sol Smith Russell, in "A Poor Relation." 87 Clark street.

AUDITORIUM—Imre Kiralfy's Spectacle "America." Congress street and Wabash avenue.

McVICKER'S THEATER—Denman Thompson, in "The Old Homestead." 82 Madison street.

CHICAGO OPERA HOUSE—American Extravaganza Company, in "Ali Baba, or Morgiana and the Forty Thieves." Eighth week. Washington and Clark streets.

SCHILLER THEATER—Chas. Frohman's Stock Company, in "The Girl I Left Behind Me." Eighth week. Randolph, near Dearborn.

HAVERLY'S CASINO—Haverly's United Minstrels. Wabash avenue, near Jackson street.

TROCADERO—Concert. Michigan avenue near Monroe street.

THE GROTTO—Vaudeville. Michigan avenue near Monroe street.

Buffalo Bill's "Wild West." 63d street. Daily at 3 and 8.30 p. m.

Pain's "Siege of Sebastopol." 60th street and Cottage Grove avenue. Tuesday, Thursday and Sunday nights.

This week Mr. Willard has returned to "The Middleman," with the exception of Saturday night, when he will appear in "The Professor's Love Story." The cast is the same as that of six weeks ago and the reappearance of this play has been greatly appreciated.

"A Poor Relation," in which Mr. Russell appears this week, has proved, it is said, to have been the most satisfactory in Mr. Russell's repertoire. This comedy was written by E. E. Kidder and has never failed to receive a large patronage.

Various scenes in "Ali Baba" seem to render it especially adapted for the summer season. The waterfalls, the forests, and caves, the moonlit garden, and other suggestions of cool retreats, seem to modify through the imagination the heat of the body. The American Extravaganza Company has done much toward making Chicago a profitable place during the summer, having opened the summer season seven years ago with the production of "Arabian Nights."

The great popularity of the spectacle "America" was shown last week when on the hottest nights of the season the theater was filled. Every World's Fair visitor comes with at least one evening reserved for seeing this spectacle, it has become so generally recognized as one of the great attractions of the World's Fair season. The Schaffers still hold the highest place in public favor. Persons present at the Wednesday matinee were fittingly reminded of the 100th performance by a handsome souvenir.

CLARK
ELECTRIC
COMPANY, NEW YORK.

192 Broadway and 11 John Street.

MANUFACTURERS OF ARC LIGHTING APPARATUS FOR EVERY PURPOSE A SPECIALTY.
The CLARK ARC LAMPS for use on EVERY CURRENT, have the reputation of being
the best and most durable of any ever made in the United States.

RAWHIDE PINIONS FOR ELECTRIC MOTORS
A SPECIALTY.
RAWHIDE DYNAMO BELTING

Greatest Adhesive Qualities   A Non-Conductor of Electricity

THE CHICAGO RAWHIDE MANUFACTURING CO.
THE ONLY MANUFACTURERS IN THE COUNTRY

LACE LEATHER ROPE
and OTHER RAWHIDE

GOODS
OF ALL KINDS
BY KRUEGER'S PATENT

75 Ohio Street,   CHICAGO, ILL.

STANDARD ELECTRIC COMPANY.

GENERAL OFFICES: 625 Home Insurance Building.

WORKS: So. Canal Street,

CHICAGO.

STANDARD SYSTEM

AT THE

WORLD'S FAIR.

MACHINERY HALL, Sec. Q, 2 Standard Arc Dynamos.
Sec. S, 20   "   "   "
ELECTRICITY BUILDING, Sec. P, Space 2, Arc Lighting Exhibit.

The Standard Lamps Light the Power Plant, Machinery Hall, Agricultural Hall, Shoe and Leather Building, and
Other Buildings and Portions of the Grounds.

See our Double Service All Night Lamp Before Buying an Old Style Two Rod Lamp.

ELECTRICITY BUILDING—EXHIBITORS AND THEIR LOCATION.

GALLERY.

MAIN FLOOR.

Exhibitor.	Section.	Exhibitor.	Section.	Exhibitor.	Section.	Exhibitor.	Section.
Austria		Electrical Review	Y	Jaeger, Chas. L	T	Reliance Gauge Co	T
Ansonia Electric Co	E	Electricity		Johns Mfg. Co., H. W	P	Rogers & Hossleter Chem. Co	S
Am. Inst. of Elec. Engrs	S	Electric Gas Co	R	Jewell Belting Co	P	Street Railway Journal	
American Battery Co	P	Electrical Engineer		Jenney Elec. Motor Co	T	Stranger Abi. Telph. Co	T
Astpid, R. M		Electrical World		Knapp Electric Works		Standard Paint Co	T
Alkg. Elec. Gesellschaft	D	Eddy Electric Motor Co		A. P. Elec. Novelty Co		Spathide, C. A	S
Bates Mfg. Co	Y	Electron Electric Co	B	Knapp & Buckley		Star Iron Tower Co	W
Bryant Electric Co		Electrical Forging Co	B	Kennedy Electric Co	L	Spoth	
Billings & Spencer	R	Equitable Brbnrc Co	P	Lawson, H. J	Y	Schlepps, Chas. A. & Co	D
Brixey, W. R	T	Eddgraw Mfg. Co	P	LeClanche Battery Co		Schomburg & Sohne	
Belknap Motor Co	E	Electrical Conduit Co		McNeal Tender Elec. Co	I	Siemens & Halske	
Bell Telephone Co	K	Eureland		Mavror, W. N		Shockert & Co	E
Brush Electric Co	L	Singer Chain Works	S	Meeker, Dr. G	N	Short Electric Co	E
Caldwell El. Cloth Cut Mch. Co	Y	Franklin Elec. Applmenit		McIntosh Bat. & opt. Co	W	Sperry Elec. Railway Co	L
Consol. Elec. Storage Co	P	Frank Piano Exhibit		Munson, C. Belting Co	D	Standard Undeg. Cable Co	L
Cutter, George		Felton & Guellraume		Mather, A. C	K	Standard Electric Co	P
Cutton Elec. Co	T	Fenner	K-P	Mather Electric Co	M	Samson Battery Co	
Chicago Elec. Wire Co	T	Ft. Wayne Elec. Co	M	Newman Clock Co		Tar-Ant. El. Signal Co	
Copenhagen Fire Alarm Co	S	Gandt & Co., N. Y		New S quote Watch Co		Todd, Applegate Co	
Central Electric Co	A	Gamewell Fire Alarm Co	N	N. Y. Insulated Wire Co	P	Taylor, Goodhue & Ames	
Commercial Cable Co	T	Genteel Electric Co B-B-N-C & J		National Carbon Co		Thomson Elec. Welding Co	
C & C Elec. Motor Co	A	Gorpenl Jewell's Air L Co		Novao Elec. Wire Co	P	Tehantogogh, Elisha Gray	
Cleveland Elec. & Mfg. Co	A	Grnly, E. S. & Co		North Am. Phonograph Co	N-P	Union Electric Co	W
Chicago Belting Co	K	Germanng		N. Y. & L. E. A		Vetter, J. C. & Co	
Daniggn Clock Co		Hubbard, Mfr. & Co		Nat. Ant. Fire Alarm Co	N	Wabb, C. F	Y
Department of Electricity	B	Hateman, C. J	N	Nat. Engraving Machine Co	N	Weston El. Instrument Co	U
Eastern Al. Industries		Hart & Hegeman Mfg. Co		Owen, D. C	S	Watchorn & Morn Mfg. Co	V
Elec. Tassell & Mac. Co		Hoge Elec. Applianc Co		Phoenix Glass Co	N	Western Union Tel. Co	
Electric Seanator	T	Hall, Chas. V		Prate, H. T		Waite & Bartlett Mfg. Co	
Edgerton, E. M	T	Holmes, S. S	W	Pulvermacher Galv. Co	T	White. S. S. Dental Mfg. Co	
Stern Telephone Co		Hermon & Bennis		Pompelly, J. S	T	Western Electrician	
Edison Elec. Mfg. Co	S	Hinman & Van Winkl		Pratt El. Med. Sup. Co		Wilder Ant. Burglar Al. Co	
Entrepreur. Elec. Co		Hoyt, J. M		Ponall, Reg. & Co		Weston Electric Co	
Eureka Trade. Supper	I	Heidtmann, E. & Co	P	Philge, A. R		Westinghouse El. & Mfg. Co	D-B
Electric Appliance Co	S	Rhode Albig Co		Page Belting Co	D	Ward & Stoughton	
Elec. Sel. & Sig't Co		Internal. Ins. L't & P't Co		Queen & Co	K	Wing, L. J. & Co	P
Electric Heat Alarm Co	Y	India Rubber Comb Co		Ringler, F. A	R	Zucker & Levett Chem. Co	F

ELECTRICAL SUPPLIES

We confine our attention entirely and strictly to electrical supplies. We are

NOT COMPETING

with our own trade in the sale of electrical machinery or in electrical contracting.

ELECTRIC APPLIANCE COMPANY,
242 Madison Street, CHICAGO.

We are thus enabled to be more in sympathy with the trade and by giving entire attention to one line of the business can secure more satisfactory results for ourselves and our customers than would be attained by trying to spread ourselves over every department of the electrical field.

ELECTRICAL SUPPLIES.

THE MATHER ELECTRIC CO,
MANCHESTER, CONN.

Dynamos, Motors, Generators,

Offices, 116 Bedford St., BOSTON.
—AND—

1002 Chamber of Commerce Bldg., CHICAGO.

THE "NOVAK" LAMP.

CLAFLIN & KIMBALL (Inc.)

General Selling Agents.

116 Bedford Street, BOSTON.

1002 Chamber of Commerce Bldg., CHICAGO.

Enterprise Electric Company

307 Dearborn Street,
Chicago ....

GENERAL WESTERN AGENTS

N.Y.I.R.

Manufacturers' Agents and Mill Representatives for

Electric Railway,
Telegraph, Telephone and
Electric Light

SUPPLIES OF EVERY DESCRIPTION

Agents for   Cedar Poles,
Cypress Poles, Oak Pins,
Locust Pins, Cross Arms, Glass
——Feeder Wire,   Insulators,

WIRES, CABLES, TAPE and TUBING

Siemens & Halske Electric Company
of America.
Chicago, Illinois.
Electrical Machinery.

Siemens & Halske,

Berlin,
Charlottenburg,
Vienna,
St. Petersburg,

Map of Chicago.

Showing Location of its Electrical and Allied Business Interests, Principal Hotels, Theatres, Depots and Transportation Lines to the World's Fair Grounds. (Index numbers refer to the black squares.)

—Lundell—
Suspended Fan Outfit

OKONITE

OKONITE

Black

Japan

Finish

Self-Oiling
....and
Self-Aligning
Bearings

Electric Fans

Electric Fans

OKONITE WIRES
OKONITE === TAPES === MANSON
INTERIOR CONDUIT.

Batteries, Bells, Push Buttons, Annunciators, Volt Meters, Ammeters, Wheatstone
Bridges, Line Wire Cross Arms, Brackets, Pins, Insulators, Tools.

GENERAL SUPPLIES.

CENTRAL ELECTRIC CO.
116-118 Franklin Street,
CHICAGO, ILLS.

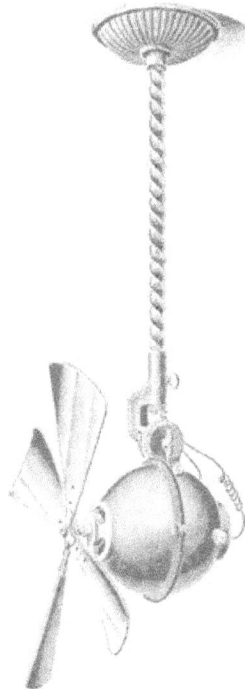

Books for Electrical Men.

THE MEASUREMENTS of ELECTRICAL CURRENTS and OTHER ADVANCED PRIMERS of ELECTRICITY by Edwin J. Houston, A. M., 429 pages, 169 illustrations, price $1. An elementary electrical treatise for students and non-technical readers giving in simple but exact terms the principles and apparatus upon which are based the practical operations of electrical measurements.

PATENTABLE INVENTION by EDW. S. RENWICK. A brief and concise summary giving the Law of Patents for inventions. An invaluable work of reference for every inventor. Bound in Sheep, price $2.00.

THE ELECTRIC RAILWAY in THEORY and PRACTICE by O. T. Crosby and Dr. Louis Bell. Second edition revised and enlarged. 416 pages, 182 illustrations. Price $2.50. A practical work on the electric railway that should be in the hands of every railway man.

ELECTRICITY and MAGNETISM by Prof. Edwin J. Houston, A. M. 306 pages, 116 illustrations. Price $1.

ANY OF THE ABOVE WORKS SENT POSTPAID ON RECEIPT OF PRICE.

Electrical Industries Publishing Co.
Monadnock Block, CHICAGO.

Western Electric Company,

CHICAGO.   NEW YORK.

Arc Lighting Apparatus
   High and Low Tension,
      Double and Single Service Lamps,
         All Night Single Lamps,
            Theater and Focusing Lamps.

ELECTRIC MOTORS

HIGH SPEED
VARIABLE SPEED
SLOW SPEED.

BUILT FOR SEVERE AND CONTINUOUS SERVICE.

SPECIAL TYPES FOR SPECIAL DUTY.

Cutter's "Boulevard" Streethood

THE LATEST NEATEST AND BEST.

GEORGE CUTTER,

851 853 855 The Rookery...CHICAGO.

SIMPLEX WIRES

INSURE
HIGH
INSULATION

SIMPLEX

Ever Onward and Upward!

Simplex Electrical Co.

620 Atlantic Ave.,

George Cutter, Chicago.   BOSTON, MASS.

XNTRIC

"That's the Switch"

And we control that movement.

H. T. PAISTE,

10 South 10th St.,

PHILADELPHIA,
PA.

Made 5 amp. S. P.
      10 amp. S. P.
      6 amp. 3 way.
      10 amp. 3 way.

P. & S.
WIRING INSULATOR,

Saves TIME
       TROUBLE
and TIE WIRE

Made only by

Pass & Seymour,
SYRACUSE, N. Y.

George Cutter,
CHICAGO.

Consolidated Electric Co.

Manufacturers and Dealers in all kinds of

ELECTRICAL . SUPPLIES,

115 Franklin Street,

CHICAGO.

GEORGE PORTER,
Contractor for All Kinds of

ELECTRICAL WORK.

Room 67, 143 La Salle St., CHICAGO.
Crary Block, BOONE, IOWA.

CHAS. A. SCHIEREN & CO.

MANUFACTURERS OF

Genuine Perforated Electric Leather Belting.

46 So. Canal Street,   -   CHICAGO.

Section 15, Dpt. F, Clm. 27,   Section D. Space 3.
MACHINERY HALL.   ELECTRICITY BUILDING.

J. HOLT GATES,
Manager Western Dept.

THE WADDELL-ENTZ CO.

MULTIPOLAR
DIRECT CONNECTED DYNAMOS.
SLOW SPEED MOTORS.

1122 Monadnock Block,   -   CHICAGO.

WAGNER ELECTRIC FAN MOTORS

For Direct or Alternating Currents.

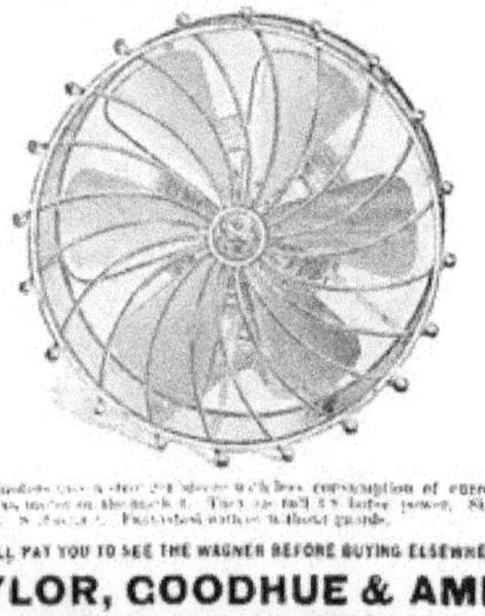

These motors give a clear and noiseless consumption of current than any other has in the market. They are full 8½ horse power. Six bladed 12 inch fan. No dust or Furnished with or without guards.

IT WILL PAY YOU TO SEE THE WAGNER BEFORE BUYING ELSEWHERE

TAYLOR, GOODHUE & AMES,
348 Dearborn Street, CHICAGO.

See Our Exhibit of # ELECTRICAL FIXTURES

IN SECTION "N", BETWEEN COLUMNS 62 AND 64, MANUFACTURES BUILDING,

GLOBE LIGHT & HEAT CO., 52 & 54 Lake St., CHICAGO.

Weekly World's Fair Supplement.

Electrical Industries

DEVOTED TO THE ELECTRICAL AND ALLIED INTERESTS OF THE WORLD'S FAIR, ITS VISITORS AND EXHIBITORS.

Vol. 1, No. 7.　　　　CHICAGO, JULY 27, 1893.　　　　FIVE MONTHS $1.00 TEN CENTS A COPY

Exhibit of the Brush Electric Company.

This exhibit, situated just west of the temple of the Bell Telephone Co., contains representative machines and appliances manufactured by The Brush Electric Company, the

The feature of the exhibit which attracts the most attention artistically is a square temple of Grecian style of architecture, at the south end of the space. The outer surface is painted white with the lightest of blue tinted trimmings. An entrance at either end, back of the row of Corinthian

FIG. 1.—THE EXHIBIT OF THE BRUSH ELECTRIC COMPANY.

earliest in the field with practical electric lighting apparatus. Although it was in the seventies that Charles Francis Brush constructed the first arc lighting dynamo bearing his name, and which by the way is exhibited here, little change is to be noted except in the substitution of a laminated armature core for one of cast iron.

columns supporting the roof, leads to a circular interior that is surmounted by a dome. At the right and left are doors leading to closets and store rooms. The central post supporting the dome is surrounded by an upholstered seat for the accommodation of visitors.

The object of the peculiar shaped interior of this temple

is to show a method for lighting theatres, churches and other large assembly rooms. Incandescent lamps are entirely concealed from view behind the frieze at the base of the dome. The light from the lamps is thrown upward on to the surface of the dome and from there disseminated throughout the room, lighting the whole interior evenly and softly.

The required effect is secured without a dazzling light tiresome to the eyes. Mr. I. R. Prentiss, who is in charge of the entire exhibit, has advocated this system of lighting for some time, and the results seem to justify him. The temple was constructed by the Henry Dibblee Company of Chicago. Without the temple on the east side of the space are displayed the eight regular sizes of Brush standard arc lighting dynamos, from one light to 65 lights capacity, each equipped with ampere meter and dial regulator. The new Brush arc dynamo, called No. 9, having a capacity of 120 to 125 10 ampere arc lamps, at 500 revolutions, will be shown coupled direct to a Willans engine, running 460

this dynamo and 20 in each half of the field. One special feature is the flexible coupling between the motor and the dynamo; while it is perfectly rigid as far as rotation is concerned it is capable of movement slightly out of line; it consists substantially of a jaw clutch with four small teeth, two on each of the large flanges, with proper recesses to receive the teeth in the opposite flange. Another direct coupled exhibit is a 50 light arc machine, connected by a flexible coupling to a 21 inch 35 horse power shunt wound crane motor, 220 volts, at 800 revolutions. An old No. 7, 16 light arc machine is accompanied by a letter from the former owners, saying it had been in use for 14 years with no repairs other than segments and brushes.

Other dynamos, etc., shown are: a 40 light arc with 10 of the original lamps sold in 1881 and used until shipped here; a 100 kilowatt 1,000 volt constant potential dynamo, such as is used at the Calumet & Hecla mines for furnishing current to motors for pumping purposes; a crane motor showing the design of motors made in sizes of from

FIG. 2.—THE EXHIBIT OF THE BRUSH ELECTRIC COMPANY.

revolutions per minute and at that speed will have a capacity of 100 or more lights.

On the west side of the temple are arranged samples of the Brush direct current, constant potential, compound wound dynamos of 20, 30, 50, and 100 kilowatts capacity, the two first being 100 volts and the two latter 110 volts. While the regular copper wire brush is used on all these dynamos it is supplemented on the constant potential machines by one carbon brush in each set, which is pushed slightly forward of the others and prevents all sparking.

Another very interesting item in this exhibit is a 150 kilowatt 2,000 volt alternating dynamo of the standard Brush type with the armature coupled up to produce 110 volts for use in the exhibit at such times as it is necessary to supply current to the three wire system current for which is furnished by the Exposition at 110 volts.

This large machine is direct coupled to a 250 horse power 220 volt brush motor, running 600 revolutions per minute, the exciter being belted from the alternating current dynamo. There are ten bobbins in the armature of

five to 35 horse power; an iron clad mining motor, 15 by 23 inches 220 volts and running 750 revolutions; a 36 kilowatt and a 60 kilowatt 2,000 volt alternating current dynamo, the armature of one being out of the machine and the armature of the other being taken apart to show the construction. Sample converters are shown of the brush types.

In switch boards the company has a very fine exhibit, including some four or five different sizes and types. The regular service board used for the exhibit is constructed of white marble on an iron frame, and has all the switches for the various motors, dynamos, lamp and power circuits used in and around the exhibit. One special feature of the board is the jointing by an iron bar of two of the regular double pole, double throw, knife switches in such a manner that when thrown one way the incandescent lamp circuits are joined onto the regular three wire system, and when thrown in the opposite direction cuts the same circuits onto the two wire system of the large brush alternating current dynamo in the exhibit.

The fuses are all placed on the back of this board and consist of copper strips cut down to a width for the proper carrying capacity. The fuses are duplex, that is, there are two, one being in circuit until blown when by putting a plug in its socket, the remaining one is cut in until the first can be replaced. There is a considerable space between

FIG. 3.—THE EXHIBIT OF THE BRUSH ELECTRIC COMPANY.

the front and back of this board, used in this case for making the circuit connections, and the end of the space between is taken up by a white marble arc-board. The 50 double carbon arc lamps used about the exhibit are connected from this board.

Machinery Hall. On the east side is another arc board for 36 dynamos and 36 circuits, with the regulation safety plug and cable. A ground detector is placed on one end of the board, consisting of a bank of lamps with a revolving switch or commutator for cutting them in or out individually in order to balance up against the resistance of a ground. These boards are all made in panels, each complete in itself and having all its contacts perfectly insulated from the slate by hard rubber bushing. The plug used has a substantial hard rubber handle and a long sleeve of the same material slips over the brass contact piece to protect the hand from an accidental contact in drawing out or pushing in the plug. The plug socket is on the back of the board and at the end of a heavy bushing of hard rubber being joined to the other sockets and circuits by wire bush bars, the several panels being connected together in this manner. The south end of the board is used to show the different styles of panels. The center one is on hinges and swings like a door, allowing access to the interior for examination of the connections.

On the west side of the quadrangle is a four unit alternating incandescent board capable of caring for 5,000 lights. Each panel is complete for one circuit and four machines, having a volt meter, ammeter magnetic cut out, which by the way is placed on but one leg of the circuit in this system. A ground detector at the end of the board serves for all the circuits. A larger board of this same type but of 10 unit size is placed just back of the Brush carbon exhibit across the aisle to the west. It has a set of instruments for the feeders, a set for each dynamo and a magnetic circuit breaker for each line. The instruments for use with the dynamos placed at either end, the feeder switches and other apparatus being on the central panels. Near the center of the exhibit space is a long table or bench

FIG. 4.—THE EXHIBIT OF THE BRUSH ELECTRIC COMPANY.

In special designs of standard boards, there are shown in the northwest corner of the space types of the arc and alternating incandescent boards of different capacities, all on black enameled slate so placed as to form the four sides of a quadrangle. On the end towards the north is a duplicate of the arc board used in the Brush service plant in

covered with olive green cloth on which are displayed a number of the smaller parts of a dynamo, such as the commutator, and a lamp unassembled. Back of this is a test rack for arc lamps, with lamps hanging in place, the sheet iron testing globe on each for allowing the arc to be seen through colored glass without injury to the eyes. The sliding volt meter for testing the lamps to a standard is also shown.

Various other detail parts are displayed on benches, among others a commutator for a 250-horse power motor. Across the aisle to the west is the exhibit of Brush carbons including all sizes from the smallest up to those huge pieces used by the Cowles Smelting Company at Lockport, being three feet long by three inches in diameter. All are tastefully arranged in geometrical figures on a platform, with statistical tables showing growth from five millions production in 1882 to twenty-six millions in 1892. Numerous pictures of the different forms of apparatus made by the Brush company are placed about the exhibit and handsome painted signs are suspended over each portion of the company's space. Festooned from post to post between the are lamps marking the outline of the space occupied are handsome wreaths of staff setting off the outlines. A great deal of credit is due to Mr. Prentiss for getting his display in so good shape in spite of the many difficulties in the way. The scheme of interior lighting mentioned above and especially advocated by him is deserving of special attention.

Electrical Exhibit of the Brazilian Government.

The Brazilian government has recently installed a very interesting exhibit of electrical apparatus in Section V, of the gallery of Electricity Building. It consists mostly of telegraph instruments made at the government workshops, Rio de Janeiro, although many pieces of electrical apparatus for fire alarm, telegraph, torpedo work, and maps of the entire system of government telegraphs have been added.

The regular Morse instruments shown are finely finished and well made. Sample relays, resistance boxes, switchboards, which, by the way have all metal surfaces tinned to prevent oxidation, galvanometers and lightning arresters are displayed about the tables in the space. The naval department shows a set of instruments after the pattern of Lieut. Bradley Fisher's range finder for locating vessels over mines, with resistance boxes, testing and firing keys, also samples of electric exploders for use with mines and torpedos.

One very large map at the south end of the exhibit gives the location of all the telegraph lines now completed and under construction. Portfolios containing maps showing every section of telegraph line in the country are displayed at the table in the center. The country now has some 8,700 miles of line in operation with 17,400 miles of wire and 300 stations. Among special features is the peculiar porcelain insulator used by this government. It consists essentially of two parts, that is the insulator itself and the lead ball used to hold the same in place.

The insulator is of hard glazed porcelain with double pettycoat and secured onto an iron pin which fits into the cross-arm. The shape is somewhat like a truncated cone. In the top is a circular depression about an inch in depth and two notches are cut in opposite sides of this depression in which the circuit wire is laid. A hole in each of the other two sides at right angles with the notches serves to hold the cross wire which is pushed through over the circuit wire and bent at right angles at either end. The principal feature, however, is the lead or tin ball, about ¾ inch in diameter, which is applied to the wire in halves and being placed over the wire at the proper spot is dipped into solder and thereby securely fastened in place. The wire and ball are then dropped into place in

the top of the insulator and the tie wire passed through the small holes over the ball preventing it from rising and holding the wire securely in place, the insulation being of the very best. These insulators were made by the Siemens & Halske Co. in Berlin; as are the cables and many of the heavier pieces of machinery.

The booth is covered with a handsome canopy in the national colors of green and yellow and is surrounded by a substantial railing.

The decorations and arrangements reflect great credit on Capt. L. M. de Lemos Basto, the Brazilian commissioner who is also director of telegraphs in Brazil. An attendant is always at hand to explain all the interesting features of the exhibit in many languages.

The Photophone.

During the past week the American Bell Telephone Co. has been making experiments with the photophone, in the transmission of sounds by a ray of light. The transmitter placed in the west gallery of the Electricity Building was used to direct the rays into the receiver placed on the steps at the north end of the telephone temple.

The apparatus is very simple and the experiment is said to be an unqualified success. The transmitter is made up of a very thin diaphragm of glass mirror set in a brass frame, with a mouthpiece facing the silvered side. The reflecting side receives the ray of light from an arc lamp directed through a strong lens. This ray is reflected from the mirror into a parabolic receiving reflector which concentrates the light on a small glass bulb filled with very dry burnt cork. Two wires are taken out of this bulb to small ear pieces similar to those used with the phonograph. The vibration of the transmitting diaphragm by the voice converges or diverges the ray of light, making it stronger or weaker in the parabolic receiver. The heating of the bulb of carbon is varied, thus causing a variation in the vibrations of the carbon which are transmitted to the ear pieces. The experiment is very neat although at present it is difficult to imagine to what practical use it can be put.

The Director General has issued an order to the departments and the department heads have issued a circular to all pass holders, ordering them to call at the department offices to have passes examined and approved. If such term passes are considered necessary and are approved by the department officials a form will be filled out for the pass holder who will then visit the office of the department of admissions either at 64th street or 62nd street, and there get his pass stamped, "Good on and after August 1, 1893." Truly the life of the pass holder is not one of unalloyed bliss.

One class of machinery seems so far to have escaped the touch of the electric motor to any extent, that is saw mill machinery; yet one would think that the high speeds at which such apparatus is run would invite the application of electric motors. A fine exhibit of wood sawing machinery is being made in a partition back of Machinery Hall, just west of the department of Mechanical and Electrical Engineering. Some of this machinery is so automatic and vigorous in its action as to appear almost human.

The Electrical World should read the World's Fair Supplement of Electrical Industries and applaud the publication of illustrations of machines that never saw the western continent, as World's Fair exhibits.

Exhibit of the Reliance Guage Company.

The Reliance Gauge Company, of Cleveland, Ohio, has one of the most attractive exhibits in the gallery of the Electricity Building, showing a full line of the well known Reliance safety alarm water columns for steam boilers. There are samples of the ordinary sizes in iron, brass and aluminum. Some very large sizes, one being shown like those used on the 1,000 horse power "Climax" boiler in the power plant in Machinery Hall.

Other sizes, ranging through all capacities and for every class of service are exhibited. The alarm whistle is usually a weak point for use in connection with high pressures of steam, but the samples shown by this company give evidence of being very strong and substantial. The Reliance safety alarm column is for use in the place of the ordinary water column and aside from having all the advantages of the plain column adds that a

patrons and friends and all others who are interested in the device, to whom he will be glad to explain the workings and good qualities. The boilers in Machinery Hall are equipped with the Reliance column, where they can be seen daily in practical operation.

Stinn & Co., of Vienna, Austria, have a small but very choice case of incandescent lamps in the northwest gallery of Electricity Building. It contains lamps of different design, candle power and color, also several styles of base.

The Star Iron Tower Co., is not to be beaten on decoration. It is painting a very handsome sign on the wall back of the sample tower in the west gallery of Electricity Building, which will be on a par with like decorations in other parts of the building.

The Newman Watchman's Clock exhibit has been moved

EXHIBIT OF THE RELIANCE GAUGE COMPANY.

safety device by which a signal whistle is blown in case the water in the boiler gets above or below a certain point. The column body is large and two oblong floats are placed in it, one at the top and one at the bottom; both join through simple levers to the valve of a small whistle placed on top of the casting. These floats hold a normal position as long as the water level is between the two pre-determined points; any abnormal change of level affects one or the other of these floats and the valve opens, allowing the steam to escape to the whistle and thus give the alarm to the fireman.

The whole device is very simple and effective. The exhibit is handsomely and tastefully arranged, the floor being covered with Brussels carpet, the stands with plush, and line olive draperies furnish the background for the exhibit. The highly polished brass and aluminum columns adding to the general luster. Mr. George B. Clark is in charge for the company and will be pleased to meet his many

to a much better location at the head of the northwest stairway of the Electricity Building, occupying the space formerly set aside for Spain.

A small but eminently practicable exhibit may be seen in the British section of the annex of the Transportation Building. It is Webb's Electric Tube Cutter, and consists of a small electric motor on the end of a four foot bar, with a guide and gauge for use in removing surplus tube after placing the same in the heads of a boiler.

So far the month of July has seen little increase in the attendance at the Fair. The total admissions for the month will barely exceed the 2,675,000 of the month of June. Even the more conservative estimates have not been realized. The warm weather, which on the whole has been very moderate, has not been conducive to sight seeing.

ELECTRICAL INDUSTRIES.

PUBLISHED EVERY THURSDAY BY THE

ELECTRICAL INDUSTRIES PUBLISHING COMPANY,

INCORPORATED 1893.

MONADNOCK BLOCK, CHICAGO.

TELEPHONE HARRISON 150.

E. L. POWERS, Pres. and Treas.   K. E. WOOD, Secretary.

E. L. POWERS,   Editor.

H. A. FOSTER,   }
W. A. REMINGTON, }   Associate Editors.

K. E. WOOD,   Eastern Manager.

FLOYD T. SHORT,   Advertising Department.

EASTERN OFFICE, WORLD BUILDING, NEW YORK.

World's Fair Headquarters, Y 27 Electricity Building.

FIVE MONTHS $1.00
SINGLE COPY   10
Advertising Rates Upon Application

News items, notes or communications of interest to World's Fair Visitors are earnestly desired for publication in these columns and will be heartily appreciated. We especially invite all visitors to call upon us or send address at once upon their arrival in city or at the grounds.
ELECTRICAL INDUSTRIES PUBLISHING CO.,
Monadnock Block, Chicago

An improvement that might be made in the Department of Electricity and also in other departments at the Exposition is the institution of more live exhibits in which electrical instruments and machines great and small are being made or assembled. The popularity of and interest taken in this class of exhibits is shown by the crowds that collect around the exhibit of the Sperry Company while they are winding armatures or some of the exhibits in the gallery where gold and silver plating is being done. A greater number of such exhibits, especially in the gallery would be an improvement.

The decision of Judge Seamans in the lamp case at Milwaukee was watched for with great interest. The defendants, the Electric Manufacturing Company of Oconto, Wis., had followed out the same line of defense as that taken by the Beacon Vacuum Pump Company at Boston, which was unsuccessful, and also followed by the Columbia Incandescent Lamp Company at St. Louis, and which in that case was successful. Thus each side victorious in a similar case had endeavored to strengthen its case by additional testimony so that the decision of the judge was difficult to predetermine. The decision which grants to the plaintiffs the injunction asked for, strengthens the decision of Judge Colt, but the next step will be watched with interest.

The Jury of Awards, Department of Electricity.

In addition to the appointments published last week one more name has been added, Adolpho Aseloff, of Brazil. Prof. W. C. Anderson, of Virginia is secretary to the executive committee.

A committee consisting of Messrs. Pope, Thomas and Ulbricht appointed to formulate rules for the guidance of judges, arranged the following schedule:

I.   Title of exhibit.
II. Description including diagrams or cuts when desirable.
III. Specific claims of exhibitor.
IV. Report (stating character, details and results of examination or tests).
V.   Finding of Judge.

1. As to originality defined by determining whether an exhibit has been in commercial use;

2. Merit.

a, Utility; b, Simplicity; c, Ingenuity; d, Reliability; e, Economic features; f, Workmanship; g, Finish; h, Design.

3. Claims sustained by examination.

VI. Recommendation as to award.

The committee suggests that there are undoubtedly many collective exhibits possessing no distinctive individual features entitled to an award; in other cases certain instruments or systems embraced in a collection may be of such special merit as to entitle them to distinct awards.

In such cases it is recommended that special awards be made for such parts of a collection as the judge may consider deserving of such honor.

It is proposed to test a number of the larger direct coupled dynamos for the purpose of ascertaining the consumption of steam and the electrical and commercial efficiencies of the apparatus under various loads. Tests of smaller apparatus of this kind may be decided upon later, but arrangements for the same are not yet completed.

Tests will be made for the purpose of helping the jury to form an estimate of the value of the exhibits. Where the jury are able to come to a conclusion without carrying out such tests they may grant awards based on an examination of the exhibits. It will only be practicable, in relation to the particular exhibits referred to herein, to test such representative apparatus as is located favorably in relation to steam supply and can be operated under a desirable load. It is, therefore, to be understood that the awards on the apparatus for which tests are herein provided are to be "granted upon specific points of excellence or advancement," the same as other exhibits, but that an abstract of the general results of the tests will be added to the report for an award, with references to the detailed report on the subject.

There are to be three classes of tests of each apparatus, to be designated A, B, and C. They are:

"A" tests: No less than one or more than three complete tests are to be made with each apparatus, to ascertain the economy in the use of steam and the relation of the indicated horse-power to the electrical output. One of such tests to be made with a load corresponding substantially with the maximum efficiency, one near the minimum allowable load, and one near the maximum allowable load.

"B" tests: Partial tests, supplementary to the above, without measuring the water evaporated, but varying the load by gradual increments through the practicable range, to be made for comparing the indicated horse-power with the electrical output.

"C" tests: These tests are designed to aid in distributing the energy developed in the engine in excess of that shown by the total output in watts to include friction tests of engine and a determination of the resistance caused by exciting the field with the main circuit open; they include also tests of electrical resistance, heating of armature, field, etc.

The order of tests provides that the apparatus shall be first operated at six-tenths of its maximum load for at least four hours before commencing the tests. The first complete or "A" test is then to be made, to be followed by the minimum "A" test, and that followed by the maximum "A" test. In case all of these tests cannot be made in succession the rule as to operating the dynamo four hours, at six-tenths load to be followed previous to each test. The same preliminary run to be made also before commencing the "B" tests.

The following sub-committees have been chosen:

Sub committee No. 1: To have charge of Groups 122,

123 and indicators, registering meters, ammeters and volt meters of Group 126. Messrs. Ayrton (chairman), Mendenhall, Rowland, Owens, Stine and Thomas.

Sub-committee No. 2: To have charge of Group 124. Messrs. Dolbear, Stine and Shrader.

Sub-committee No. 3. To have charge of Groups 125, 127 and 128, and j. of 138 a. Messrs. Carhart, Emery, Forbes, Jackson, Reber, Ryan and Rathenau.

Sub-committee No. 4: To have charge of Groups No. 126 and 129. Messrs. Thomas (chairman), Ayres, Owens (sec'y), O'Dea, Thompkins and Ulbricht.

Sub-committee No. 5: To have charge of Groups No. 130, 131 and 133 and L. M. of 138 a. Messrs. Barker, Ayres, Rathenau and Warner.

Sub-committee No. 6: To have charge of Groups 133 and 134. Messrs. Ayres, O'Dea (sec'y), Pope (chairman), and Ulbricht.

Sub-committee No. 7: To have charge of Group No. 135. Not yet appointed.

Sub-committee No. 8: To have charge of Groups 136, 137, 138 and 138 a. Not yet appointed.

Space has been secured on the ground floor of the Electricity Building just west of the Jenney Motor Company's exhibit for testing incandescent lamps; this space will be entirely enclosed by wire netting so the tests can be witnessed by visitors, no attempt being made to conceal any part except the photometer tests which necessarily have to be dark.

Space for battery testing has been secured on the west side of the building in that marked off for the Equitable Dynamo Co. The room at the northwest corner of the building and now used as an office by the Wellington Catering Co. will be fitted up for testing instruments.

Rooms for meeting of committees have been provided at the south end of the gallery between those of the Department and the American Institute of Electrical Engineers.

General meetings of the jury take place every day at 12 o'clock noon, the sub-committees meeting directly afterwards.

WORLD'S FAIR NOTES.

The new Wood 3,000 light alternator arrived early last week and was put in place Saturday; it seems to fulfill as far as appearance goes all the promises made as to its merits and as it is to be supplied with power one can soon test its running qualities.

A meeting of the Gallery Exhibitors Club was held last Saturday. Another committee was appointed, consisting of Messrs. Newman, Clark and Eckert, with power to provide signs, elevators, souvenirs, and such other devices as may in the opinion of the committee be desirable.

John Stephenson, the oldest and best known street car builder in the world, has an exhibit in the Transportation Building annex, showing his latest style of truck for electric street cars, a closed car with electric truck and a sample of the closed cable car, as made for the Broadway road, New York.

A very interesting exhibit at this day is that of A. S. Hallidie, of San Francisco, Cal., consisting of the first dummy or grip car, the first trailer used for the purpose and a section of first cable tube and grip, as originally used on the Clay street road in San Francisco, August 1, 1873. As compared with the present heavy and solid construction, the wooden braced casting used for the cable conduit, together with the wooden sleepers for the rails, makes one wonder how the cable road ever reached its present condition. A very fine collection of more modern cable grips is also shown.

Another instance of the liberality of the chief of the Department of Electricity is shown in the caps worn by the department officials. Recognizing that much trouble and annoyance would be saved the public if some way of designating an officer or employe of the department could be had, and knowing that the Directory would spend no money for the purpose, he supplied the caps at his own expense.

The perpetual motion crank has at last arrived. He reported at the Institute headquarters last Friday and talked one of the members into a fit that would enable him to get his brain composed again. He (the crank) said he had offered to explain his device to the Germans and French but they were so prejudiced that they said they would not go across the street to see it.

C. S. Van Nuis, of 136 Liberty St., New York, is showing a line of "Ajax" specialties in the annex to the Transportation Building, samples of the Ajax lightning arrester, which might be called a magazine fuse box, switchboards of enameled slate with Weston instruments, Ajax switches and fuses, also a glass show case containing various other samples and specialties. Albert & J. M. Anderson show in the same space the Boston pivotal trolley and a show case with numerous samples of line material, insulators, overhead frogs and switches, and a lot of other material for trolley line work.

In the present day of question as to the style and weight of rail to be used for electric roads, not to mention the kind of joint, one can do no better than to visit the very exhaustive and comprehensive exhibit made by the Gelenc Museum in the annex to the Transportation Building. From the old plank road and first short sections of cast iron rail laid on stone blocks up through the various styles of sleepers and with lap joints and very heavy fish or joint plates, nearly everything is shown as used on the European continent. Our American railway engineers may criticise, but will doubtless get some valuable points.

The Chicago Rawhide Manufacturing Company, Chicago, has a very nice exhibit in Section 15 J, 28 and 29, Machinery Hall, consisting of rawhide belting, both flat and round, together with samples of all the various goods of its manufacture. The flat belts shown consist of several rolls from one inch to twenty-four inches in width, while the round belting includes samples from 1-32 of an inch to 1½ inches in diameter. The flat belts include everything from small thin belts to dynamo and heavy belts. Rawhide lace leather is also shown in sides and cut laces, together with harness leather halters and straps. Rawhide pinions for electric cars, of which the company makes a specialty, are exhibited, together with a large roll of hydraulic packing. The company has, in addition to its display of manufactured goods, a large 38 inch belt in operation in Machinery Hall, running a Willans high speed engine in the British section; also an 18 inch belt running a smaller Willans engine. These belts attract a great amount of attention from their extreme smooth running. A large number of other smaller belts made by the company is also in operation on various parts of the grounds.

PERSONAL.

Prof. George Forbes was called to Niagara Falls last week on business.

Prof. Ulbricht, of Dresden, Germany, and a member of the jury of awards, has arrived.

Mr. E. E. Wood, eastern manager of the ELECTRICAL INDUSTRIES has been visiting the Fair for a few days.

Prof. N. M. Terry, of the U. S. Naval Academy of Annapolis, has returned home after a two weeks' visit to the Exposition.

Mr. W. B. Grimes, of Grimes Bros., electrical engineers and contractors. Grand Bend, Kas., is in the city visiting the Fair.

Prof. G. B. Owens, of the University of Nebraska, arrived last week and expects to remain in Chicago for a couple of months.

Mr. A. J. Martin, superintendent of the West End Light Company, of Philadelphia, spent several days at the Exposition last week.

Prof. A. E. Dolbear, of Tufts College, Mass., arrived in Chicago last week and is devoting himself to the work of the jury of awards.

Prof. Brown Ayres, of Tulane University, is now in the city and attending the meetings of the jury of awards, of which he is a member.

Mr. James J. Wood was at the Fair a couple of days last week looking after the erection of the 3,000 light alternator which had arrived.

Mr. Marcellus Reid, formerly with the Short Electric Railway Co., Cleveland, has been in the city visiting the Exposition for the past few weeks.

Mr. F. B. Starr, Jr., manager of the Gonzales Light and Power Company, Gonzales, Texas, accompanied by his father, recently made an extended visit to the Fair.

Mr. Aaron C. Wright, of the Hope Electrical Appliance Co., Providence, R. I., has completed arrangements for the handling of his company's goods by local agents in Chicago and other western points.

DEPARTMENT OF ELECTRICITY.

OFFICES: SECTION E, ELECTRICITY BUILDING.

Chief, JOHN P. BARRETT.
Assistant Chief, J. ALLEN HORNSBY.
General Superintendent, J. W. BLAISDELL.
Electrical Engineer, W. W. PENN.

DEPARTMENT OF MECHANICAL AND ELECTRICAL ENGINEERING.

OFFICES SOUTH OF MACHINERY HALL.

Mechanical Engineer, C. F. FOSTER.
Electrical Engineer, G. H. PIERCE.
First Asst. Mechanical Engineer, JOHN MEADEN.
First Asst. Electrical Engineer, S. G. NEILE.

AMERICAN INSTITUTE OF ELECTRICAL ENGINEERS.

World's Fair Headquarters,
SECTION S. ELECTRICITY BUILDING.

RALPH W. POPE, Secretary.

Open from 9 a.m. to 5 p.m.

CHICAGO WORLD'S CONGRESS OF ELECTRICIANS.

OPENING SESSION, MONDAY, AUGUST 21st, &c. &c.

ADVISORY COUNCIL.

President, Dr. ELISHA GRAY, Highland Park, Ill.
Secretary, Prof. H. S. CARHART, Ann Arbor, Mich.

EXECUTIVE COMMITTEE.

Chairman, Prof. ELIHU THOMSON, Lynn, Mass.

COMMITTEE ON INVITATIONS.

Chairman, T. COMMERFORD MARTIN, 253 Broadway, New York.

COMMITTEE ON PROGRAM.

Chairman, Prof. T. C. MENDENHALL, Washington, D. C.

COMMITTEE ON FINANCE.

Chairman, B. E. SUNNY, 115 Adams Street, Chicago.

BUSINESS NOTES.

Mr. W. L. ABBOTT, 419 La Salle street, Chicago, has been retained by the council of Rochelle, Ill., as consulting engineer for the municipal plant to be installed.

L. K. COMSTOCK, 1449 Monadnock Block, Chicago, has just completed several large wiring contracts, among which are the Columbus Memorial building, the New Era building, and the Polk St. depot.

QUEEN & COMPANY (Inc.) has in its exhibit of high grade instruments much that attracts attention. Some $3,000 worth of the instruments in the exhibit have been reserved for the Armour Institute, Chicago.

THE JENNEY ELECTRIC MOTOR COMPANY, Indianapolis, Ind., has had a very busy season at its western office, Chicago. A number of large orders and a steady stream of small ones has kept this office busy.

THE ELECTRIC APPLIANCE COMPANY, 242 Madison St., Chicago, reports a continued large demand for the swinging ball lightning arrester. It is a frequent sight to see a line of poles with some shaped boxes on every fifth or tenth pole, as 10 or 12 of the swinging ball lightning arresters are used on a circuit.

THE ANSONIA ELECTRIC COMPANY, Michigan Ave. and Randolph streets, Chicago, finds the warm weather has so stimulated the sale of fans that it has been almost unable to supply the demand. It makes a specialty of fans operated by primary batteries and this summer there has been a large demand for them. The Edison motor, operated by the Edison Lalande batteries has been found very popular and a large stock is carried.

THE BERGMANN SMOKE COMPANY, Cleveland, Ohio, report among recent sales the following: American Straw Board Company, Circleville, Ohio, 16 machines; Pittsburgh Water Works, 30 machines; Rocket River Paper Company, Potsdam, N. Y., 2 machines; Brown & Co. (Inc.), Pittsburgh, Pa., 8 machines; Bailey-Farrell Manufacturing Company, 7, and W. H. Birge & Sons, Buffalo, N. Y., 2 machines. The works of this company have been very busy this season and the outlook is very bright for the later summer and fall.

Amusements.

HOOLEY'S THEATER— Mr. E. S. Willard, in "The Middleman." Saturday matinee, "The Professor's Love Story." 149 Randolph street.

COLUMBIA THEATER— Miss Lillian Russell in "La Cigale." 108 Monroe street.

GRAND OPERA HOUSE — Sol Smith Russell, in "A Poor Relation." 87 Clark street."

AUDITORIUM—Imre Kiralfy's Spectacle "America." Congress street and Wabash avenue.

MCVICKER'S THEATER—Denman Thompson, in "The Old Homestead." 82 Madison street.

CHICAGO OPERA HOUSE—American Extravaganza Company, in "Ali Baba, or Morgiana and the Forty Thieves." Washington and Clark streets.

SCHILLER THEATER—Chas. Frohman's Stock Company, in "The Girl I Left Behind Me." Randolph, near Dearborn.

HAVERLY'S CASINO—Haverly's United Minstrels. Wabash avenue, near Jackson street.

TROCADERO Concert. Michigan avenue near Monroe street.

THE GROTTO—Vaudeville. Michigan avenue near Monroe street.

Buffalo Bill's "Wild West," 63d street. Daily at 3 and 8.30 p.m.

Pain's "Siege of Sebastopol," 69th street and Cottage Grove avenue. Tuesday, Thursday and Sunday nights.

The season of America, which ends in October, is about half over and still the attendance is only limited by the capacity of the Auditorium, which bids fair to continue to the end of the season. The trapeze combines, Basco and Roberts continue to afford unrestrained fun by their inimitable act in the Merry-mount scene.

The ninth week of the second successful run of Ali Baba began at the Chicago Opera House last Sunday night. It is certainly one of the most popular entertainments of the World's Fair season. While the spectacular, musical, comic and terpsichorean features please the adults, the amusing donkey, the extraordinary lion and the wonderful dragon, constructed from designs not known to naturalists, afford the greatest amusement to the younger members of the family.

Weatherproof Wire.

We are the largest manufacturers of weatherproof wire in the west.

We guarantee the quality of our wire to be equal, if not superior, to any other wire of like character in the market.

Special prices on application.

WESTERN ELECTRIC COMPANY,

227-275 South Clinton Street,

CHICAGO.

ELECTRICITY BUILDING—EXHIBITORS AND THEIR LOCATION.

GALLERY.

MAIN FLOOR.

HIGH GRADE ONLY.

This applies not only to our specialties but to our general line of supplies. Send for a sample of anything we carry and we will guarantee that you will take no exception to the above claim. Remember that our

"O. K." AND PARANITE

wires are the leaders in their two classes.

PACKARD LAMPS continue to burn as brightly as ever and the

MESTON MOTOR revolves with its accustomed effectiveness.

ELECTRIC APPLIANCE COMPANY,

ELECTRICAL SUPPLIES,

242 Madison Street, CHICAGO.

THE MATHER ELECTRIC CO.

MANCHESTER, CONN.

Dynamos, Motors, Generators,

Offices, 116 Bedford St., BOSTON.

— AND —

1002 Chamber of Commerce Bldg., CHICAGO.

THE "NOVAK" LAMP.

CLAFLIN & KIMBALL (Inc.)

General Selling Agents.

116 Bedford Street, BOSTON.

1002 Chamber of Commerce Bldg., CHICAGO.

Enterprise Electric Company

307 Dearborn Street, Chicago ....

GENERAL WESTERN AGENTS

N.Y.R.

Manufacturers' Agents and Mill Representatives for

Electric Railway, Telegraph, Telephone and Electric Light

SUPPLIES OF EVERY DESCRIPTION

Agents for Cedar Poles, Cypress Poles, Oak Pins, Locust Pins, Cross Arms, Glass Insulators, Feeder Wire,

WIRES, CABLES, TAPE and TUBING

Siemens & Halske Electric Company

of America.

Chicago, Illinois.

Electrical Machinery.

Siemens & Halske.

Berlin,
Charlottenburg,
Vienna,
St. Petersburg.

Map of Chicago.

Showing Location of its Electrical and Allied Business Interests, Principal Hotels, Theatres, Depots and Transportation Lines to the World's Fair Grounds. (Index numbers refer to the black squares.)

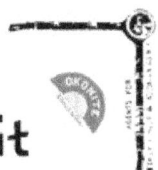

~Lundell~
Suspended Fan Outfit

Black

Japan

Finish

Self-Oiling
....and
Self-Aligning
Bearings

Electric Fans   **Electric Fans**

OKONITE WIRES
OKONITE ═══ TAPES ═══ MANSON
INTERIOR CONDUIT.

Batteries, Bells, Push Buttons, Annunciators, Volt Meters, Ammeters, Wheatstone
Bridges, Line Wire Cross Arms, Brackets, Pins, Insulators, Tools.

GENERAL SUPPLIES.
CENTRAL ELECTRIC CO.
116-118 Franklin Street,
CHICAGO, ILLS.

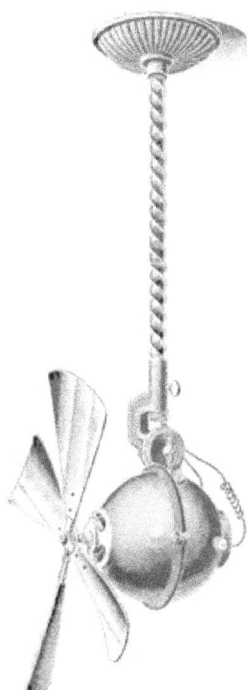

THE BRUSH ELECTRIC COMPANY,

CLEVELAND, OHIO.

World's Fair Exhibit, Section L, Electricity Building.

CLARK ELECTRIC COMPANY, NEW YORK.

192 Broadway and II John Street.

MANUFACTURERS OF ARC LIGHTING APPARATUS FOR EVERY PURPOSE A SPECIALTY.
The CLARK AR : LAMPS for use on EVERY CURRENT. have the reputation of being
the best and most durable of any ever made in the United States.

RAWHIDE PINIONS FOR ELECTRIC MOTORS
A SPECIALTY.
RAWHIDE DYNAMO BELTING

Greatest Adhesive Qualities   A Non-Conductor of Electricity
Carries Less Friction than any other Belt.

THE CHICAGO RAWHIDE MANUFACTURING CO.
THE ONLY MANUFACTURERS IN THE COUNTRY.

LACE LEATHER ROPE
and OTHER RAWHIDE

GOODS
OF ALL KINDS
BY KRUEGER'S PATENT

This Belting and Lace Leather is
not selected by steam or dampness,
never becomes hard, is stronger,
more durable and the most economical Belting made. The Rawhide Rings for Round Belting
Transmission is superior to all
others

75 Ohio Street,   CHICAGO, ILL.

STANDARD ELECTRIC COMPANY.

GENERAL OFFICES: 625 Home Insurance Building.

WORKS: So. Canal Street,

CHICAGO.

STANDARD SYSTEM

AT THE

WORLD'S FAIR.

MACHINERY HALL, Sec. Q, 2 Standard Arc Dynamos.
Sec. S, 20   "   "   "
ELECTRICITY BUILDING, Sec. P, Space 2, Arc Lighting Exhibit.

The Standard Lamps Light the Power Plant, Machinery Hall, Agricultural Hall, Shoe and Leather Building, and
Other Buildings and Portions of the Grounds.

See our Double Service All Night Lamp Before Buying an Old Style Two Rod Lamp.

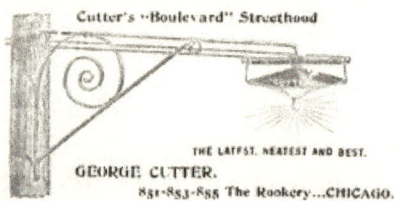

Cutter's "Boulevard" Streethood

THE LATEST, NEATEST AND BEST.
GEORGE CUTTER.
851-853-855 The Rookery...CHICAGO.

SIMPLEX WIRES

SIMPLEX
Ever Onward and Upward!

INSURE
HIGH
INSULATION

Simplex Electrical Co.
620 Atlantic Ave.,

George Cutter, Chicago.   BOSTON, MASS.

XNTRIC

"That's the Switch"

And we control that movement.

H. T. PAISTE,
10 South 19th St.,
PHILADELPHIA,
PA.

Made 5 amp. S. P.
10 amp. S. P.
5 amp. 3 way.
10 amp. 3 way.

P. & S.
WIRING INSULATOR,
Saves TiME
TROUBLE
and TIE WIRE

Made only by
Pass & Seymour,
SYRACUSE, N. Y.

George Cutter,
CHICAGO.

Consolidated Electric Co.

Manufacturers and Dealers in all kinds of

ELECTRICAL . SUPPLIES,

115 Franklin Street,

CHICAGO.

GEORGE PORTER,

Contractor for All Kinds of

ELECTRICAL WORK.

Room 07, 143 La Salle St., CHICAGO.
Crary Block, BOONE, IOWA.

CHAS. A. SCHIEREN & CO.

MANUFACTURERS OF

Genuine Perforated Electric Leather Belting.

46 So. Canal Street,   -   CHICAGO.

Section 15, Dpt. F, Clm. 27.   Section D, Space 3
MACHINERY HALL.   ELECTRICITY BUILDING

BEAR IN MIND

that the regular monthly issue of ELECTRICAL IN-
DUSTRIES contains the most complete and correct
directories published of the electric light central stations
and the electric railways in North America.

World's Fair Headquarters Y 27 Electricity Building.

CITY OFFICES, Monadnock Block.

WAGNER ELECTRIC FAN MOTORS

For Direct or Alternating Currents.

These motors give a steadier breeze with less consumption of current than
any other fan motor on the market. They are but 1-5 horse power. Six kinds.
15-inch fan. Self-oiling. Furnished with or without guards.

IT WILL PAY YOU TO SEE THE WAGNER BEFORE BUYING ELSEWHERE.

TAYLOR, GOODHUE & AMES,

348 Dearborn Street, CHICAGO.

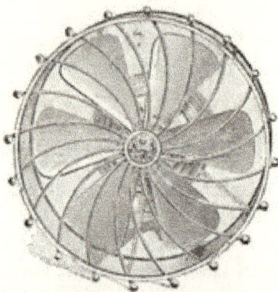

WEEKLY WORLD'S FAIR

ELECTRICAL INDUSTRIES

DEVOTED TO THE ELECTRICAL AND ALLIED INTERESTS OF THE WORLD'S FAIR, ITS VISITORS AND EXHIBITORS.

Vol. I, No. 8.   CHICAGO, AUGUST 3, 1893.   FIVE MONTHS $1.00
TEN CENTS A COPY

De Laval's Steam Turbine.

The exhibit of the De Laval steam turbine in connection with electric lighting is located in the Swedish section, K, 22, Machinery Hall. In these days of the large compound and triple compound steam engine, a novelty of this sort is sure to be interesting to steam engineers and users.

Steam turbines have been in use for some time past, and of late, after the development that comes with use, their economy has been made something great for such small machines. Instead of using a number of turbine wheels, one exhausting into the next and so on until the pressure of steam is wholly used, this machine has but one wheel

and in place of taking steam all around its circumference of blades at the same time, it uses a limited number of nozzles directed tangentially against one side of the wheel, admitting steam at great velocity against a portion of the blades. As the speed of this wheel is something terrible,

THE EXHIBIT OF DE LAVAL'S STEAM TURBINE.

being 30,000 revolutions per minute for the five-horse power machine; and about 22,000 for the 20-horse power. A unique style of gear has been invented to accomplish a reduction in the speed. This gear reduces the speed to one-tenth that of the turbine.

In the larger engines, as many as eight nozzles are used, four being arranged with valves that may be controlled from outside by hand, so varying the power and speed as two to one.

Great difficulties were met with in overcoming faults and troubles developed by the very high speed. However carefully the turbine may be manufactured, it is impossible, on account of unevenness of the material, to get its center of gravity to correspond exactly to its geometrical axle of revolution; and however small this difference may be, it becomes very noticeable at such high velocities. De Laval has nevertheless succeeded in solving the problem, by providing the turbine with a flexible shaft, the wheel disc being placed on the thin portion of the shaft and at a considerable distance from the bearings, the shaft for 20-horse power being about three-eighths of an inch in diameter for a distance of about four or five inches each side of the disc.

A yielding shaft allows the turbine at the high rate of speed to adjust itself and revolve around its true center of gravity, the center line of the shaft meanwhile describing a surface of revolution. If the shaft were stiff, the vibrations of the turbine wheel would be communicated to the bearings, which would then run warm. The purpose of allowing an adjustment of the wheel can also be accomplished by making the journals yielding; but this arrangement requires complicated and expensive details of construction. The bearing of the free end of the flexible shaft is provided with a ball fastening, in order to allow it to adjust itself to the shaft. Moreover, all the journals are provided with lubricating grooves and anti-friction metal, thus being specially adapted for effective lubrication.

The gear for the 20-horse power engine is about one and a quarter inches in diameter, with a double face for teeth, about two inches long over all. There are 23 teeth, there being screw threads of that number and of very steep pitch, thus forming a multiple spiral gear, the teeth on the two faces running in opposite directions, to prevent any end thrust. The teeth on the large gear are of like pattern, and as some 20 or 30 of them mesh at the same time and enter with a long sliding motion, the teeth meet with no shock and need not be excessively strong. These gears run with practically no noise. The governor is a simple, centrifugal device, not unlike many others, but of small and neat appearance. It acts on the throttle valve in the main steam inlet.

A copy of a certificate is shown, stating the results of an economy test made at Stockholm, Sweden, on the 13th of May last. In an eight hours run, the consumption of water is said to have been 19.68 pounds and of coal 2.67 pounds each per brake horse power. In connection with this engine, and to provide for the different sizes shown, there are several constant potential dynamos directly coupled to the engines. They are made after the Manchester type, but with different winding, owing to the high speed at which they are run. One is a dynamo with two armatures in the same frame, with the turbine gear meshing in between the armature gears. Leads from this machine can be coupled in multiple or series, as desired. A bank of incandescent lamps is provided for load, and may be cut on or off, as desired, to show the regulation of both dynamo and engine.

The exhibit, though small, is very finely arranged, the engines being placed on pedestals, to show that heavy foundations are not necessary.

A brass rail surrounds the space and in one corner is exhibited, in a handsome frame, medals and decorations presented to Gustaf De Laval, the inventor.

The machine is creating as much interest among engineers as did the other steam turbines when they came out.

Although it has been in use in Sweden for some three years, it has not previously been shown to the American public.

The International Electrical Congress.

Preparations for the meeting of the Congress of Electricians in this city on the 21st of August are progressing rapidly. Many inquiries being received regarding railroad facilities, etc. Mr. C. O. Baker, Jr., has been appointed chairman of the transportation committee by Dr. Elisha Gray; this alone guarantees the comfort of guests from the east. Mr. Baker will have his headquarters at the office of the National Electric Light Association, 136 Liberty street, New York, Secretary Porter having placed it at his disposal. Mr. Baker will soon issue a circular for the benefit of the eastern members and in all probability a special congress train will be run.

The Congress committee on invitations has already finished the larger part of its work but will probably be busy up to the very opening day. Upwards of 950 individual invitations have already been issued of which 900 are American and Canadian, including some of the foreign electricians attached to the World's Fair. General invitations have also been issued to twelve foreign electrical and scientific societies, of which the membership is upwards of 3,000, only a small proportion of which may be expected to attend.

It is believed that the attendance at the Congress will be at least 500. The individual invitations have all been canvassed by Dr. Gray and the committee on invitations and it is safe to say that a more representative body of scientific, professional and practical men has never before met in this country. Many very distinguished foreigners will attend. The program for papers and topics is in preparation and will be issued shortly before the meeting of the Congress.

Exhibit of the Sperry Electric Railway Company.

This exhibit, located just north of the Brush carbon display and across the aisle from the large exhibit of the Brush Electric Company, is unique in showing the company's ideas of what is the proper design of gearing for street car motors. Although the idea of beveled gears with one motor in the center is perhaps not new with this company, the designs shown are of recent construction. Trouble has heretofore been experienced from the very rigid connections necessary in order to keep gear teeth in good bearing. This trouble has been attacked by the Sperry Electric Railway Company and the result is shown on the McGuire truck equipped with a new type of Sperry motor and jacked up from the floor so that the motor may be shown running.

The special feature in this gearing, in which it differs from other similar types, is the flexible connection between the motor shaft and the gears at each end. This connection is formed of a series of links so arranged that while the shafts are freely movable out of line in every direction the rotating movement is as solid as though it were a straight clutch coupling. The pinion and gear are arranged in a solid casting in such manner that it is impossible for them to become separated in the least. A solid cast iron case encloses the gear and pinion against all dirt and dust and with a liberal amount of grease the gearing runs with very little noise. The motor itself is suspended from the side bars of the truck with its shaft at right angles with the axles and from this position is attached to the motors on either axle. This method of suspension allows the motor free play over rough ground and prevents many of the small evils due to the heavy jar from the usual methods of suspension.

Another truck near the first, having no platform shows the arrangement of motor, gears and connections. The first design was to have the link flexible connection encased to keep out dirt and other obstructions but by improving the design of the links they are now left open and free and are easily examined for defects. At the north end of the exhibit on a wooden frame is shown one of the company's street car motors with one half of the field thrown back to show the armature and arrangement of the fields. The type of motor is after the style of the old Wenstrom dynamo, having two field coils and four poles, two of them being consequent poles. The special feature of the machine which removed many of the difficulties attending this special design is the dividing of the casting at diagonal corners instead of in the centers either top or bottom. In this way about two-thirds of the surface of the armature is

commutator. This avoids much trouble due to the old methods of cross connections. Numerous sample armature cores are exhibited; some bare, some partially insulated and others partly wound.

One of the most interesting features of the exhibit and the one which attracts the most attention from both the lay and the professional public is the armature winders in one corner of the exhibit who are at work insulating and winding armatures for this railway motor. So much interest is manifested in this part of the work that it is astonishing that there is not more of it done in the Fair. On the west side of the space is a table on which are displayed various parts of the motor and accessories. The exhibit is surrounded by a handsome railing and a large sign in white and gold is suspended over the center of the space. This company claims to have received many orders for its particular style of car equipments and to have met with very satisfactory results on such roads as have so far been equipped. Special attention is called to the flexible connection

EXHIBIT OF THE SPERRY ELECTRIC RAILWAY COMPANY.

exposed to view for examination. One field coil can be entirely removed and all parts of the machine are as readily accessible as any type of motor so far examined.

The machine is very simple and much surprise is expressed by most visiting electricians to see it run with the top half of the machine entirely removed as it is here. The armature is of the iron clad type wound with ribbons of wire about three-eighths of an inch wide in radial slots cut in the surface of the core and secured in place after the usual manner of the day by wedge shaped hard wood sticks driven in under the converging edges of the slot over the coil of wire after it is wound. The connections between the armature coils and the commutator are made by flexible copper wire heavily insulated with cotton and the cross connections are accomplished by a heavy hard wood ring placed over the armature shaft and next to the commutator. Insulated flexible copper connections are wound on this ring before it is put in place so that all that remains is to connect to the ends of the armature windings one of the ends of these ring connections and the other end of the connections comes in proper position to connect directly to the

between the gearing and armature shaft as explained above and a thorough examination will amply repay anyone interested in this class of apparatus.

Exhibit of the Telegraph Construction Maintenance Company.

The problem of electrical communication between the shore and light ships anchored outside has been before naval boards for a long time. Direct cable connection has never been accomplished to the entire satisfaction of all concerned, owing principally to the fact that tides twist the ship around and entangle the cable and anchor chains.

An exhibit model shown by the Telegraph Construction and Maintenance Company, limited, 38 Old Broad street, London, near the center of the Transportation Building, seems to have eliminated most of the objectionable features and is claimed to have been in use from 1885 to 1889 on a light ship, nine miles off the east coast of England, until a committee of the Board of Trade decided that the expenditure involved was not commensurate with the advantages gained

The ship is anchored by a chain from the bow, which at a short distance below the surface of the water is attached to a heavy swivel having a hollow bolt. The chain divides here and two chains are led in opposite directions to anchors. The swivel forms a point, around which the ship swings.

The cable from the shore runs up through the hollow bolt of the swivel, and following the chain to near the ship enters a hole provided for the purpose, and following along the deck is led through an aperture to a drum below. This drum, holding a considerable length of extra cable, is so hung as to be capable of revolving not only on its own axis, but at right angles thereto, permitting twists to be taken out, in case the ship swings around often.

It was found that the regular stranded conductor was easily broken when the ship strained at the anchors in a storm, and a spiral conductor was substituted that is said to have given no further trouble in that line.

allowing considerable movement of the axle and shaft, avoiding any jar to the motor, which is suspended wholly from the arc truck.

A number of diagrams are shown stating the efficiency of this motor as compared with others with single and double reduction gearing.

A Dorner & Dutton improved flexible truck on one side of the exhibit is equipped with two of the latest type of 20-horse power gearless motors. It is supported above the floor leaving the wheels free so the motors may be run. Along side of this truck is a stand supporting the motor controller and rheostat so that both may be easily and closely examined. Other samples of the rheostat are also shown. It is made entirely of iron and asbestos and is most thoroughly fire proof.

One each of the 20 and 30-horse power motors with single reduction gearing are shown.

Among the larger exhibits is a 500-horse power, 500-

EXHIBIT OF THE SHORT ELECTRIC RAILWAY COMPANY.

In the United States, the Light-house Establishment has tried numerous methods, but never to their own complete satisfaction. With some few improvements, that could be easily added, it seems as if this system might be made successful.

Exhibit of the Short Electric Railway Co.

This exhibit is located just across the aisle to the west of the Greek temple of the Brush company. It embodies everything in the line of electrical apparatus for railway work, from the generator to the motor.

This company was one of the very earliest in the field with a practicable gearless motor and its faith still clings to it as is shown by the special exhibition made of that style of apparatus. The original type is exhibited and the different styles designed since are represented, up to and including the latest, which is constructed with three field coils and six poles half of them being consequent poles. This last motor embodies many new improvements, runs at 100 to 175 revolutions per minute, and has flexible link connections between the axle and hollow armature shaft

volt generator having six field coils on a side and a capacity of 300 kilowatts at 300 revolutions. A special feature of this machine is a device attached to one of the main bearings of the armature for moving that part endwise so as to center the armature windings in the fields. An extra armature for this large generator is exhibited and attached to it is one of the bearings fitted with the above mentioned shifting device.

The casting to which the bearing is attached is square, having surfaces parallel with the axis of the journal, across the bottom of this casting at right angles to the axis of the bearing is a wedge, fitting into a recess of the main casting having another loose wedge. A screw attached to the loose wedge and extending through the side of the main casting may be turned by the hand wheel attached shifting the journal bearing one way or the other as may be found necessary. The usual rings found on all Brush armature shafts are located on the shaft in this bearing and prevent side play of the armature.

These machines are very compact, well finished and run at a very low rate of speed.

On a cloth covered bench at one side are displayed

numerous parts of the armatures and other apparatus, including automatic cut-outs, switches, commutators and various parts of the motor fields.

A very fine black enameled slate switchboard in panels like all other well arranged boards of today, is exhibited fully equipped with all the instruments, apparatus and devices necessary for controlling the railway currents of two generators. In addition to the circuit breakers provided for each machine, very long fuses are supplied, mounted on the backs of long single insulated black switches, so that when a fuse blows there is no trouble in replacing it.

The exhibit covers the field in a very comprehensive manner, shows good judgment in its selection, and is standard all the way through, embodying as it does only such types of apparatus as are advocated for practical every day use. Numerous pictures of Shoct appliances and apparatus are distributed about the space, and a large sign in white and gold is suspended over the center of the exhibit.

Exhibit of Hart & Hageman.

The exhibit of switches by Hart & Hageman M'f'g Co., of Hartford, Conn., located in Sec. 8, space 5, Electricity building, is in charge of George S. Searing, the western

EXHIBIT OF HART & HAGEMAN.

manager. Since it was first installed, more space has been secured, and the display somewhat enlarged. Rugs, table and chairs give an air of comfort to the place. A brass rail on oak posts surrounds the space with gates on both sides. The real exhibit of the company consists of various sizes and styles of the Hart switch arranged in geometrical figures on small panels in mahogany, white and ebonized wood. The two mahogany boards support the various sizes of the standard three point, single and double pole Hart switches, which are well known to the trade. The flush switches, of which this company makes a specialty are shown in various styles and finish on the white and ebonized board, the different colors of finish showing well in contrast to the white and the black of the panels. These flush switches are made in single units or in any multiple desired. The principal styles of finish are nickel plated, dark copper, mottled steel and bronze. Special orders are filled in any color the purchaser may desire. Colored banners, in two corners, with the words "Hart Switches," and a sign across one end with the name of the company, attract the attention of visitors. Mr. Searing is present

most of the time, and takes pleasure in showing the Hart switch to visitors.

Brewster Electrolytic Disinfectant Plant.

The recent installation of an electrical disinfecting plant at Brewster, New York, gives evidence of a new field for the use of the electric current. The Electrical Engineer recently described this plant which contains one 15-horse power engine coupled to a Zucker & Levitt dynamo with a capacity of 500 amperes at five volts. Near the dynamo is an electrolyzing tank with a capacity of 1,000 gallons, in the bottom of which are placed three copper plates plated with platinum alternating with four carbon plates. This tank is fed from a larger tank and the supply is so regulated that the water is electrolyzed and overflows into the sewer.

As a result of the threatened invasion of cholera, examinations were made of New York's water supply, and it was found that the sewerage from Brewster which drained into a marsh situated on an elevation, percolated through the soil and finally reached one of the streams forming the water supply of New York. As these marshes were also a a nuisance to the health of the citizens of Brewster, the authorities of both places were interested in means for relief.

The ordinary methods employed for the purification of sewerage and the destruction of the germs to which cholera is due, and by which it is disseminated in drinking water, involve the use of chemicals containing hypochlorites and chlorides more or less expensive. Dr. Edison was led by the great difference in cost and other advantages to adopt the system proposed by Mr. Albert E. Woolf.

The chlorides, bromides, etc., in sea water are converted by the passage of the electric current into hypochlorites, hypobromites and various other compounds. When a solution of hypochlorite of sodium is brought into contact with organic matter decomposition takes place and if the solution is strong enough the matter is completely disinfected.

The use of this system at Brewster is said to be entirely satisfactory. The offensive odors are absent from the marshes and the green algae and other vegetable matter have become bleached since this plant has been operated. Tests of the Woolf disinfectant show that it equals in strength a one per cent solution of chloride of lime. A one per cent solution of chloride of lime costs about 1.4 cents per gallon with lime at six cents per pound, while the estimated cost of the electrolyzed sea water is only 10 cents per 1,000 gallons. The electrolyzed sea water is harmless, while many of the chemical disinfectants are harmful in the hands of inexperienced parties.

The Babcock & Wilcox Company has an exhibit in Machinery Hall, section 25, M 24. The space is nicely carpeted and provided with desk, chairs and table. On the table are displayed samples of the tubes used by the company bent in a great variety of ways to show the quality of the metal. Samples of the cast mud drum, of the cast and wrought iron heads are shown. A very neat railing surrounds the exhibit one side being made of brass tube while the other three sides are made of boiler tubes with headers. In a glass cabinet is shown a very handsome model of a single drum boiler. It has all the attachments of a full sized boiler. On the Exposition grounds there are installed 6,000 horse power of Babcock & Wilcox boilers, 3,000 of which are in the power house of the Intramural Railway and 3,000 in the Machinery Hall boiler plant.

ELECTRICAL INDUSTRIES.

PUBLISHED EVERY SATURDAY BY THE

ELECTRICAL INDUSTRIES PUBLISHING COMPANY,

INCORPORATED 1893.

MONADNOCK BLOCK, CHICAGO.

TELEPHONE HARRISON 198.

E. L. POWERS, Pres. and Treas.   E. E. WOOD, Secretary.

E. L. POWERS, - - - - - - - - - - - - - - - Editor.
H. A. FOSTER,   } - Associate Editors.
W. A. REMINGTON, }
E. E. WOOD, - - - - - - - - - - - - Eastern Manager.
FLOYD T. SHORT, - - - - - Advertising Department.

EASTERN OFFICE, WORLD BUILDING, NEW YORK.

World's Fair Headquarters, Y 27 Electricity Building.

SUBSCRIPTION.

FIVE MONTHS, - - - - - - - - - - - $1.00
SINGLE COPY - - - - - - - - - - - - 10
Advertising Rates Upon Application.

News items, notes of communications of interest to World's Fair
Visitors are earnestly desired for publication in these columns and will
be heartily appreciated. We especially invite all visitors to call upon us
or send address at once upon their arrival in city or at the grounds.
ELECTRICAL INDUSTRIES PUBLISHING CO.,
Monadnock Block, Chicago

Monday, July 31, was mechanical engineers' day at the
Fair, and the American Society of Mechanical Engineers
visited the Fair in a body, going at 1:30 p.m., on the
Illinois Central R. R. A regular program had been provided by Mr. H. F. J. Porter for their entertainment. They
inspected the Multi-platform Railway on the Pier, after
which a reception by the engineers of the Exposition was
held in Music Hall at which the engineering features of the
Exposition were described. After a trip to the Krupp Pa-
villion, where they were entertained by the representatives
of Fried Krupp, and to the pumping station of Henry R.
Worthington, they separated to a number of the more
amusing features of the Exposition.

As proposed some time ago it has now been decided to
close a majority of the Exposition buildings at 7 p.m. The
Electricity Building, Machinery Hall and some other building will remain open. On Monday evening Horticultural
Building will be open; on Tuesday Manufactures and
Liberal Arts Building; Wednesday, Transportation Building; Thursday, Art Gallery, Anthropological, Forestry and
Shoe and Leather Building; Friday, Agricultural and
Fisheries; Saturday, Mines and Mining. This order goes
into effect immediately. By this arrangement exhibitors
can easily keep their exhibits open all the time the
building is open, and the visitor will be assured of seeing a
majority of the exhibits in the buildings on their open
night. Electricity Building has no reason to complain of
lack of visitors at night; so far the attendance has been very
good. With its interesting exhibits, its handsome decorations, and brilliant illumination it is one of the greatest
attractions of the Fair.

A very interesting type of engine is that shown by the
Dake Engine Manufacturing Company of Grand Haven,
Mich., in Column G, 1-37, Machinery Hall annex. It is
exceedingly simple in construction, having but two moving
parts outside of the crank itself, and consists of a thin,
oblong rectangular box containing the piston. The piston
is double, the outside part sliding from end to end of the
box; the other part, being inside the first, slides up and
down in a direction at right angles to that of the first.
Steam is admitted through ports cast in the casing to the
center of the shell and exhausts through a circular port
surrounding the central admission port. Owing to the
double piston and double action, there are no dead points
and the engine will start from any part of the stroke. It
is said that the engine may be run at any speed up to a
thousand revolutions per minute.

The Baltimore Car Wheel Company has an exhibit in the
annex to Transportation Building, a sample truck for electric
street railway motors. It is equipped with two Baxter
motors, and has oil and dust tight axle boxes. As is usual
in electric car trucks at the present time, the car bearings
overhang, giving all the effect of a long wheel base.

Chicago World's Congress of Electricians.

As the time approaches for the assembling of the World's
Congress of Electricians, August 21, something of its history may be of interest to our readers. The congress is
held under the direction of the World's Congress Auxiliary
of the World's Columbian Exposition, the Electrical Congress Committee having the matter in charge. In the work
of preparation the congress committee of the American
Institute of Electrical Engineers has assisted. As early as
1889, the Institute appointed a committee to prepare for a
congress during the World's Fair. The adjournment of
the Frankfort Congress in 1891 to meet during the World's
Fair in 1893, seemed to confirm the work of the Institute.
Having had the advantage of experience in two electrical
congresses, the Institute proceeded in a systematic manner
in the work of preparation. After the appointment of the
Electrical Congress committee of the World's Congress
Auxiliary, with Prof. Elisha Gray at its head, the two committees have co-operated. Electricians in all parts of the
world have become interested in the congress, and in the
technical press have appeared many discussions of subjects
which will be brought up at the congress. As many new
units have appeared in connection with the growth of electrical knowledge and the use of electricity, the necessity of
adopting universal terms which would be international has
lead to a general discussion.

As already announced, the congress will be divided in
three sections, which will meet at 10 o'clock a.m., on
Tuesday, Wednesday, Thursday and Friday of congress
week. Prof. H. A. Rowland has been appointed temporary
presiding officer of section (A) of pure theory, including
electric waves, theories of electrolysis, electric conductors,
magnetism, etc.; Prof. Charles R. Cross, of section (B) of
theory and practice, including studies of dynamos,
motors, storage batteries, measuring instruments, materials
for standards, etc.; and Prof. A. Graham Bell, of the section
(C) of pure practice including telegraphy and telephony,
electric signalling, electric traction, transmission of power,
systems of illumination, etc.

At one o'clock p.m., Friday, the congress will assemble as
a whole to hear reports, etc., and will adjourn at three
o'clock p.m. Several prominent electricians have signified
their willingness to deliver lectures of a popular character
which will be delivered on Tuesday, Thursday and Friday
evenings of congress week. While the public can attend
the meetings of the congress only those who hold cards of
invitation from the committee on invitations will be per-
mitted to take part in the deliberations of the congress.

WORLD'S FAIR NOTES.

The Ferris Wheel was completely lighted for the first time last week. The rows of lights on the sides of the rim of the wheel and the rows on the towers supporting the axle produce a striking scene at night.

J. M. Jones & Sons, of West Troy, N. Y., show two styles of electric street cars in the annex to Transportation Building. One is an open car, the other of the closed type. Both exhibit fine design and workmanship.

A considerable amount of interest is taken in the vertical marine engine of the General Electric Company in Machinery Hall. It is fitted with Corliss valve gear. In the same space are shown two dynamos of the type constructed especially for steamships.

The Westinghouse company has recently started in the southeast corner of Machinery Hall a large continuous current multipolar generator coupled direct to a cross compound horizontal Alles Corliss engine. The armature is on the shaft between the two main bearings, and a fly wheel is provided to steady the speed.

The electrical apparatus of Prof. Galileo Ferraris of Turin, Italy, is expected to arrive about August 10th. This apparatus has been very unfortunate on the way having been sunk in the harbor of Genoa. Afterwards recovered, repaired and started again for the World's Fair on July 23d.

The scenic theater of the Western Electric Company is again running in its new quarters in the southeast corner of Electricity Building. A number of additions have been made in the stage equipment, and chairs have been provided for the audience. A piano adds to the attractiveness of the entertainment. The storm and rainbow effects are especially applauded. It requires 20 minutes for the production of the different light effects, and it is intended to give performances every half hour.

The Electric Appliance Company has recently added to its World's Fair exhibit a lot of twenty finely finished fancy reels of Paranite wire. The reels are stacked in the four corners of the space in such a way as to form four pyramids of reels tapering from very large at the bottom to small at the top. These reels thus serve the double purpose of helping to mark out the space and making an attractive exhibit of Paranite. This addition just about fills up the Electric Appliance Company's space, and seems to be about all that was wanting to make it symmetrical in arrangement and complete in detail.

A grate bar which will be found of interest to steam users will be found in an exhibit in the British section of Machinery Hall. It is made by Caddy & Company, Limited, Daybrook Iron Works, near Nottingham, England. The bar is a hollow piece of chilled cast iron. The top edge is harder and wider than the lower. The rear end enters a chamber in the bridge wall at the back of the furnace while the front end projects in front. Thus a current of air passes through the bar. The bars are said to wear much longer on account of the air keeping them cooler and the air entering the back of the furnace the gases are consumed that would be otherwise passed out as smoke.

Just beyond the door at the right of the south entrance to the Government Building is a very interesting exhibit of photographs of nearly all the electrical apparatus used by Joseph Henry in his electrical experiments. The Sturgeon magnet, the quantity magnet, Henry's intensity magnet made in 1829, pole changers and the induction coils used

by him in the discovery of the laws of induction are shown. There is also a picture of the first electro-magnetic motor made by Henry in 1831, and called the father of all electro-motors. As the originals themselves are nearly all on exhibition in the Princeton College exhibit these photographs are interesting only as a matter of record at the Fair.

In the Manufactures and Liberal Arts Building, just north of Tiffany's beautiful exhibit, is that of the Self-Winding Clock Company, of New York. The booth is a large square room, with one side open. The walls are hung with clocks of all styles, sizes and purposes, each being provided with the electric winding arrangement owned by this company. The standard time clocks rented all over the country by the Western Union Telegraph Company are of this make, and are synchronized with the standard time of the Naval Observatory at Washington, D. C., at noon each day. The company also shows samples of devices, for use with clocks, to strike a starting bell for street car lines and for any purpose where it is desirable to have a bell struck at a given minute of time. The large clocks and chimes in the central clock tower in the Manufactures and Liberal Arts Building are all furnished and run by this company, the bells being struck from a desk located in the front of its exhibit, and connected with it by electric wires and the proper electrical devices.

H. T. Paiste, of Philadelphia, has a very attractive exhibit at the south end of the gallery in Electricity Building. The space which is next to the wall is surrounded by a very neat railing, is carpeted and furnished with chairs and desk. At the ends are easels, one of which supports a polished oak board on which are switches illustrating the evolution of the Paiste switch. On the wall on a very handsome background are arranged main and branch cutouts, switches, etc. In the center of the background cutouts are arranged showing the outlines of the "Antric" movement of the switch, with the words, "Antric; that's the switch." The exhibit is very tastily arranged and reflects great credit on the designer. Mr. E. A. Jenkins is in charge of the exhibit.

Cutter's Economic Door Switch.

Among the most novel devices which are on exhibition at the electricity building are some which George Cutter has been developing, and which are only now ready for the market. One of these is a switch for controlling the current by the opening and closing of a door, so as to keep the cost of the lighting at a minimum. Most housekeepers know how expensive it is to light closets, store rooms and the like as the hired help will persist in burning the lamps continuously. Such a waste of current is easily avoided by setting an economic door switch into the casing of the door jamb, so that the door itself will operate the switch. Then the light will burn only so long as the door is open, and the saving in current and in the life of lamps will soon pay for the controlling device.

Our cut shows the general appearance of this new switch which has a spring action, is well built and carefully insulated. Having a cylindrical casing, it is very easy to fit to the woodwork, no expert being required. Other uses for this switch will soon suggest itself, and it looks as if Mr. Cutter ought to find a wide market for it.

PERSONAL.

Prof. George Forbes has returned from Niagara Falls and is now at the Fair.

Mr. Frederick A. Scheffler, of The Sterling Co., New York, is visiting the Exposition.

C. C. McNutt, president of the Warren Electrical and Specialty Co., Warren, Ohio, is visiting the Fair this week.

Prof. N. M. Terry of the U. S. Naval Academy at Annapolis, has returned home after an extended visit to the Fair.

Prof. D. C. Jackson, of the University of Wisconsin, will remain at the Fair for some time on work connected with the awards.

Mr. W. R. Brixey of Day's Kerite, is again at the Fair extending his already large acquaintance among the electrical fraternity.

Mr. Leo Daft, one of the earliest pioneers in electric railway work, and now settled at Seattle, Washington, is in the city stopping at the Great Northern.

Mr. W. C. Cheney superintendent and engineer of the Portland (Oregon) General Electric Company made a short visit to the Fair last week while on his way east.

Prof. E. Hospitalier of Paris, delegate of the French Government to Chicago World's Congress of Electricians arrived in Chicago on the 27th after a two weeks stay in New York.

Prof. H. A. Rowland, of Johns Hopkins University, arrived in the city Monday, July 31, and will attend the meetings of the Jury of the Department of Electricity, of which he is a member.

Prof. Baurath Ulbricht of Dresden, Germany, has been elected first vice-president and Prof. W. E. Ayrton of England second vice-president of the Jury of Awards Department of Electricity.

Mr. Harold B. Smith, superintendent of the draughting department of the Elektron Manufacturing Co., has accepted the professorship of electrical engineering in Purdue University, Lafayette, Ind.

Mr. Alfred A. Cobb, of the India Rubber Comb Company, sales agent for the Chicago Electric Wire Company, designed and arranged the company's exhibit of wire. He has shown excellent taste in the arrangement as he has in conducting other affairs of his company.

The many friends of Mr. L. W. Barnham, manager of the Electric Gas Lighting Company, Boston, will be pained to learn that just as he was ready to start on a visit to the Fair last week he was taken with a sudden illness. He is, however, improving slowly and will, we hope, be able to visit the Fair a little later in the season.

DEPARTMENT OF ELECTRICITY.

OFFICES: SECTION E. ELECTRICITY BUILDING.

Chief, JOHN P. BARRETT.
Assistant Chief, J. ALLEN HORNSBY.
General Superintendent, J. W. BLAISDELL.
Electrical Engineer, W. W. PRIMM.

DEPARTMENT OF MECHANICAL AND ELECTRICAL ENGINEERING.

OFFICES SOUTH OF MACHINERY HALL.

Mechanical Engineer, C. F. FOSTER.
Electrical Engineer, E. B. PRESS.
First Asst. Mechanical Engineer, JOHN MEADS.
First Asst. Electrical Engineer, S. G. NEILER.

AMERICAN INSTITUTE OF ELECTRICAL ENGINEERS.

World's Fair Headquarters.
SECTION S. ELECTRICITY BUILDING.

RALPH W. POPE, Secretary.

Open from 9 a. m. to 5 p. m.

CHICAGO WORLD'S CONGRESS OF ELECTRICIANS.

OPENING SESSION, MONDAY, AUGUST 21ST, NO. 9

ADVISORY COUNCIL.

President, DR. ELISHA GRAY, Highland Park, Ill.
Secretary, PROF. H. S. CARHART, Ann Arbor, Mich.

EXECUTIVE COMMITTEE.

Chairman, PROF. ELIHU THOMSON, Lynn, Mass.

COMMITTEE ON INVITATIONS.

Chairman, T. COMMERFORD MARTIN, 253 Broadway, New York.

COMMITTEE ON PROGRAM.

Chairman, PROF. T. C. MENDENHALL, Washington, D. C.

COMMITTEE ON FINANCE.

Chairman, B. E. SUNNY, 115 Adams Street, Chicago.

BUSINESS NOTES.

WM N. MARCUS, of 218 North Second street, Philadelphia, has a small exhibit of his patent auxiliary mouth piece for telephones in section 8 Electricity Building, next to the Hart & Hegeman exhibit. This device is applied to the Blake transmitter of the ordinary commercial telephone and is adjustable for persons of any height.

CUTLER & HAMMER, Chicago, manufacturers of electrical goods, are pushing rapidly to the front. Their shop is receiving orders for all the work it can turn out and instead of being obliged to lay off some of their men, as many of the older shops are doing, they have been obliged to increase their force. The new "C. & H." snap knife switches are meeting with a large sale.

THE C. H. STOELTING MANUFACTURING COMPANY, Chicago, are manufacturing several electrical specialties, among others the Blair Lamp Adjuster. This adjuster has been improved lately in construction by using a porcelain base, so made that wires can be run under it and the metal cover locks on to the base, giving easy access to the roller. The roller is now being so constructed that it will a much larger amount of cord.

Amusements.

HOOLEY'S THEATER—Mr. E. S. Willard, in "The Middleman." Saturday matinee, "The Professor's Love Story." 149 Randolph street.

COLUMBIA THEATER—Miss Lillian Russell, in "La Cigale." 108 Monroe street.

GRAND OPERA HOUSE — Sol Smith Russell, in "A Poor Relation. 87 Clark street."

AUDITORIUM—Imre Kiralfy's Spectacle "America." Congress street and Wabash avenue.

McVICKER'S THEATER—Denman Thomson, in "The Old Homestead." 82 Madison street.

CHICAGO OPERA HOUSE—American Extravaganza Company, in "Ali Baba, or Morgiana and the Forty Thieves." Washington and Clark streets.

SCHILLER THEATER—Chas. Frohman's Stock Company, in "The Girl I Left Behind Me." Randolph, near Dearborn.

HAVERLY'S CASINO—Haverly's United Minstrels. Wabash avenue, near Jackson street.

TROCADERO—Concert. Michigan avenue near Monroe street.

THE GIOTTO —Vaudeville. Michigan avenue near Monroe street.

Buffalo Bill's "Wild West." 63d street. Daily at 3 and 8.30 p.m.

Pain's "Siege of Sebastopol," 60th street and Cottage Grove avenue. Tuesday, Thursday and Sunday nights.

"The Girl I Left Behind Me" has had a very successful run at the Schiller during the last ten weeks, and the advance sale of seats seems to promise a continuance of the success. On August 14th, the 250th performance, handsome souvenirs will be presented to the ladies in attendance.

This week is the closing week of Mr. E. S. Willard's engagement at Hooley's. His season has been most successful. Thursday and Saturday evenings he appears in "The Middleman," Friday evening in "The Professor's Love Story."

The special attractions of the Trocadero, Ammani, the mimic, Paquerette, the electric fire works, and the music of the Von Bulow band and the Rosenbecker orchestra are specially attractive, while the dancers, the acrobats, the jugglers and Sandow who is making his first appearance in the city are intensely interesting and amusing to all lovers of vaudeville.

"Ali Baba" is keeping up its reputation for genuine success so well established in its previous productions. The 33rd week it has been played at the Chicago Opera House began last Sunday night. Several new people with special parts have been engaged and will shortly appear. The production has grown so large that at the close of the season in Chicago it will be taken to only the larger cities.

In "America" is found a genuinely cosmopolitan entertainment. It is not an uncommon sight to see groups of Turks, Japanese, Singalese, and representatives from other nations in the audience applauding the different parts of the spectacle. The vivid scenes and the pantomimic action makes the piece understood by everyone. The play is now at its zenith.

WEATHERPROOF
WIRE

We Are The Largest Manufacturers of

WEATHERPROOF WIRE...IN THE WEST

WE GUARANTEE THE QUALITY OF OUR WIRE TO BE EQUAL IF NOT SUPERIOR TO ANY OTHER WIRE OF LIKE CHARACTER IN THE MARKET.

SPECIAL PRICES ON APPLICATION.

WESTERN
ELECTRIC
COMPANY,

227-275 South Clinton Street,

CHICAGO.

ELECTRICITY BUILDING EXHIBITORS AND THEIR LOCATION.

GALLERY.

MAIN FLOOR.

SLOW MONEY ....Makes Low Prices

Now is the time to begin to lay in your supplies for Fall Extensions.

PRICES AWAY DOWN....QUALITY AWAY UP

WE ARE STILL SUPPLYING

Meston Alternating Current Fan Motors in large quantities to those who know a good thing when they see it. Are you going to get in line before it is too late?

ELECTRIC APPLIANCE COMPANY,
ELECTRICAL SUPPLIES.
242 Madison Street,   CHICAGO.

THE MATHER ELECTRIC CO.
MANCHESTER, CONN.

Dynamos, Motors, Generators,

Offices, 116 Bedford St., BOSTON.

—AND—

1002 Chamber of Commerce Bldg., CHICAGO.

THE "NOVAK" LAMP.

CLAFLIN & KIMBALL (Inc.)

General Selling Agents.

116 Bedford Street, BOSTON.

1002 Chamber of Commerce Bldg., CHICAGO.

Siemens & Halske Electric Company
of America.
Chicago, Illinois.

Electrical Machinery.

Siemens & Halske;

Berlin,
Charlottenburg,
Vienna,
St. Petersburg.

Enterprise Electric Company

307 Dearborn Street,
Chicago . . . .

N. I. R.

GENERAL WESTERN AGENTS

Manufacturers' Agents and Mill Representatives for

Electric Railway,
Telegraph, Telephone and
Electric Light

SUPPLIES OF EVERY DESCRIPTION

Agents for   Cedar Poles,
Cypress Poles, Oak Pins,
Locust Pins, Cross Arms, Glass
Feeder Wire.   Insulators,

WIRES, CABLES, TAPE and TUBING

Map of Chicago.

Showing Location of its Electrical and Allied Business Interests, Principal Hotels, Theatres, Depots and Transportation Lines to the World's Fair Grounds. (Index numbers refer to the black squares.)

~Lundell~
Suspended Fan Outfit

Black

Japan

Finish

Self-Oiling

....and

Self-Aligning

Bearings

Electric Fans

Electric Fans

OKONITE WIRES
OKONITE === TAPES === MANSON
INTERIOR CONDUIT.

Batteries, Bells, Push Buttons, Annunciators, Volt Meters, Ammeters, Wheatstone Bridges, Line Wire Cross Arms, Brackets, Pins, Insulators, Tools.

GENERAL SUPPLIES.
CENTRAL ELECTRIC CO.
116-118 Franklin Street,
CHICAGO, ILLS.

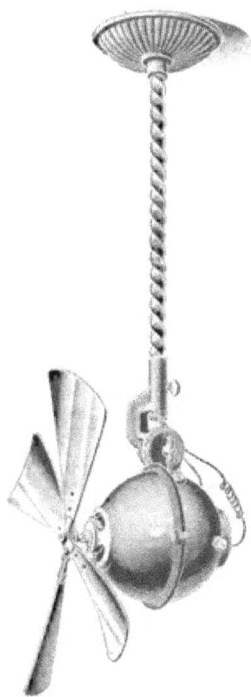

VISITORS SHOULD NOT FAIL TO SEE THE

First Souvenir Half Dollar

...AT THE EXHIBIT OF...

Remington Typewriters

in the N. E. Corner of the Main Gallery of the Manufactures and Liberal Arts Building.

———————————

$10,000 was paid for this coin, making it the most valuable piece of silver in the world.

THE MONTHLY ISSUE FOR AUGUST

ELECTRICAL INDUSTRIES

Should be read by everyone interested in electrical matters.   In its table of contents is the following:

"Incandescent Lighting at the World's Fair."
"The Electric Power Plant of the Chicago City Railway."
"Steam Engine Efficiency—Its Possibilities and Limitations" by Wm. H. Bryan.
"Alternating Arc Lighting for Central Stations" by H. S. Putnam.
"Hard Rubber as an Insulator in Street Railway Work" by W. R. Mason.
"A Brief Review."
Together with illustrations of the recent applications of electricity.
The paper also contains regularly
A Buyer's Directory of Manufacturers and Dealers in Electrical Supplies and Appliances.
A Complete Directory of Electric Light Stations in North America and a Complete Directory of Electric Railways in North America.
These directories are revised each issue to the date of going to press and are to be found in no other electrical journal in the World. Its articles are read carefully and its directories used constantly by all the buyers in the trade. These facts make it without a superior as an advertising medium. Sample copies and rates sent on application.
Subscription price $3 per year. Six months trial $1, if ordered during the next 30 days.

ELECTRICAL INDUSTRIES PUB. CO.,
Monadnock Block, CHICAGO.

CLARK
ELECTRIC COMPANY, NEW YORK.

192 Broadway and 11 John Street.

MANUFACTURERS OF ARC LIGHTING APPARATUS FOR EVERY PURPOSE A SPECIALTY.
The CLARK ARC LAMPS for use on EVERY CURRENT, have the reputation of being the best and most durable of any ever made in the United States.

RAWHIDE PINIONS FOR ELECTRIC MOTORS
A SPECIALTY.
RAWHIDE DYNAMO BELTING

Greatest Adhesive Qualities   A Non-Conductor of Electricity
Causes Less Friction than any other Belt.

THE CHICAGO RAWHIDE MANUFACTURING CO.
THE ONLY MANUFACTURERS IN THE COUNTRY.

LACE LEATHER ROPE
and OTHER RAWHIDE

GOODS
OF ALL KINDS
BY KRUEGER'S PATENT

This Belting and Lace Leather is not affected by steam or dampness; never becomes hard; is stronger, more durable and the most economical belting made. The Rawhide Rope for Round Belting Transmission is superior to all others.

75 Ohio Street,   CHICAGO, ILL.

STANDARD ELECTRIC COMPANY.

GENERAL OFFICES: 625 Home Insurance Building.
WORKS: So. Canal Street,

CHICAGO,

STANDARD SYSTEM

AT THE

WORLD'S FAIR.

MACHINERY HALL, Sec. Q, 2 Standard Arc Dynamos.
Sec. S. 20 "   "   "
ELECTRICITY BUILDING, Sec. P, Space 2, Arc Lighting Exhibit.

The Standard Lamps Light the Power Plant, Machinery Hall, Agricultural Hall, Shoe and Leather Building, an Other Buildings and Portions of the Grounds.

See our Double Service All Night Lamp Before Buying an Old Style Two Rod Lamp.

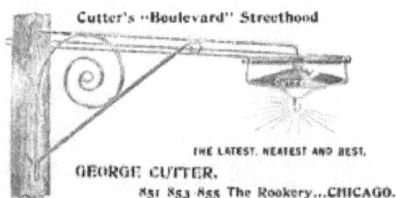

Cutter's "Boulevard" Streethood

THE LATEST, NEATEST AND BEST.

GEORGE CUTTER.

851 853 855 The Rookery...CHICAGO.

SIMPLEX WIRES

INSURE
HIGH
INSULATION

SIMPLEX

Ever Onward and Upward!

Simplex Electrical Co.

620 Atlantic Ave.,

George Cutter, Chicago.   BOSTON, MASS.

XNTRIC

"That's the Switch"

And we control that movement.

H. T. PAISTE,

10 South 16th St.,
PHILADELPHIA,
PA.

Made 5 amp. S. P.
      10 amp. S. P.
      5 amp. 3 way.
      10 amp. 3 way.

P. & S.
WIRING INSULATOR,

Saves TIME
        TROUBLE
and TIE WIRE_____

Made only by

Pass & Seymour,
SYRACUSE, N. Y.

George Cutter,
                    CHICAGO.

P.&S.

Consolidated Electric Co.

Manufacturers and Dealers in all kinds of

ELECTRICAL . SUPPLIES,

115 Franklin Street,

CHICAGO.

GEORGE PORTER,

Contractor for All Kinds of

ELECTRICAL WORK.

Room 67, 143 La Salle St., CHICAGO.

Crary Block, BOONE, IOWA.

CHAS. A. SCHIEREN & CO.

MANUFACTURERS OF

Genuine Perforated Electric Leather Belting.

46 So. Canal Street,   -   CHICAGO.

Section 15, Dpt. F, Clm. 27.   Section D, Space 3
MACHINERY HALL.   ELECTRICITY BUILDING

BEAR IN MIND

that the regular monthly issue of ELECTRICAL IN-
DUSTRIES contains the most complete and correct
directories published of the electric light central stations
and the electric railways in North America.

World's Fair Headquarters Y 27 Electricity Building.

CITY OFFICES, Monadnock Block.

WAGNER ELECTRIC FAN MOTORS

For Direct or Alternating Currents.

These motors give a stronger breeze with less consumption of current than
any other fan motor on the market. They are full 18 horse power. Six bladed
12-inch fan. Self-oiling. Furnished with or without guards.

IT WILL PAY YOU TO SEE THE WAGNER BEFORE BUYING ELSEWHERE.

TAYLOR, GOODHUE & AMES,

348 Dearborn Street, CHICAGO.

WEEKLY WORLD'S FAIR

ELECTRICAL INDUSTRIES

DEVOTED TO THE ELECTRICAL AND ALLIED INTERESTS OF THE WORLD'S FAIR, ITS VISITORS AND EXHIBITORS.

Vol. I, No. 9.   CHICAGO, AUGUST 10, 1893.   FIVE MONTHS $1.00
TEN CENTS A COPY

Exhibit of the Westinghouse Electric & Manufacturing Company.

While the lighting system of the Westinghouse company is best shown in the electric plant of the Exposition, its ex-

the latest system of long distance power transmission of this company.

Another section occupying a similar space but toward the south of the building, contains an exhibit of railway power machinery. Directly to the east of this exhibit is the

FIG. 1.—EXHIBIT OF THE WESTINGHOUSE ELECTRIC & MANUFACTURING COMPANY.

hibit in Electricity Building better represents the varied products of its factories. The exhibit is divided into sections, each representing a particular branch of the industry. The section located near the center of the building between the Barbier display of lighthouse lenses and the Edison In candescent lamp exhibit, contains an exhibit representing

display of both arc and incandescent lighting. The Shal lenberger meter, lighting arresters and apparatus showing some novel effects of the two phase alternating currents are also displayed in this section. In the corner of the space a room has been built in which are shown a number of startling experiments in high tension currents. Mr.

Tesla expects to be present to utilize this room for exhibitions for invited guests during congress week. To the east of this section is the space in which representative machines from the Newark factory of the company are shown. In addition to the letter and horizontal type several machines of the Manchester type are shown. Also a 50-horse power multipolar direct current machine made in Pittsburg. The different sections are surrounded by substantial and ornamental railings open at the corners which are closed by brass chains.

The exhibit of long distance power transmission system consists of a complete plant with all the machines and apparatus necessary for the transmission and the reception of the electric current. The plant consists of a 500-horse power, two phase alternating current generator directly connected to a pelton waterwheel as illustrating the source of power while the real power is furnished by a 500-horse power two phase Tesla motor which is belted to the generator by a 24 inch belt. The prime generator is of the new type now known as the rotary transformer, although why it is so named is hard to say as it is simply a commutating

sion plant proper. The greater number of the switches mounted on the board are used for controlling the Tesla motors supplied from the lighting circuits and used in this case to furnish power and exciting current to the transmission plant. All the connections and contacts are made on the back of the board, only the handles of the specially designed switches and the fuse blocks are on the front of the board. All the wires have Okonite insulation and are fastened on porcelain. All terminals for the instruments are insulated so that it is almost impossible to come in contact with the current on the front of the board. The fuse blocks are specially designed single pole blocks and the fuses are of a non-oxidizing metal requiring but a small mass of metal in comparison with the ordinary fuse. The block is eight inches long by four wide, of two pieces an inch thick. The pieces are recessed for the fuse and the outer block has a hole above the fuse through which the melted fuse may pass when blown out.

The bearings of the fuses are two long brass plugs an inch wide by three-eights inch thick and about six inches long extending out at right angles from one side and fit in

FIG. 2.—EXHIBIT OF THE WESTINGHOUSE ELECTRIC & MANUFACTURING COMPANY.

device for an alternating current and might better be called a commutating machine. As used in this case the full horse power output can be taken off either in alternating or direct current or any proportion of both currents at the same time.

The machine is separately excited by a five-horse power direct current dynamo directly coupled to a Tesla motor of the same capacity. The dynamo may be easily made self exciting. The primary voltage is of course provided for according to distance and load, in this case the voltage is 580. From the generator the current is conveyed to the switch board by the wires which are concealed beneath the floor. From the switch board the current is conveyed to the step up transformers for long distance transmission. This switch board is of white marble mounted on an iron frame and is arranged after a design specially made for this work. It is provided with the necessary switches for controlling the 500-horse power motor and the current from the generator so arranged that the current may be thrown from the step up transformers to any part of the exhibit.

The switch board contains much more apparatus than is required by the 500-horse power generator of the transmis-

to suitable contacts on the back of the board, holes in the marble permitting the placing of the block from the front. Other instruments on the board are as follows: Four 300 ampere ammeters; two 100 ampere ammeters; four 120 volt volt meters; two switch board transformers; one rheostat; three double pole switches for exciters; two fuses for direct current lines and eight for alternating current lines. The back of the board is protected on the back by a closet about three feet deep, with doors at the ends allowing access to all connections.

Just back of the switch board on glazed tiling are the six step up transformers. A frame is built over them to support the four primary and four secondary wires. The 360 volt primaries of the three transformers are connected to one pair of wires from the switch board and generator and the primaries of the other three to the remaining two wires. The high tension secondaries, in this case 1,200 volts, are connected in the same grouping to the two pairs of transmission wires which run on a couple of poles to the step down transformers at the opposite side of the space. The insulators used are of the type employed by the Westinghouse Electric & Manufacturing Company in the plant now operating at Pomona and at San Bernardino, California. Usually the transmission wires would be conducted directly to the receiving switch board first, before reaching the transformers so in case there were trouble with the transformers, current could be entire cut off from them.

The connections of the transmission wires to the step down transformers are precisely like those of the same at the power end of the line. The board varying only from the fact that connections, instruments, switches and other appliances are provided for control of all the various apparatus used to illustrate the methods of using the power. The principal machine shown at this end of the transmission is another 500-horse power motor or rotary transformer which takes the two phase current into the four rings at one end of the armature from the switch board and at 360 volts, and running as a motor supplies power to several machines belted from its pulley and direct current from the commutator on the opposite end of the armature for use in a number of direct current motors used for various purposes, thus illustrating the great flexibility of the system. The voltage from the direct current end is 550, being somewhat greater than the alternating current voltage supplied to the motor.

current from this machine. Samples of the Ingersoll-Sargeant rock drill, and coal cutter are also shown. Connected to the alternating current switch board, current for which is taken from the secondaries of the step down transformers, are coupled direct to a 1,000 volt constant potential alternating dynamo; the other being employed simply as a rotating transformer giving out a continuous current of 50 volts, showing that low voltage can be had for electrolytic or electroplating, and electric tanning, etc. It is intended to employ current from this machine to operate one of the large Schuckert search lights in Electricity Building.

This covers the mechanical and electric part of the section and fully illustrates the development today of the two-phase long distance electrical transmission of power. In the center of the space on a platform slightly raised above the surrounding floor is a very handsome kiosk or booth, that is used as a reception room and office, tables and chairs are provided and files of the electrical journals are supplied for the visitor.

This booth is oblong in shape, made in wood and staff, the lower or supporting part consisting of columns at either end, the roof curving over and a square tower rising from

FIG. 3.—EXHIBIT OF THE WESTINGHOUSE ELECTRIC & MANUFACTURING COMPANY.

Driven by belt from the pulley of this machine is a large Worthington pump which supplies water to the Pelton wheel which is coupled to the shaft of the prime generator, and is done to show the regulator of the Pelton wheel which naturally should form a part of the power end of the exhibit.

Besides the pump, from another pulley on the same shaft is also driven by belt a 10 light alternating arc dynamo running 1,000 revolutions per minute, supplying a 10 ampere current for 25 arc lamps suspended in a frame on the side of the space. A small switch board for controlling the output of this machine is provided near the dynamo.

Leads from the commutator end of the rotary transformer are taken to a portion of the receiving switch board where all the necessary bus bars and switches are provided for utilizing the current for running a pair of Westinghouse 30-horse power railway motors, mounted on a Dorner & Dutton truck, the object being to show the ease with which this style of long distance transmission apparatus can be connected to the existing railway lines, which might be located in towns sufficiently near to the plant.

An Ingersoll Sargeant air compressor equipped with a 60-horse power 500-volt direct current motor is also run by

it to the height of several feet, all being highly ornamented in design and painted in cream color, with the points touched out with gold. Flags and curtains are draped around the sides and ends, the margin of the cornices are ornamented and lighted by numerous incandescent lamps with globes of various designs and the interior is lighted by three highly ornamented short arc lamps running on the alternating current. These are lamps are of the constant potential A. C. type. A large number of these lamps are supplied with current from the incandescent lighting circuits throughout the Exposition grounds.

The name Westinghouse Electric & Manufacturing Company Tesla Polyphase System is in large letters on the north and south sides of the square tower on top of the pavilion. This is one of the handsomest exhibits in the Fair and is attracting a great deal of attention, as the system is new and not yet generally understood by the public.

The section of the Westinghouse company's exhibit coming next in point of interest is the railway department, located between the two main aisles just north of the Bell Telephone temple. This display is intended to include a complete list of the apparatus necessary for use in connection with street railway power, with samples of such types as are considered embodying the most points of excellence. A very handsome octagonal kiosk occupies the center of the space furnished with tables and chairs and files of several trade journals. Curtains are draped across the openings, and it is illuminated by arc and incandescent lamps from the alternating circuits.

Ornamented wrought iron brackets are raised around the roof dome, from which are suspended eight short arc lamps, very highly decorated in Berlin black finish. These serve to light the exhibit space as well as to ornament the booth. Suspended over the pavilion is a sign in black and gold displaying the Westinghouse company's name.

FIG. 4.—EXHIBIT OF THE WESTINGHOUSE ELECTRIC & MANUFACTURING COMPANY.

The most noticeable feature of the exhibit is the "Kodak," or 270-horse power, 500-volt compound wound railway generator on the same base with and directly coupled to a Westinghouse compound engine, with cylinders 16 by 16 and 27 by 16. This generator is over compounded about eight per cent, is very compact and said to be highly efficient. Another heavy piece of apparatus is a 400-horse power, 500-volt compound wound belt driven railway generator located on the opposite side of the pavilion from the "Kodak." Both of these machines have the special windings and all the latest features of the Westinghouse systems and are mechanically and electrically well balanced. On proper frames for the purpose are two generator armatures, one for a belt driven dynamo of 250-horse power, and another for a 400-horse power direct coupled machine.

In motors there are shown, one of 20-horse power and one of 25-horse power on frames, with the fields thrown back to show the manner of construction and repair; another of 30-horse power capacity is closed just as it goes on the truck. A "Three Rivers" equalizing truck on the south end of the space is equipped with a pair of the standard 30-horse power motors, which may be run as the truck is raised from the floor. The motor armatures here shown are of the toothed type in which the wire is first wound on a form, then applied to the core in coils, the connections being made after the coils are placed.

On polished oak tables at the west side of the space are shown a sample field coil such as is used on the large compound wound railway generators; rheostats for the shunt fields of generators, these rheostats being thoroughly fireproof, the front of marble, the box of iron and the insulation as far as possible of porcelain. A pair of bearings and a complete set of brush holders with the yoke and attachments, a trolley stand and pole are also shown.

Another table displays the company's well-known series

parallel controller for street cars, with miscellaneous parts of motors, controllers, commutators, canopy switches, connecting wires from motors to the controller, which by the way are run in a light canvas hose that is easily fastened to the underside of the car floor, motor fields, a diverter or resistance for use with the series parallel controller, and another for use with the plain rheostat.

A set of motor gears are shown of the proportion of three and one-half to one, motor bearings, trolley wheels, carbon brushes, lightning arresters, which are of the air gap type with carbon contacts in an asbestos lined box.

The fuse shown for use in connection with street car work is of peculiar form. A double lignum vitae block is made, one part having a slot cut in it into which fits the other part. A small copper wire is used for a fuse which is doubled around the ends of one part of the block, and entirely protected from all other parts of the connections.

On a rack placed in the front part of the exhibit and made to imitate part of a skeleton switch board, is a Wurtz tank lightning arrester. As is well known this is a device for connecting onto railway circuits only during heavy storms, and at such times current flows directly from the circuit wire into a tank of running fresh water in connection with the ground. The special device for diverting the lightning from the machines is a series of simple copper wire coils or helices, the inductive or reactive effect of which is enough to turn the current off into the water ground provided for it. Another section of frame shows the method of making connections between circuit or terminal wires and bus bars.

Arranged on one of the sloping sides of a strong bench made for the purpose are samples of the automatic carbon

FIG. 5.—EXHIBIT OF THE WESTINGHOUSE ELECTRIC & MANUFACTURING COMPANY.

shunt circuit breaker, running from 150 to 1,500 ampere capacity, and having the trigger adjustable by small weights or washers, each one representing 50 amperes capacity.

Another well designed switch is for use on feeders, and like the automatic switch it has the carbon shunt but is operated by hand.

Samples of the various sizes of current and potential indicators are shown running in capacity from 100 to 4,000 amperes. These instruments are all on bases of white marble, with sides and front made of plate glass with metallic fastenings. Double pole knife switches are shown as well as single pole, both in capacities from 30 to 1,000 amperes. On another table alongside the last are shown the heavy triple point knife switches used on generator switch boards, and in capacities from 200 to 1,000 amperes. On a track over a pit running the whole length of the north end of this space are two street cars, one a Brownell 20-foot accelerator and the other a Stephenson 18-foot car with Tackaberry truck, both having 6½-foot wheel base. Both these cars are fitted throughout with the Westinghouse standard 30-horse power equipment, with the series parallel controllers, and all the lamps and other devices that go with a car, complete and ready for the road. The New Haven fare register is used, and is quite neat in appearance as compared with the old style of circular dials in common use on old lines. The machines are connected to a 500 volt circuit so that their actual operation is shown to visitors.

Nothing seems to be missing from this exhibit that can be needed or useful in a railway power station, and with the courteous attendants always present to explain matters, visitors should be well pleased and get a fair knowledge of what the Westinghouse company has in this most interesting line.

In the section to the east are shown the various machines and apparatus used in electric lighting. Sample dynamos, converters, switchboard instruments and devices, lightning arrestors, etc., are exhibited. The transformers and meters are shown complete, and their unassembled parts showing the construction. The exhibit of the Wurtz non arcing lightning arrester is neatly arranged in the center of the space. The exhibit of dynamos includes sizes ranging from a 25 light to a 7,500 light, of both continuous and alternating current dynamos. Sample field rheostats, ground detectors, theater regulators with numerous other devices are shown.

While the Westinghouse company have arranged the exhibits of a practical nature of the greatest interest to the electrician, the mechanic and the capitalist, the decorative features have not been overlooked. One of the most noticeable features at night is the large illuminated sign on which the name of the company, the dates 1492 and 1892 with the head of Columbus in the center are thrown out in the artistically arranged incandescent lights.

A very interesting novelty is the Columbus egg, as it is called, shown in the lighting exhibit. On a table on the west side of the space are placed a pair of large induction coils for exhibiting the effects of the two-phase rotary current. A wooden table is placed over these on which metal objects commenced to spin around as soon as placed upon it. Two copper eggs, one small, the other about eight inches long, when placed over these coils commence whirling and soon turn up on the end and continue to whirl. In the room provided for the exhibition of high tension currents a series of tranformers and Leyden jars are so arranged as to give heavy discharges over glass and rubber plates.

The Cook Elevated Electric Railway.

Near the Westinghouse Air Brake exhibit in the Transportation Building is a working model of the Cook Elevated Electric Railway. A single row of truss iron pillars, placed from 20 to 50 feet apart and of the desired height, supports the girders, which are constructed in the most approved style of brace work. The top of the girder has an outward and upward flange. The track runs the entire length of the upper right hand side of the car, the weight resting on two anti-friction wheels placed near either end of the track, and running on the inside of the flange.

The driving wheel, which is directly connected with the armature shaft and made with bevelled edge, travels on the track at the lower part of the girder, which is made to slope out and down. As the speed increases the driving wheel climbs the track and takes a portion of the load from the truck wheels, which ordinarily support four-fifths of the weight of the car. The driving wheel and the truck wheels are compensating, thus lessening noise and vibration.

A guide wheel runs on the under side of the lower flange and serves the double purpose of preventing the truck from leaving the upper track and of increasing the traction of the driving wheel when desired, being controlled by a lever in the motor room. A shoe brake, also controlled by lever, bears on the outer edge of the lower flange and is wonderfully efficient in conjunction with the under running guide wheel in making quick stops. In practical tests, made last fall at Tacoma on a 600 foot eliptical wooden track, a speed of 42 miles per hour was developed, and stops were made inside of 600 feet while running at this speed, and without inconvenience to the passengers.

The current is applied through a flat copper conductor placed in a perfectly insulated trough on the under side of the upper track, where it is necessarily protected from rain, sleet or snow. A metal brush trolley is used, which will be perfected later so as to allow the car to move in either direction. The motor is of the ordinary street railway type, that used in the Tacoma tests being of the Sperry manufacture. The ground is made by the contact of the driving wheel with its track.

Among the claims made for this system by Mr. Cook, who is personally in charge of the exhibit, are the following: A speed of 200 miles per hour; an absence of any jerking or surging motion in starting or stopping; minimum possibility of accidents to passengers or the public generally; low cost of construction; small amount of ground space required for the single line of pillars; one above the other can be constructed on the same line of supports; a modification of the system can be used for express or parcel transportation.

A company has been formed and it is expected that at least two miles of track will be in operation in Chicago before the end of the present year. Such an eminent specialist on the rapid transit problem as Prof. Haupt, of the University of Pennsylvania, unhesitatingly pronounces in favor of the Cook system.

A well known electrician was heard to remark the other day that electricity could show the only notable progress of any thing at the World's Fair. In art, science, manufacture and nearly every branch of industry, he said, the advance made could not be compared with electricity. The Jumbo dynamo was one of the first incandescent machines built, and is now on exhibition at the World's Fair. Compare that machine with those of to-day and a person sees the improvement at once. This cannot be shown of the steam engine or any other class of machinery in so marked a degree. In transportation the progress made has been that of electric traction. In fact, electricity is the one feature of the Fair that is always interesting and constantly reveals some new wonder to the thousands of people who visit the Exposition both day and night.

England, Germany, Belgium, Brazil and Turkey are now represented on the jury of awards: Ahmed Fahri Bey, an electrical engineer of Constantinople, having recently been appointed.

ELECTRICAL INDUSTRIES.

PUBLISHED EVERY THURSDAY BY THE

ELECTRICAL INDUSTRIES PUBLISHING COMPANY,

INCORPORATED 1889.

MONADNOCK BLOCK, CHICAGO.

TELEPHONE HARRISON 398.

R. L. POWERS, Pres. and Treas.   E. E. WOOD, Secretary.

R. L. POWERS, · · · ·   Editor.

H. A. FOSTER, }
W. A. REMINGTON, }   · Associate Editors,
E. E. WOOD, }   · Eastern Manager,
FLOYD T. SHORT, · · ·   · Advertising Department.

EASTERN OFFICE, WORLD BUILDING, NEW YORK.

World's Fair Headquarters, Y 27 Electricity Building.

SUBSCRIPTION:	
FIVE MONTHS,	$1.00
SINGLE COPY,	10

Advertising Rates Upon Application.

News items, notes or communications of interest to World's Fair Visitors are earnestly desired for publication in these columns and will be heartily appreciated. We cordially invite all visitors to call upon us or send address at once upon their arrival in city or at the grounds.
ELECTRICAL INDUSTRIES PUBLISHING CO.,
Monadnock Block, Chicago.

THE interest taken in the World's Congress of Electricians by foreign nations and societies is a matter of commendation. The Electrotechnischer Verein of Vienna has just appointed the following of its members delegates to represent it at the meetings of the congress: Nikola Tesla, A. Prosch, inspector of the Austrian State Railways; Ernst Egger, Dr. Johann Sahulka, constructor at the Imperial High School, Vienna, Fred W. Tischendorfer and Joseph Wetzler.

THIS month will be the crowning one to all interested in electricity who are able to visit Chicago and the World's Fair. The number of electricians now here is being constantly increased. Prof. Helmholtz, whose name is familiar to everyone, is expected to arrive in Chicago in a few days. The educational advantages of this meeting can hardly be estimated. To all who are in any way interested in electricity, the congress and the Fair will be found of the greatest interest and benefit. Electricity Building affords unexcelled advantages for educational improvement. The exhibits, with the attendants in charge who are ready to explain any part of the exhibit; the various features installed, not so much to show the practical side of electricity as to show what can be done with the electric current; the popular lectures and other attractions arranged by the department afford opportunities never equalled. The congress, with the men prominent in arts and science assembled from distant cities, will also afford to all interested in electricity, no matter in what branch, an opportunity of gathering a mass of information such as would otherwise require years to gather.

Popular Lectures in Electricity Building.

The Department of Electricity desiring to create as much general interest as possible in the exhibits in Electricity Building, has mapped out a course of free lectures to be given every Tuesday, Thursday and Saturday afternoon from now on until the close of the Fair. Mr. Hawley, of the Department of Electricity has the matter in charge.

The regular course will begin next week and various subjects pertaining to the uses and application of electricity will be treated under the following heads: Lighting (a), Arc (b), Incandescent; Power Transmission; Experiments in High Potential, High Frequency and Alternating Current; Electrical Signals; Electrical Railways; Electricity applied to Mining and Milling Machinery; Wires and Insulation; Ocean Cables; The Telegraph; The Telautograph; The Phonograph; Metal Working; Scientific Instruments; Patent Law Applied to Electricity.

Preliminary to the regular course on Tuesday this week, Mr. C. P. Frey, electrician, with E. S. Greeley & Company, New York City, gave a talk on "Electrical Test Instruments," and on Thursday, Mr. Fred W. Tischendorfer, representative of Schuckert & Company, Nuremberg, Germany, will speak on "Search Lights," and on Saturday of this week "The Fire Alarm and Police Telegraph" will be discussed by Mr. E. Bruce Chandler, of the Gamewell Fire Alarm & Telegraph Company of New York.

The lectures will all be given in the new scenic theatre of the Western Electric Company in the southeastern corner of the Electricity Building, and will begin promptly at 2 o'clock P. M. It is expected that they will not last more than an hour. As soon as the program for the entire course is completed it will be announced in full in ELECTRICAL INDUSTRIES.

The Importance of Practical Experience.

The importance of practical experience in the training of electrical engineers is emphasized by Prof. S. P. Thompson in a letter to the editor of Electrical Plant and Industry, London, in which he says: "The education of a student of electrical engineering, whether liberal or illiberal, must be an education obtained from things and men, in the laboratory, the workshops, and the drawing-office, can be aided by books and lectures; but these do not and cannot constitute the education of an electrical engineer. From that point of view there is more educational value in a half-yearly volume of one of the weekly technical journals than in a ton of text-books. Give the student of electrical engineering the run of a laboratory well equipped with modern electrical machinery and instruments, and a good library of books of reference, and bring him into daily contact with men who are themselves both electricians and engineers, he will want few books beyond his own note-books and a pocket-book of numerical data. It is a real disaster to the electrical industry, and to the thousands of young men who are just now swarming into it on all sides, that the idea should prevail that so essentially practical a subject can be crammed up by mastering a certain list of books. It is mainly because we are convinced of the evil wrought by such misinformed ideas that Mr. Knapp and I, who have now for two years been colleagues as examiners in this subject in the City and Guilds Technological examinations, have determined to set such questions as shall be impossible to be satisfactorily answered by the mere paper electrician, whilst they shall be comparatively easy to answer by the candidate who has really mastered his subject by becoming practically acquainted with it in the laboratory, the drawing-office, and the workshop, the lighting station, or the testing-room. Electrical engineering can no more be learned from books than can railway engineering, or hydraulic engineering, or any other branch of the great electrical industry.

WORLD'S FAIR NOTES.

The exhibit of the Hicks (Troy) Electrical Door Operator on the west gallery, near the Enterprise Electric Co., is nearing completion.

Dr. N. S. Keith, of San Francisco, expects to shortly install an exhibit of constant current motors, similar to those used largely on the Pacific Coast.

Sidney Smith & Sons of the Basford Brass Works, Nottingham, Eng., have displayed in a neat case in Machinery Hall a line of brass goods, including steam boiler alarms, gauges, valves, water columns, etc.

The Yale & Town traveling crane with a band of music and a number of visitors, traveling back and forth the length of Machinery Hall and Annex over the heads of visitors and exhibitors was watched with a considerable of interest last week.

The new scenic theater of the Western Electric Co. is "playing to large houses." People can be seen crowding around the entrance awaiting the next performance from the time the theater is opened till it is closed in the afternoon.

A large jeweled sign, 24 feet 4 inches in length, bearing the words, "Western Electric Company," is to be hung in front of the Egyptian Temple, near the southern entrance to Electricity Building. Ten are lamps will be placed at the back of the sign to give the jewel effects.

The Association of Edison Illuminating Companies, of which Mr. M. J. Jenks, of the legal department of the General Electric Company, is Secretary, are holding their annual meeting this week at the Fair. The opening session was held in the Wisconsin State Building on Tuesday, August 8th.

In Machinery Hall the ornamental posts in the railing which surrounds the service plant have attracted a good deal of interest. The tops which support incandescent lamps in very neatly designed shades are very handsome. The different colors of the shades which were made by the Phoenix Glass Company give a variety to the combination.

The register of the American Institute of Electrical Engineers contains much interesting data. On August 4th there were nine entries only, but these nine are literally from the four corners of the globe. We give the addresses in order: San Francisco, Constantinople, Vienna, Stockholm, Darmstadt, Cornwall-on-Hudson, Morgantown, W. Va., Lafayette, Ind., and New York City.

Preparations for the California Midwinter International Exposition is progressing rapidly at Golden Gate Park at San Francisco, Cal. The Park covers some 100 acres and will contain five principal buildings, viz: Manufactures and Liberal Arts, Agricultural and Horticultural, Fine and Decorative Arts, Mechanical Art and Administration buildings. Besides these buildings there will be numerous smaller buildings occupied by various concessions.

McIntosh, Seymour & Company of Auburn, N.Y., have in use in the service plant in Machinery Hall a 1,200 horse power engine of the double tandem compound type belted to one of the large Westinghouse alternators. Among the special features of the construction of this engine are the water jackets for the guides and main bearings, the latter being of ball and socket pattern with oil settling chambers and pumps for continuous oiling, copper heating coils in the receiver fed from high pressure cylinder jacket, and a drag link shaft which gives motion to the governor placed on the outside of the frame. Engineers will find many things about this engine of interest.

What to see in the Electricity Building in the Daytime.

"There is nothing worth seeing in there," is a remark one quite frequently hears made in reference to the Electricity Building. Of course it is always made by visitors who are not well acquainted with the grounds and having no special interest in electrical matters. This is to be very much regretted, for there certainly are many exhibits of interest to the general public in the Department of Electricity.

Entering the Electricity Building at the north door the visitor should take the electric elevator directly in front of the entrance, or the one in front of the French section on the west side of the building, to the gallery. At the north center of the gallery is the exhibit of the Ansonia Electric Company, and in the east end of its pavilion is the exhibit of the American Electrical Heating Company, where the visitor may witness the exceedingly novel and interesting sight of cooking by electricity. This, of course, is more especially interesting to ladies.

Going west from the Ansonia exhibit and then south past the French piano exhibit the visitor will come to the booth of the Commercial Cable Company, where the sending and the receiving of cablegrams is something that the visitor should not go away without seeing.

Opposite to this booth is Prof. Gray's telautograph and everyone of course wishes to carry away from the Fair a souvenir telautogram as well as to have explained the workings of the mysterious little instrument that not only duplicates a message hundreds of miles away, but produces it in an exact facsimile of the sender's hand writing.

Continuing down the main aisle to the extreme southern end of the gallery the visitor will arrive at the exhibit of the North American Phonograph Company, where will be found the Edison Phonograph applied to all its various uses. Directly opposite the phonograph exhibit is one of the most interesting pieces of mechanism in the Fair, the National engraving machine, a machine capable of engraving one's name in characters so fine that a magnifying glass is necessary to decipher them.

Descending to the main floor the Grecian temple of the Brush Electric Company is directly in the foreground and opposite it is the smaller but no less attractive temple of the Jenney Electric Motor Company. Opposite the main southern entrance is the exhibit of the Bell Telephone Company with all of its historical apparatus. Here is to be seen the photophone by means of which sounds are transmitted through the medium of a ray of light.

Directly east of the Bell telephone exhibit is that of the Western Electric Company. If the day is warm the visitor should pass through the Egyptian temple and enjoy the cool breeze produced by the large exhaust fans, then if the scenic theatre is open a visit should be paid to it. The automatic writing machines at the northern end of this exhibit are extremely novel. Near by a magnetic lift offers an opportunity for the "strong man" to exercise his muscle.

Farther north past the main eastern entrance is the exhibit of the Electrical Forging Company, which, it is to be regretted, is too seldom in operation. Next north of this exhibit is that of the Belknap Motor Company where the visitor will get another refreshing breeze from a large exhaust fan and have an opportunity of witnessing the amusing spectacle of "fishes swimming in the air."

PERSONAL.

Mr. Wm. Stanley has been spending a few days at the Fair and about Chicago.

Mr. Luther Stieringer arrived in the city last Thursday and will spend a few days at the Fair.

Dr. Chas. E. Emery is now in New York but will return later when the active work of the jury commences.

Mr. E. A. Falconer, of the Falconer Mfg. Co., Boston, was a caller at the booth of ELECTRICAL INDUSTRIES this week.

Prof. George F. Barker, of the University of Pennsylvania, one of the judges of electrical exhibits, is now in the city.

Mr. Richard O. Heinrich, of the Weston Electrical Instrument Co., Newark, N. J., is at the Fair again for a few days.

Mr. J. T. Burke, secretary of the Western Electrical Supply Company, Omaha, Neb., has been in the city for some days past.

Mr. John M. Marvin, of Milwaukee, the inventor of Marvin's Electrical Brick Baker, has been a visitor at the Fair the past week.

Prof. Henry A. Rowland, of Johns Hopkins University, Baltimore, is attending the jury meetings of the department of electricity.

Mr. Phillip H. Campbell, general manager of the India Rubber Comb Co., New York, is spending a few days in Chicago and at the Fair.

Prof. Horace S. L. Verney, from the Stevens Institute, New York, is registered at the office of the American Institute of Electrical Engineers.

Mr. Adolpho Ashoff, of Brazil, has been appointed on the jury of awards of the Department of Electricity as one of the foreign members. He is now in attendance.

Prof. George F. Barker, of the University of Pennsylvania, is now in attendance at the Exposition. He is on committees No. 1 and 2 of the jury of awards of the department of electricity.

Dr. N. S. Keith, of San Francisco, Cal., arrived in Chicago on Friday last and will spend a few weeks at the Fair. Dr. Keith was one of the originators of the American Institute of Electrical Engineers.

Mr. S. D. Greene, general manager General Electric Co., New York; Mr. A. D. Page, assistant manager of the Edison Lamp Works, Harrison, N. J.; and Mr. Jno. W. Howell of the Edison Lamp Works, Newark, N. J., are in Chicago this week.

DEPARTMENT OF ELECTRICITY.

OFFICES: SECTION 3, ELECTRICITY BUILDING.

Chief, JOHN P. BARRETT,
Assistant Chief, J. ALLEN HORNSBY.
General Superintendent, J. W. BLAISDELL.
Electrical Engineer, W. W. PRIMM.

DEPARTMENT OF MECHANICAL AND ELECTRICAL ENGINEERING.

OFFICES SOUTH OF MACHINERY HALL.

Mechanical Engineer, C. F. FOSTER.
Electrical Engineer, R. H. PIERCE.
First Asst. Mechanical Engineer, JOHN MEADEN.
First Asst. Electrical Engineer, S. G. NEILER.

AMERICAN INSTITUTE OF ELECTRICAL ENGINEERS

World's Fair Headquarters.
SECTION 8, ELECTRICITY BUILDING.

RALPH W. POPE, Secretary.

Open from 9 a.m. to 5 p.m.

CHICAGO WORLD'S CONGRESS OF ELECTRICIANS.

OPENING SESSION, MONDAY, AUGUST 21st, 10 A. M.

ADVISORY COUNCIL.

President, DR. ELISHA GRAY, Highland Park, Ill.
Secretary, PROF. H. S. CARHART, Ann Arbor, Mich.

EXECUTIVE COMMITTEE.

Chairman, PROF. ELIHU THOMSON, Lynn, Mass.

COMMITTEE ON INVITATIONS.

Chairman, T. COMMERFORD MARTIN, 393 Broadway, New York.

COMMITTEE ON PROGRAM.

Chairman, PROF. T. C. MENDENHALL, Washington, D. C.

COMMITTEE ON FINANCE.

Chairman, B. E. SUNNY, 175 Adams Street, Chicago.

BUSINESS NOTES.

The ELECTRIC APPLIANCE COMPANY, 242 Madison St. Chicago, reports that it has a surprise in store in the shape of something new and progressive in the transformer line and advise intending purchasers to correspond with them before making any converter contracts or placing any transformer orders.

THE CENTRAL ELECTRIC COMPANY, Chicago, reports having just secured the contract for the complete installation with Okonite wire of the new Y. M. C. A. Building on LaSalle St. The company also secured the contract for the complete equipment of the great Ferris wheel with Okonite wire. The structure of this wheel being entirely of iron it was found necessary to have a very high grade of insulation and the fact that this brand was selected speaks volumes for that popular wire.

ELECTRICAL ENGINEERING COMPANY, 249 Second Ave., south, Minneapolis, Minn., formerly the Electrical Engineering & Supply Company, of St. Paul, is handling a large line of electrical machines and supplies, including National incandescent and Standard arc dynamos, Eddy and Holtzer-Cabot motors, Paranite, Grimshaw and O. K. Weatherproof wires and electrical supplies and specialties of all kinds. The company just completed the installation of a 2,500 light alternator with the necessary transformers, etc., at Brainard, Minn., and has several large contracts on hand.

Amusements.

HOOLEY'S THEATER—Mr. Nat C. Goodwin, in "Mizzoura," 149 Randolph street.

COLUMBIA THEATER—Miss Lillian Russell, in "The Mountebanks." 108 Monroe street.

GRAND OPERA HOUSE — Sol Smith Russell, in "A Poor Relation." 87 Clark street.

AUDITORIUM—Imre Kiralfy's Spectacle "America." Congress street and Wabash avenue.

McVICKER'S THEATER—Denman Thomson, in "The Old Homestead." 82 Madison street.

CHICAGO OPERA HOUSE—American Extravaganza Company, in "Ali Baba, or Morgiana and the Forty Thieves." Washington and Clark streets.

SCHILLER THEATER—Chas. Frohman's Stock Company, in "The Girl I Left Behind Me." Randolph, near Dearborn.

HAVERLY'S CASINO—Haverly's United Minstrels. Wabash avenue, near Jackson street.

TROCADERO—Vaudeville. Michigan avenue near Monroe street.

THE GROTTO—Vaudeville. Michigan avenue near Monroe street.

Buffalo Bill's "Wild West." 63d street. Daily at 3 and 8.30 p.m.

Pain's "Siege of Sebastopol," 60th street and Cottage Grove avenue. Tuesday, Thursday and Sunday nights.

The Alhambra Theater opens Saturday evening with Corinne in "Henrick Hudson."

The principal attraction at the Trocadero this week is the appearance of the strong Prussian, Sandow, whose feats requiring great strength are a surprise to all.

"The Mountebanks" by Gilbert and Cellier, is being presented this week at the Columbia by The Lillian Russell Opera Comique Company, Miss Russell appearing as Teresa.

'America" is enjoying its sixteenth week at the Auditorium. This fascinating spectacle once seen leaves an impression that cannot be effaced, and repeated visits are made with increased pleasure. The Schaffers are still one of the great attractions of the piece.

The principal event of this week has been the appearance of Mr. Nat C. Goodwin at Hooley's in "In Mizzoura," a play of which Augustus Thomas is the author. The play has been favorably received by good houses, which seem to be more enthusiastic on each succeeding night.

At the Chicago Opera House "All Baba" is having the same crowded houses. New ideas and new hits keep the play fresh. The jokes of the players keep pace with the times and some of the local hits are especially amusing. Norman's "Midway Plaisance" is a decided hit. It is announced that the piece is soon to be taken from the boards and will make a tour of the principal cities.

Batteries

Disque La Cianche Battery.

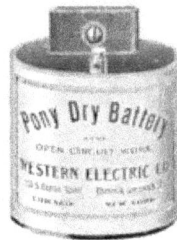

Pony Dry Battery

Phoenix Dry Battery.

Quad Battery.

Battery Exhibit Electricity Bldg. World's Fair.

Weco Carbon Battery.

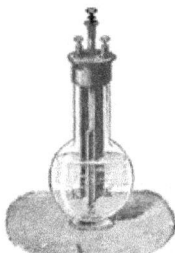

Crenet Battery.

Gravity Battery.

Smee Battery.

Carbon Battery.

Western Electric Company,
CHICAGO.   NEW YORK.

ELECTRICITY BUILDING—EXHIBITORS AND THEIR LOCATION.

GALLERY.

MAIN FLOOR.

Exhibitor.	Section	Exhibitor.	Section	Exhibitor.	Section	Exhibitor.	Section
Austin	Y	Electrical Review	V	Jaeger, Chas. L.	T	Helmers Gauge Co.	
Ansonia Electric Co.	Z	Electricity	H	Johns Mfg. Co., H. W.	T	Roessler & Hasslacher Chem. Co.	N
Am. Inst. of Elec. Eng.		Electric Gas Co.	M	Jewell Belting Co.	P	Street Railway Journal	
American Battery Co.	S	Electrical Engineer	Y	Jancy Elec. Motor Co.	L	Stromeyer Aut. Telph. Co.	
Alford, D. M.		Electrical World	Y	Knapp Electrical Works	V	Standard Paint Co.	
Allg. Elec. Gesellschaft	D	Eddy Electric Motor Co.	B	K. A. P. Elec. Novelty Co.	V	Sombolt C. L.	
Bulow Mfg. Co.		Excelsior Electric Co.	B	Knapp & Buckley	S	Star Iron Tower Co.	
Bryant Electric Co.	R	Electrical Forging Co.	H	Kennedy Electric Co.	S	Spain	
Billings & Spencer	R	Reynolds Dynamo Co.	P	Lawton, H. A.	U	Schiereu, Chas. A. & Co.	
Bricey, W. H.		Elektron Mfg. Co.	U	Le Clanche Battery Co.	U	Schoenberg & Sohne	
Belknap Motor Co.	S	Electrical Conduit Co.	P	McNeil Tinfoil Elec. Co.		Menozen & Hahlo	
Bell Telephone Co.	E–G	England		Marene, W. N.		Schuckert & Co.	
Brush Electric Co.	L	Emerer Glass Works	N	Meeker, Dr. G.	S	Short Electric Co.	
Caldwell El. Cloth Cut. Mch. Co.	Y	Franklin Elec. Appliances	N	McIntosh Bat. & Opt. Co.	W	Sperry Elec. Railway Co.	
Cutout, Elec. Storage Co.	T	French Piano Exhibit		Munson, C., Belting Co.	B	Standard Underg. Cable Co.	
Cutru, George	T	Fulton & Gallahouse		Matter, A. C.		Standard Electric Co.	
Cutlers Elec. Co.	Y	France	K–P–G	Mather Electric Co.	M	Samson Battery Co.	
Chicago Elec. Wire Co.		Ft. Wayne Elec. Co.	M	Newman Clock Co.	Y	The Am. El. Signal Co.	
Copenhagen Fire Alarm Co.	S	Gerill & Co. S. C.		Non Magnetic Watch Co.	M	Todd, Appleton Co.	
Crystal Electric Co.		Gamewell Fire Alarm Co.	S	N. Y. Insulated Wire Co.	V	Taylor, Goodhue & Ames	
Commonwealth Sup Co.		General Electric Co.	B–H–N–C–K–J	National Carbon Co.		Thomson Elec. Welding Co.	
C. & C. Elec. Motor Co.	A	General Incand't Arc El't Co.	J	Norwich Wire Works		Transbargel, Eliska Gray	
Cleveland Elec. & Mfg. Co.	A	Gisuley, E. S. & Co.		North Am. Phonograph Co.	S–V	Union Electric Co.	
Chicago Belting Co.		Germany	C	N. Y. & L. E. A.	Y	Vetter J.C. & Co.	W
Dulaney Clock Co.		Hubbard, Wm. & Co.	T	Nat. Aut. Fire Alarm Co.	S	Volk, G. F.	
Department of Electricity	H	Hodgson, C. J.		Nat. Engraving Machine Co.	U	Western El. Instrument Co.	
Electrical Engineers	Y	Hart & Hegman Mfg. Co.	S	Owen, Dr. A.		Washburn & Moen Mfg. Co.	
Elec. Launch & Nav. Co.	X	Hap. Elec. Appliance Co.	S	Phoenix Glass Co.		Western Union Tel. Co.	
Electric Separator		Hall, Chas T		Paist, H. T.		Waite & Bartlett Mfg. Co.	
Edgerton, E. M.	T	Holmes, W. S.	W	Polsermacher Galv. Co.	T	White, S. S. Dental Mfg. Co.	
Elgin Telephone Co.		Hartman & Braun		Pumpelly, J. S.	T	Western Electrician	
Edison Elec. Mfg. Co.		Hanson & Van Winkle		Perl El. Nod. Sup. Co.	T	Wilder Aut. Burglar Al. Co.	
Enterprise Elec. Co.		Hush, J. W.		Powell, Wm. & Co.		Western Electric Co.	
Frank Fruge Copper's		Handtranch, F. & Co.	W	Phelps, A. H.	U	Westinghouse El. & Mfg. Co.	B–H–J
Electric Appliance Co.		International Art Lt. & Pw Co.		Page Belting Co.		Weller & Sombolt	
Elec. Sol. & Ins't Co.		India Rubber Comb Co.	S	Queen & Co.		Wing, L. J., & Co.	
Electric Heat Alarm Co.	Y			Riegler, F. A.	M	Zucher & Levett Chem. Co.	P

"DULL TIMES"

Get in your orders before prices are sent skyward by favorable legislation in

Prices will probably prevail for the balance of this month, which will make it expedient for intending purchasers to anticipate their requirements as much as possible.

EXTRA SESSION.

ELECTRIC APPLIANCE COMPANY,
ELECTRICAL SUPPLIES AND SPECIALTIES,
EXTRAORDINARY,
242 Madison Street, - - CHICAGO.

THE MATHER ELECTRIC CO.
MANCHESTER, CONN.

Dynamos, Motors, Generators,

Offices, 116 Bedford St., BOSTON.

—AND—

1002 Chamber of Commerce Bldg., CHICAGO.

THE "NOVAK" LAMP.

CLAFLIN & KIMBALL (Inc.)

General Selling Agents.

116 Bedford Street, BOSTON.

1002 Chamber of Commerce Bldg., CHICAGO.

Siemens & Halske Electric Company
of America:
Chicago, Illinois:
Electrical Machinery.

Siemens & Halske,

Berlin.
Charlottenburg.
Vienna:
St. Petersburg.

Enterprise Electric Company

307 Dearborn Street, Chicago....

GENERAL WESTERN AGENTS

N. I. R.

Manufacturers' Agents and Mill Representatives for

Electric Railway,
Telegraph, Telephone and Electric Light
SUPPLIES OF EVERY DESCRIPTION

Agents for Cedar Poles,
Cypress Poles, Oak Pins,
Locust Pins, Cross Arms, Glass
Feeder Wire, Insulators,
WIRES, CABLES, TAPE and TUBING

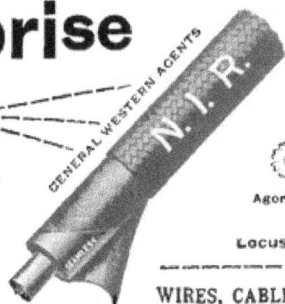

Map of Chicago.

Showing Location of its Electrical and Allied Business Interests, Principal Hotels, Theatres, Depots and Transportation Lines to the World's Fair Grounds. (Index numbers refer to the black squares.)

THE
FERRIS WHEEL

When you visit the World's Fair, you will naturally take a ride on the FERRIS WHEEL and be interested in the ELECTRIC LIGHT INSTALLATION. which is wired throughout with

OKONITE WIRE

FURNISHED BY THE

CENTRAL ELECTRIC COMPANY,

116-118 Franklin Street,

CHICAGO, ILLS.

VISITORS SHOULD NOT FAIL TO SEE THE

First Souvenir Half Dollar

...AT THE EXHIBIT OF...

Remington Typewriters

in the N. E. Corner of the Main Gallery of the Manufactures and Liberal Arts Building.

$10,000 was paid for this coin, making it the most valuable piece of silver in the world.

THE MONTHLY ISSUE FOR AUGUST

Should be read by everyone interested in electrical matters. In its table of contents is the following:

"Incandescent Lighting at the World's Fair."

"The Electric Power Plant of the Chicago City Railway."

"Steam Engine Efficiency- Its Possibilities and Limitations" by Wm. H. Bryan.

"Alternating Arc Lighting for Central Stations" by H. S. Putnam.

"Hard Rubber as an Insulator in Street Railway Work" by W. R. Mason.

"A Brief Review."

Together with illustrations of the recent applications of electricity.

The paper also contains regularly

A Buyer's Directory of Manufacturers and Dealers in Electrical Supplies and Appliances.

A Complete Directory of Electric Light Stations in North America and a Complete Directory of Electric Railways in North America.

These directories are revised each issue to the date of going to press and are to be found in no other electrical journal in the World. Its articles are read carefully and its directories used constantly by all the buyers in the trade. These facts make it without a superior as an advertising medium. Sample copies and rates sent on application.

Subscription price $3 per year. Six months trial $1, if ordered during the next 30 days.

ELECTRICAL INDUSTRIES PUB. CO.,
Monadnock Block, CHICAGO.

CLARK ELECTRIC COMPANY, NEW YORK.

192 Broadway and 11 John Street.

MANUFACTURERS OF ARC LIGHTING APPARATUS FOR EVERY PURPOSE A SPECIALTY.
The CLARK ARC LAMPS for use on EVERY CURRENT, have the reputation of being the best and most durable of any ever made in the United States.

RAWHIDE PINIONS FOR ELECTRIC MOTORS
A SPECIALTY.
RAWHIDE DYNAMO BELTING

Greatest Adhesive Qualities.   A Non-Conductor of Electricity.
Causes Less Friction than any other Belt.

THE CHICAGO RAWHIDE MANUFACTURING CO.
THE ONLY MANUFACTURERS IN THE COUNTRY

LACE LEATHER ROPE
AND OTHER RAWHIDE
GOODS
OF ALL KINDS
BY KRUEGER'S PATENT

This belting and Lace Leather is not affected by steam or dampness; never becomes hard; is stronger, more durable and the most economical belting made. The Rawhide Rope for Round Belting Transmission is superior to all others

75 Ohio Street,   CHICAGO, ILL.

STANDARD ELECTRIC COMPANY.

GENERAL OFFICES: 625 Home Insurance Building.

WORKS: So. Canal Street,

CHICAGO.

STANDARD SYSTEM

AT THE

WORLD'S FAIR.

MACHINERY HALL, Sec. Q, 2 Standard Arc Dynamos.
Sec. S, 20   "   "   "
ELECTRICITY BUILDING, Sec. P, Space 2, Arc Lighting Exhibit.

The Standard Lamps Light the Power Plant, Machinery Hall, Agricultural Hall, Shoe and Leather Building, and Other Buildings and Portions of the Grounds.

See our Double Service All Night Lamp Before Buying an Old Style Two Rod Lamp.

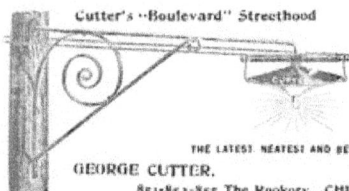

Cutter's "Boulevard" Streethood

THE LATEST, NEATEST AND BEST.

GEORGE CUTTER.

851-853-855 The Rookery...CHICAGO.

SIMPLEX WIRES

**INSURE
HICH
INSULATION**

SIMPLEX

Ever Onward and Upward!

Simplex Electrical Co.

620 Atlantic Ave.,

George Cutter, Chicago.   BOSTON, MASS.

XNTRIC

"That's the Switch"

And we control that movement.

H. T. PAISTE,

10 South 11th St.,

**PHILADELPHIA,
PA.**

Made 5 amp. S. P.
     10 amp. S. P.
      5 amp. 3 way.
     10 amp. 3 way.

P. & S.

WIRING INSULATOR,

Saves TIME
        TROUBLE
and TIE WIRE

Made only by

Pass & Seymour,

SYRACUSE, N. Y.

George Cutter,

CHICAGO.

P.&S

Consolidated Electric Co.

Manufacturers and Dealers in all kinds of

ELECTRICAL . SUPPLIES,

115 Franklin Street,

CHICAGO.

GEORGE PORTER,

Contractor for All Kinds of

ELECTRICAL WORK.

Room 87, 143 La Salle St., CHICAGO.
Crary Block, BOONE, IOWA.

CHAS. A. SCHIEREN & CO.

MANUFACTURERS OF

Genuine Perforated Electric Leather Belting.

46 So. Canal Street,   -   CHICAGO.

Section 15, Dpt. F. Clm. 27.   Section D, Space 3.
MACHINERY HALL.   ELECTRICITY BUILDING.

BEAR IN MIND

that the regular monthly issue of ELECTRICAL IN-
DUSTRIES contains the most complete and correct
directories published of the electric light central stations
and the electric railways in North America.

World's Fair Headquarters Y 27 Electricity Building.

CITY OFFICES, Monadnock Block.

WAGNER ELECTRIC FAN MOTORS

For Direct or Alternating Currents.

These motors give a stronger breeze with less consumption of current than
any other fan motor on the market. They are full 1-6 horse power. Six bladed
12 inch fan. Self-oiling. Furnished with or without guards.

IT WILL PAY YOU TO SEE THE WAGNER BEFORE BUYING ELSEWHERE.

TAYLOR, GOODHUE & AMES,

348 Dearborn Street, CHICAGO.

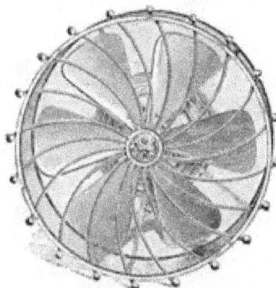

WEEKLY WORLD'S FAIR

ELECTRICAL INDUSTRIES

DEVOTED TO THE ELECTRICAL AND ALLIED INTERESTS OF THE WORLD'S FAIR,
ITS VISITORS AND EXHIBITORS.

Vol. I, No. 10.   CHICAGO, AUGUST 17, 1893.

FIVE MONTHS $1.00
TEN CENTS A COPY

Exhibit of George Cutter in Electricity Building.

An exhibit that has been arranged with a view to the comfort of visitors as well as the display of electrical goods

The exhibit is interesting to electrical people in many different ways and shows the variety of devices which the electrical specialist of the west has put on the market. Occupying a central position is a large panel on which are

EXHIBIT OF GEORGE CUTTER IN ELECTRICITY BUILDING.

is that of George Cutter in the southeastern part of the gallery of Electricity Building. The goods are arranged in attractive figures on panels about the space which is fitted with a handsome carpet, comfortable chairs and other furniture. A brass railing marks the front of the space.

displayed Simplex wires in a variety of sizes and kinds which Mr. Cutter has been pushing in the west. Large cables plain and armored form an artistic border to the panel. In the center are shown different simplex wires from the smallest to the largest with the insulation and armor slipped back to show its character. At the sides of these samples of wire are shown the wire on spools. The arrangement on the antique oak background makes an interesting display.

Part of the space has also been given to the display of the Pass & Seymour specialties. Insulators of almost every kind. Several sizes of the self-fastening porcelain insulator, outlet insulators, etc., cut-outs in numerous styles and sizes are shown. The line of switches is one that has only recently been developed and contains some novel features. The most striking point is the abandonment of the inner contact, the current being carried through the shaft as well as through a series of conducting ribbons. The ribbons increase the carrying capacity of the switch without adding to the power required to throw the switch, a point specially noticeable in the large sizes. The display includes sizes up to 1,000 amperes in single, double and three pole styles. The arrangement of the snap spring in these switches is also novel.

On the same board are shown several of the new economic door switches as well as the voltmeter and in the center of the board an eight point switch. This switch is a special four way switch designed for alternating current work. Arc light cut-outs are next shown in a number of styles, among which is the Dow hanger board shown in several sizes. The exhibit would hardly be complete without the Cutter lamp supporting pulleys used so largely in central station lines. One special form is shown in connection with the Cutter mast arm that surmounts the exhibits. On the same mast is also shown Cutter's boulevard street-hood. A large variety of smaller devices and specialties are shown, among which are mining sockets, tree insulators and lightning arresters. The goods shown in this exhibit of Mr. Cutter's are noticeable for their originality as well as for their practical design.

The Baker Gas Engine.

Joseph Baker & Sons of 58 City Road, London, England, have an extensive display of gas and oil engines in the east end of Machinery Hall. Several types are shown, some running with, others without, load; some as regular gas engines and others of the same type with the necessary parts added running as oil engines.

The special claims made for it are, that it can be changed from an oil to a gas engine or vice versa in a very short period of time; will use any commercial oil of flash test between 140 and 300 degrees; is very economical of fuel and can be started in from 7 to 10 minutes. The engine as a gas engine is of the common Otto Cycle type with exhaust and air valves, gas flame, water jacket and pump. To make the same machine an oil engine, there is bolted onto the rear end of the gas cylinder a vaporizing cylinder which being first heated to a proper degree by the flame from a blow lamp vaporizing the oil flowing into it and the heat is maintained by the explosions of the oil in the cylinder. A lamp on the outside is furnished with oil and a wick, the air blast from a pump blowing the flame into the igniting tube where it lights the gas. A double pump on the side of the cylinder actuated by an eccentric on the engine shaft, pumps water to the cylinder jacket and furnishes air to the lamp and cylinder. A very small pump attached to the under side of the vaporizing cylinder and actuated by the same shaft that moves the gas valve levers, pumps oil a few drops at a time into the vaporizer. The supply is controlled by the governor which is remarkably simple.

A crank on the end of the valve shaft moving in a slot in the upper end of a vertical lever that is pivoted below moves the lever back and forward for a space of about two and a half inches. Attached to the end of this lever at its upper end by a pin, is a flat finger about six inches long by three-fourths of an inch wide. On the end of the lever next to where it rests on the pin is an arm extending downward at an angle and sliding on this is a ball of iron which can be fastened at any point on the rod by a set screw.

This weight almost exactly counter-balances that of the flat arm and at slow speed the arm impinges against a notch in the end of the oil pump lever, pushing it forward and pumping a few drops of oil into the vaporizer. In case the speed increases the momentum or inertia of the ball being slow the lever falls and does not catch on the pump rod so no fuel is supplied and the engine slows down. A very slight variation in speed causes the governor to act. Common commercial burning oil is used, a small supply being kept in a tank at the side of the engine and conducted by very small copper pipes to the pumps.

THE BAKER GAS ENGINE.

The amount of fuel required per horse power per hour is said to be about one half pint. The cut herewith shows the three cylinder vertical oil engine displayed in the front part of the exhibit. The cranks are set at 120 degrees from each other to provide smooth running, and each cylinder is a complete engine in itself, being provided with running gear so that any one or two or all engines may be run at once. About the only difference is the change of governor to meet different conditions. A special feature on the machine not often seen in this country, is the style of gear used to run a shaft at right angles to the main or crank shaft; the gears have diagonal teeth cut in their peripheries and mesh at right angles, one acting as an exaggerated worm.

The application of oil engines seems to have been quite extensive both in England and on the continent and it would seem as if they might be used to advantage in this country if once properly introduced. It is probable that with gas at present prices the gas engines will not be universally used, but with the large supply of oil, the oil engine would be used economically.

The Exhibit of the Elektron Manufacturing Company.

In the northwestern part of the Electricity building opposite the French section is located the exhibit of the Elektron Manufacturing Company. The space has aisles on three sides allowing the machinery arranged about the space to be easily examined.

The electric elevator in the space forms a very attractive part of the exhibit as well as a most convenient way of reaching the gallery. The Elektron company has had an extended experience in the applications of electric power to elevator service and this elevator with the accompanying machines and apparatus demonstrates in an excellent manner the adaptability of electricity to this purpose.

A wooden frame with guides was erected on the platform in the space to the proper height above the gallery floor. The lower part is surrounded with an ornamented iron network for protection. Running in this frame is an ordinary elevator car rather prettily ornamented. The regular counter-balance weights are applied over pulleys at the top as usual. The elevator machinery consists of a 10-horse

thrown by means of the controlling rope and sheave which at the same time raises the brake. The starting switch turns the current into the motor and automatic rheostat and before admitting the current sets the reversing switch in the proper position for ascending or descending. The automatic rheostat switch always stands at the point of all resistance in except when the car is in motion and thus protects the motor from any rush of current in case the primary current is cut off for any purpose and the starting devices left on. When current is turned on, a strong solenoid magnet attached to the rheostat lever working against the dash pot very gradually cuts the resistance out and the motor gets up to speed slowly and safely. As this automatic rheostat has no mechanical connection with any part of the elevator machinery it can be set in any convenient place and wires run from the motor.

The main brake is a strap of iron almost surrounding the brake flange on the worm shaft. The compound lever actuating this brake terminates in a roller engaging with a cam on the shaft that turns the switch, thus whenever the switch is moved the brake is applied by the falling of the

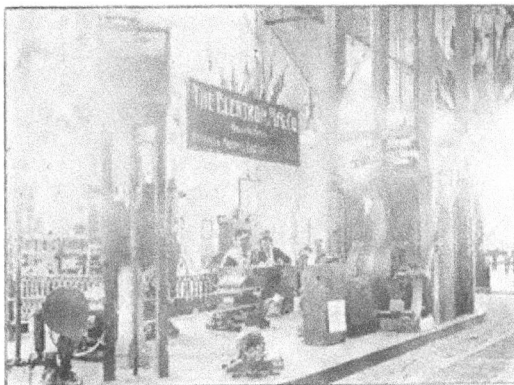

THE EXHIBIT OF THE ELEKTRON MANUFACTURING COMPANY.

power Perret six pole motor directly connected through an insulated universal coupling to the worm and gear attached to the cable drum. The whole machine was designed and constructed by the Elektron company.

The gear on the drum shaft has bronze teeth and cast iron hub on which is a good sized flange to prevent any danger from the keys between drum and shaft working loose. This hub flange is bolted to another similar flange on the head of the drum itself. A strap brake is provided for the drum and is automatically applied in a number of ways; for instance, if the car should happen to be very much overheated when coming down and attempt to race, a couple of weights inside the drum spread apart by centrifugal force and through connections to the outside, trip a lever supporting the brake weight, allowing it to fall and instantly apply the safety brake, gradually bringing the car to a stop. Again, in case the car comes to a standstill or meets with an obstruction in its descent, the consequent slacking of the cable causes a pulley riding against the cable to fall as the slack gains, tripping the brake and shutting off the power.

The motor is equipped with a starting switch which is

weight at the end of the lever; if the switch is off, the brake is on and vice versa. The entire machine is well designed, compact and seems to embody all the safety devices necessary to meet any emergency that can occur to such a machine. This particular machine has a lifting capacity of 1,500 at 150 feet per minute.

In the exhibit of electric motors for stationary work there is, first, a 15-horse power 500 volt Perret motor of the six pole type receiving current from Machinery Hall and running at 600 revolutions per minute. This motor is belted to another similar machine which as a generator produces 75 amperes of current at 110 volts for some of the motors in the exhibit. Current from this generator is conducted to a marbleized slate switch board and from there distributed to the various motors and other apparatus. The board is provided with Queen & Company's magnetic vane instruments, a rheostat built into the board, and all the necessary switches and cut-outs.

In order to show the adaptability of electric power to pumping water for domestic use, a one horse power 110 volt motor is belted to a Gould triplex pump which raises water from a tank under the floor to another raised on supports a considerable height above the floor. In the upper tank is a copper float which by a chain attached to a snap switch cuts off the current when the water reaches a certain height and cuts it on again when the water level is lowered. an automatic throttle similar to the one used with the elevator prevents rush of current through the motor.

Glass in the side of the tank permits inspection of the action. Several electric fan motors are shown: one, a quarter horse power, is of the new inverted type and carries a fan 15 inches in diameter; uses pencil carbon brushes and has ring coiled bearings. Other small motors are displayed, two motor generators, one giving out continuous current at 60 volts and the other alternating current at 100 volts, one armature is used with two windings in these machines.

Four Mosher constant potential arc lamps, run two in series are used to light up the exhibit at night. There is also a four horse power motor belted to a 50 ampere 110 volt dynamo. As a display of stationary motor work and incidentally of constant potential generators this exhibit is comprehensive, well arranged and enables the lay visitor to learn just what is being accomplished today in this line.

with brass body, up to very large sizes, with iron bodies, as used in the heaviest steam work. Samples are shown of large brass valves of this make as used by the naval department on the new cruisers. Others with quick opening handles are also shown.

Numerous samples of globe valves with regrinding seat are shown. In this valve there is a small hole in the stem and a corresponding one in the socket of the disc by placing a pin in the two when in line, they are joined together, so that the process of grinding may be carried on.

A simple type of sight feed lubricator for use on threshing machine engines and in other similar places, is exhibited. This instrument has the condenser bulb attached to a short piece of pipe extending from the upper side, and requires but one opening to the steam pipe for the entire apparatus, and but one valve is used to turn oil on or off after the first adjustment.

Dynamo oil cups, with patent feed, set so that oil may always be started instantly and feed the same. Grease cups, both with automatic compression feed and plain screw feed, the latter with a spring inside, so arranged as to prevent the

EXHIBIT OF THE LUNKENHEIMER COMPANY.

Mr. M. H. Robbins, Jr., in charge of the exhibit, will be found constantly on hand and ready to explain all the apparatus.

Exhibit of the Lunkenheimer Company.

The Lunkenheimer Company, of Cincinnati, Ohio, has an exhibition of valves and other similar material in Sec. 25, Col. Q 24, Machinery Hall. Broad platforms have been built and covered with yellow and black velvet, forming three terraces for the display of the company's product.

The company is making a specialty of the Lunken gate valve, which has an renewable seat, renewable disc, removable top, and is self balancing for steam pressure by the simple process of releasing the pressure on the handle, thus opening a small by pass hole through the gate. The top containing the screw, nuts and stem is secured to the valve body by a U shaped clasp surrounding the body and passing up through holes in the top casing, packing being inserted between the two parts. The removal of the two nuts holding this clasp enables one to take the entire valve to pieces.

These valves are exhibited in all sizes, from half inch

cover coming loose, glass sight gauges for ring oiled bearings, so constructed as to keep oil in easy and plain sight; and various other sight feed devices for attaching to points where bearings are oiled from a general system of piping, are shown in many sizes and styles.

The company makes water columns, pop valves, and a specialty of steam whistles, both the regular type and the famous mocking-bird whistle, such as is furnished to the Fair for signaling purposes, this one being an eight inch cylinder and run on 125 lbs. of steam. The exhibit is well arranged, quite comprehensive in the material shown and receives much attention from visitors.

The large engine manufactured by the Sioux City Iron Works seems to attract attention, placed as it is near the center of the Machinery Hall annex among looms, printing presses, lath's, etc. It is finished in nickel plate and white, of the Corliss pattern, 100 horse power and 24 by 48 in dimensions. The fly wheel is 16 feet by 44 inches and weighs 16 tons. It is belted by two Schleren electric belts running side by side to the two lines of shafting by which power is distributed throughout the annex.

The Western Electric Company's Exhibit of Primary Batteries.

In the exhibit of primary batteries made by the Western Electric Company in Electricity Building will be found bat teries of every desired size and type. There are batteries displayed from small dry batteries that can be placed in a wine glass to the largest used. Arranged in a pyramid are twelve hundred dry batteries and porous cups, and next to this pyramid is another pyramid of wet batteries.

The dry batteries shown, the Phoenix and Pony, are the results of the extended experience of the manufacturers, and on tests show excellent results. The resistance is very low and the electro motive force is as high as the Le Clanche. The Phoenix battery is shown in the standard size adopted of eight inches high by three and three-quarters in diameter, while the Pony battery is but one half as high. These batteries are largely used on electric bell circuits.

Several crowfoot gravity batteries of the size used on the telegraph lines are shown. The zinc and copper of these batteries are made of specially refined metal free from impurities. Several sizes of Grenet batteries, finely finished and

Exhibit of the Chapman Valve Manufacturing Company.

The Chapman Valve Manufacturing Company, of Boston, U.S.A., whose works are at Indian Orchard (Springfield), Mass., have a fine and extensive exhibit in Machinery Hall at K 28, center aisle.

The question of valves has become of such paramount importance in the consideration of the economic generation of power, that this company has taken advantage of the World's Columbian Exposition to give visiting managers and engineers a chance to see for themselves the great advance made by them in this line, especially calling attention to the valves recently designed for high pressure steam service.

The high pressure, renewable bronze seat valves, just mentioned, are shown with the bypass arrangement, the outside screw and yoke, the self packing spindle, by means of which the valve may be repacked when wide open with the steam on the line; and with the bodies heavily ribbed so that the valve is not affected by the expansion of the pipe line. In addition to these features all kinds of end

THE WESTERN ELECTRIC CO'S EXHIBIT OF PRIMARY BATTERIES.

EXHIBIT OF THE CHAPMAN VALVE MANUFACTURING COMPANY.

designed for laboratory use, are shown; also the Smee battery, which is very simple and plain in construction, using dilute sulphuric acid for the liquid. Daniell, Fuller's mercury bicromate and other batteries are shown. Several of LeClanche batteries, which have been so successfully used on electric gas lighting, bell and other circuits, are exhibited.

The various kinds of carbon batteries made by the Western Electric Company, in which the carbon used is of such a form that a proportionately large surface of carbon is presented to the acid are shown. The acid and manganese used by the company are remarkably free from iron and other impurities. To secure pure material has required a number of years of persistent effort

The porous cups used by the company are made from special clay found in but few places in the United States. In connection with the exhibit are shown the different parts of the batteries and the liquids used in them. This part of the exhibit is especially interesting as showing how each battery is made up. A number of specially designed porous cups in which by different means the walls are decreased in thickness are shown, it being the object to lower the resistance of the cells.

connections indicating devices and methods of opening are shown.

The display is arranged in a very appropriate manner, valves for like purpose being bolted or screwed to brass or iron masts in tiers, or arranged in columns ranging from very large at the bottom to quite small at the top.

The line of flanges shown are intended for use in connection with high pressure steam, and are shown, both plain, tongued and grooved. These may be used either bolted to valve or as flange unions.

All kinds of brass and iron valves are shown, adapted to feed and circulating systems, exhaust and low pressure steam; also ammonia valves for mechanical refrigeration, water valves and fire hydrants for use on cast iron pipe.

Besides the Machinery Hall exhibit, working displays may be seen at the Intramural Electric Railway power house, on the Heine boilers in the boiler house annex to Machinery Hall, and upon the big engine in Machinery Hall itself, which was made by the E. P. Allis Manufacturing Company, of Milwaukee, Wis.

Edward L. Ross, M. E., is representing the company, and gives a hearty welcome to visitors at the space in Machinery Hall.

ELECTRICAL INDUSTRIES.

PUBLISHED EVERY MONDAY BY THE

ELECTRICAL INDUSTRIES PUBLISHING COMPANY,

INCORPORATED 1893.

MONADNOCK BLOCK, CHICAGO.

TELEPHONE HARRISON 395.

E. L. POWERS, PRES. AND TREAS.   E. E. WOOD, SECRETARY.

E. L. POWERS, ...   EDITOR.

H. A. FOSTER,   }
W. A. REMINGTON,   }   .   .   .   ASSOCIATE EDITORS.

E. E. WOOD, ...   EASTERN MANAGER.

FLOYD T. SHORT, ...   ADVERTISING DEPARTMENT.

EASTERN OFFICE, WORLD BUILDING, NEW YORK.

World's Fair Headquarters, Y 27 Electricity Building.

SUBSCRIPTION:

FIVE MONTHS, ... $1.00
SINGLE COPY, ...   10

Advertising Rates Upon Application.

News items, notes or communications of interest to World's Fair Visitors are earnestly desired for publication in these columns and will be heartily appreciated. We especially invite all visitors to call upon us or send address at once upon their arrival in city or at the grounds.
ELECTRICAL INDUSTRIES PUBLISHING CO.,
Monadnock Block, Chicago

The total paid admissions to the Fair to date reaches nearly eight millions. The pleasant weather and lower railroad rates have brought more people during the last few days. The interest shown in the popular amusements arranged by the management should encourage their continuance. They not only furnish amusement for the visitors but a diversion for the exhibitors and attaches of the Fair, while the participants in the swimming and canoe contests seem to enjoy them greatly.

The Electric Committee of Underwriters meet this week at 157 LaSalle street, Chicago, to discuss the recently adopted international rules and will themselves formally adopt rules and regulations for electric light and railway circuits. The members of the committee in the city are Mr. F. E. Cabot, Boston, chairman; Mr. LeLoup, New Orleans; Mr. Devitt, Philadelphia; Mr. E. A. Fitzgerald, Syracuse; Mr. Geo. P. Low, San Francisco, and Mr. E. A. Van Geisen. While in session the committee hope to devise some means for standardizing municipal and automatic fire alarm systems.

The World's Congress of Electricians.

About one thousand invitations have been sent out and from the number of acceptances received, it looks as though the congress would be well attended. At the meeting of the Advisory Council held January 17th., a committee was appointed to decide on the number of members that each foreign country should be invited to send to sit in the Chamber of Delegates. Their apportionment was as follows:

Five each for England, France, Germany, Austro-Hungary and the United States.

Three each for Belgium, Italy and Switzerland.

Two each for Norway and Sweden, Holland, Denmark, Russia and Spain.

One each for Portugal, British North America, Australian Colonies, India, Japan, China, Mexico, Brazil, Chili, Peru and the Argentine Republic. Making fifty-five in all.

Up to the present moment eight of the more important countries have officially responded to the call. We give below the names of the delegates in the order of their official appointments.

England.—W. H. Preece, F. R. S.; Prof. W. E. Ayrton, F. R. S.; Prof. S. P. Thompson, D. Sc., F. R. S.; Alex. Siemens, Major Carden, R. E.

France.—M. Mascart, M. Hospitalier, M. Violle, M. de la Touanne.

Germany will be represented by Dr. H. von Helmholtz who requests that the following gentlemen of the Physikalische-Technische Reichsanstalt of Charlottenburg, be allowed to attend the meetings of the Chamber of Delegates, namely: Dr. Feussner, Dr. Kurlbaum, Dr. Leman, Dr. Lindeck, Dr. Lummer and Dr. Pringsheim.

United States.—Prof. H. A. Rowland, Prof. T. C. Mendenhall, Dr. H. S. Carhart, Prof. Elihu Thomson, Prof. Edward L. Nichols.

Switzerland.—M. le Dr. A. Palaz, M. Thury, M. le Dr. Weber.

Italy.—Prof. Gallelio Ferraris.

Mexico.—Senor Don A. M. Chavez.

China.—Mr. Peng Kuang-Yu, Mr. Teng Shen, Mr. Shou Yeo.

The reply of the Electrotechnische Verein of Vienna published in our last issue will be referred to the committee on credentials at the opening of the congress.

Program of Popular Lectures.

The hour for the lecture course in the scenic theatre of the Western Electric Company has been changed from 2 P.M. to 11 A.M. The course, under the supervision of Mr. Hawley of the Department of Electricity, is exciting interest in the exhibits of electrical wares, and the idea so far has been extremely well received. The program, for this week and next, will be as follows: Tuesday, Mr. George C. Holland, representative of the North American Phonograph Company, lectured on the Edison phonograph. In another column will be found a summary of his address.

On Saturday Mr. R. O. Heinrich, electrical engineer for the Weston Electrical Instrument Company, will talk of electrical testing apparatus.

On Thursday, Mr. E. J. Jenness, representing the Western Electric Company, will speak on the subject of arc lighting.

Tuesday, August 22, Mr. Wm. Carroll, superintendent of city construction (Chicago), will deliver an address on municipal arc lighting.

Thursday, August 24, the address will be on safety alarm water columns, and Mr. Geo. B. Clark, manager of the Reliance Gauge Company, will be the lecturer.

Saturday, August 26, Mr. Geo. D. Benton, president of the Electrical Forging Company (Boston), will speak of electric bath heating and electric heating and welding.

Obituary.

The many friends of Mr. Louis W. Burnham, vice-president and general manager of the Electric Gas Lighting Company, Boston, were pained to hear of his death on Wednesday the 9th instant. His recovery was confidently expected up to within a very short time of his death and the sudden termination of his illness was a shock to his family and friends.

Funeral services were held at his late residence in Dorchester Friday noon and the burial took place in his family lot at Palmer, Mass.

WORLD'S FAIR NOTES.

Director Emil Rothenan of the German exhibit and Prof. Ulbricht started on their return to Berlin Wednesday the 16th.

Prof. Louis C. Hill of the School of Mines, Colo., Dr. W. J. Hershman of the University of Michigan and Dr. Louis Duncan have recently been appointed on the jury of awards.

The General Electric Company installed an additional exhibit on the experimental track at the side of the Terminal Station last week in the shape of a 30 ton high speed electrical passenger locomotive of the gearless type. It will soon be seen in operation.

The exhibit of the Mather Electric Company, Electricity Building is at last completed, and is exciting much admiration and favorable comment both from attaches of the building and from visitors.

The ventilating exhaust fans just placed in the boiler annex of Machinery Hall will be greatly appreciated by the visitors as well as the attendants. The Jenney Electric Motor Company has installed 3 three-horse power motors to operate them, which are of the well known Jenny type.

The Hoppes Manufacturing Company has an exhibit of feed water heaters in the boiler annex showing close together a new heater and one that has been used with the pans incrusted with scale. In a case are shown numerous samples taken from heaters which are interesting to steam users.

Major Gen. A. Benton, vice-president of the Standard Signal Company, Rochester, N. Y., accompanied by Mr. W. W. Hibbard, the inventor and electrician for the company, is in Chicago this week, showing a working model of the company's fire and auxiliary alarm system at the rooms of the Fire Underwriters Association in the N. Y. Life Building.

The Maharajah of Kapurthala reviewed the military organization at the Fair at 5 o'clock p. m. on Tuesday. The band pavilion near the north entrance of Electricity Building was used as the reviewing stand. The Maharajah was escorted from the boat landing to the pavilion by Director General Davis and the official representatives of various foreign countries in full uniform.

An informal meeting of the American Institute of Electrical Engineers has been called for Wednesday evening, August 23d, at the World's Fair headquarters. It is to be of a social character for extending personal acquaintance between members. A good attendance is expected as a large number of members are in the city attending the Congress and Fair.

Some exceedingly novel electroliers are shown in connection with the exhibit of earthquake instruments from Japan. The electroliers are constructed of bamboo bent in a very artistic manner. The sockets are concealed within leaf shaped pieces made of small pieces of bamboo. The different parts are well put together and make a very ornamental electrolier.

The elevators at the north end of Electricity Building put in by the Frisbie Elevator & Manufacturing Company, of New Haven, Conn., are greatly appreciated by visitors to whom the stairs seem like mountains after walking around the grounds. This easy and quick way of reaching the gallery largely increases the number of visitors to that part of the building. This elevator is very appropriately operated by electricity. The motion is easily regulated; the

car stopped or started. It is also fitted with safety devices by which the car would be stopped automatically should the cable break or the speed of the car exceed the limit for which the device is adjusted. A 10 horse power Crocker & Wheeler motor furnishes the motive power. The motor runs in one direction and is not reversed. The reversal being effected by a friction clutch which when thrown in different directions changes the direction of the drums.

The various special attractions at the Exposition arranged by the management have drawn large crowds to the part of the grounds where they took place. This was especially true of the swimming contests of last Friday when this course was lined with interested spectators. These races were entered by the foreigners at the Fair, the prizes offered being five dollar gold pieces. The race of the Turks won by Comaste and the Dahomey race won by Byotoglan were particularly amusing.

In response to an invitation from Mr. W. A. Grant, of Niagara Falls, general manager of the Niagara Falls Park & River Railway, the eastern members and friends of the Electrical Congress will take a ride over the above road Sunday morning August 20th. The special train will stop at Niagara Falls giving time for all to see the beautiful scenery about Niagara. This electric road runs for 12 miles in view of the falls and rapids. The ride will take about two hours. Special cars will be provided.

The American merchant might learn a thing or two from His Highness of Kapurthala. The Maharajah wished to purchase a supply of electrical ventilating apparatus. He wrote Mr. L. J. Auerbacher, of the E. S. Greeley & Co., to call on him at his hotel. Mr. Auerbacher called, and His Highness invited him to lunch with himself and his wife, and after luncheon was over he placed a thousand dollar order for goods, and we venture to suggest he paid less for them than he would have paid had he been wined and dined at the expense of the firm that secured the order.

Exhibit of the Joseph Dixon Crucible Company.

The extent to which graphite is used in some form or other in nearly every branch of business makes the exhibit of the Joseph Dixon Crucible Company of interest to every one.

In the electrical industry graphite is largely used and its production is consequently of great importance to the trade. The two exhibits of the company, one in the northeastern part of the gallery of Mining Building, and the other in the northeast gallery of Manufactures and Liberal Arts Building, show the graphite in both the crude form and the manufactured product.

The space in Mining Building, 25 by 28 feet is very handsomely finished and arranged. In upright cabinets are displayed numerous articles made of graphite, crucibles, retorts, ladles, stopper heads, graphite boxes, resistance rods, incandescent filaments, moulds and other goods. The process of electrotyping in which graphite is used is illustrated in a case placed in the center of the room.

In other cases are shown graphite lubricants, stove polish, foundry facings, paints, etc., etc. Distributed about the room are samples from all the different sources from which that article is obtained. One sample from the island of Ceylon weighing nearly 300 pounds seems to attract attention. The space is handsomely carpeted and furnished with comfortable chairs, tables, etc.

PERSONAL.

Mr. E. Egger, Vienna, is paying the Fair a visit.

Mr. Ch. Lotter, Zurich, was seen in the Electricity Building last week.

Dr. Martin Kellmann, Berlin, called at the office of the A. I. E. E. this week.

Mr. Geo. E. Metcalfe, editor of Electricity, New York City, is at the Exposition this week.

J. M. Rich, electrical engineer, New York, was a caller at the office of ELECTRICAL INDUSTRIES this week.

T. C. Rafferty & Co., Chicago, have located their offices in the Como block, and are now ready for business.

Mr. W. H. St. John, electrician of the Lazoo Company, Lazoo, Miss., called at the World's Fair office of ELECTRICAL INDUSTRIES.

Mr. Stephen L. Coles, the associate editor of the Electrical Review, is looking over the Fair, and the Electricity Building especially.

Mr. M. J. Sullivan, of the Electrical World, was married on Tuesday of this week to Miss Frances Read at Freeport, Ill. A number of Chicago friends attended. ELECTRICAL INDUSTRIES extends congratulations.

Mr. Geo. B. Ellison, formerly engineer with the North Western Thomson-Houston Company, of Portland, Ore., has just accepted a position as engineer for the western department of the Waddell-Entz Company, with offices at 1140 Monadnock building.

Mr. Geo. P. Low, the well-known insurance inspector of the Pacific coast, is in the city accompanied by Mr. W. F. C. Hasson. Both gentlemen are the delegates from California to the Electrical Congress, and while here will visit the Fair. Mr. Hasson is the special agent of the electrical department of the Fair to be held in San Francisco next winter. It is hoped to have an electricity building at this Fair, and while in Chicago Mr. Hasson will do what he can to secure exhibits for his department.

DEPARTMENT OF ELECTRICITY.

OFFICES SECTION R, ELECTRICITY BUILDING.

Chief, JOHN P. BARRETT.
Assistant Chief, J. ALLEN HORNSBY.
General Superintendent, J. W. BLAISDELL.
Electrical Engineer, W. W. FRIMM.

DEPARTMENT OF MECHANICAL AND ELECTRICAL ENGINEERING.

OFFICES SOUTH OF MACHINERY HALL.

Mechanical Engineer, C. F. FOSTER.
Electrical Engineer, R. H. PIERCE.
First Asst. Mechanical Engineer, JOHN MEADEN.
First Asst. Electrical Engineer, S. G. NEILER.

AMERICAN INSTITUTE OF ELECTRICAL ENGINEERS.

World's Fair Headquarters,
SECTION S, ELECTRICITY BUILDING.

RALPH W. POPE, Secretary.

Open from 9 a. m. to 5 p. m.

CHICAGO WORLD'S CONGRESS OF ELECTRICIANS.

OPENING SESSION MONDAY, AUGUST 21ST, 3 P. M.
ADVISORY COUNCIL.

President, Mr. ELISHA GRAY, Highland Park, Ill.
Secretary, Prof. H. S. CARHART, Ann Arbor, Mich.
EXECUTIVE COMMITTEE.
Chairman, Prof. ELIHU THOMSON, Lynn, Mass.
COMMITTEE ON INVITATIONS.
Chairman, T. COMMERFORD MARTIN, 203 Broadway, New York.
COMMITTEE ON PROGRAM.
Chairman, Prof. T. C. MENDENHALL, Washington, D. C.
COMMITTEE ON FINANCE.
Chairman, R. E. SUNNY, 175 Adams Street, Chicago.

BUSINESS NOTES.

THE CHICAGO RAWHIDE MANUFACTURING COMPANY, Chicago, reports that the foreign demand for its rawhide products is increasing in spite of the present dull times. The company has just received an order by cable for 5,000 feet of belting. It also made a large shipment of goods to Australia last week.

THE MOSHER ELECTRIC COMPANY, Chicago, is having many inquiries and making a number of shipments of its street railway arc lamps. Among the orders recently filled are those to the South Chicago Railway; Paducah, Ky., and the Los Angeles Street Railway. The alternating arc lamp recently brought out by this company is also selling well.

THE WESTERN ELECTRIC COMPANY, Chicago, has just secured the contract for the complete installation of a electric lighting central station for the city of Rochester, Minn. The plant will consist of 80 Western Electric Company arc lights and 1,000 alternating incandescent Westinghouse. The entire construction of the plant will be done by the Western company.

THE ENTERPRISE ELECTRIC COMPANY, Chicago, has just received contract for the entire electric equipment for the Medina Temple, corner Jackson St. and Fifth Ave. The McFell Electric Construction Company is doing the work of installation. The entire plant is to be wired throughout with N. I. E. wire, for which the Enterprise Company is the general western agent.

ELECTRIC APPLIANCE COMPANY report that their business in electrical house goods does not seem to feel the depression that exist in other lines of electrical trade and is proving a valuable line to fall back on at a time that in other departments is comparatively quiet. The Electric Appliance Company are building up a splendid trade in this line of goods, which has been comparatively neglected for the past few years in the interest of electric light supplies, and their electrical house goods have already established for themselves a first-class reputation among the trade of this material.

Amusements.

HOOLEY'S THEATER—Mr. Nat C. Goodwin, in "Mizzoura." 149 Randolph street.

COLUMBIA THEATER—Miss Lillian Russell, in "The Mountebanks." 108 Monroe street.

GRAND OPERA HOUSE — Sol Smith Russell, in "A Poor Relation." 87 Clark street.

AUDITORIUM— Imre Kiralfy's Spectacle "America." Congress street and Wabash avenue.

McVICKER'S THEATER—Denman Thompson, in "The Old Homestead." 82 Madison street.

CHICAGO OPERA HOUSE—American Extravaganza Company, in "Ali Baba, or Morgiana and the Forty Thieves." Washington and Clark streets.

SCHILLER THEATER—Chas. Frohman's Stock Company, in "The Girl I Left Behind Me." Randolph, near Dearborn.

HAVERLY'S CASINO—Haverly's United Minstrels. Wabash avenue, near Jackson street.

TROCADERO —Vaudeville. Michigan avenue near Monroe street.

THE GROTTO—Vaudeville. Michigan avenue near Monroe street.

Buffalo Bill's "Wild West." 63d street. Daily at 3 and 8.30 p.m.

Pain's "Siege of Sebastopol," 60th street and Cottage Grove avenue. Tuesday, Thursday and Sunday nights.

Sol Smith Russell, at the Grand Opera House, continues to have good houses. This is the sixteenth week of his present engagement, and the present piece "A Poor Relation," in which Mr. Russell has been such a favorite, will be continued some time longer.

"Ali-Baba" has but three more weeks to run, when "Sinbad," the revival of which has been postponed several times on account of the continued success of "Ali-Baba," will be again placed on the stage. The various productions of the American Extravaganza Company form a remarkably successful list. The first one, "The Arabian Nights" was presented at the Chicago Opera House 96 times, "Sinbad" ran 122 times; "Bluebeard Jr.," 125; "The Crystal Slipper," in two engagements, 266 times, and "Ali-Baba," when it closes its present engagement, September 2, will have been played 331 times in Chicago.

Batteries

Disque La Clanche Battery.

Phoenix Dry Battery.

Quad Battery.

Battery Exhibit Electricity Bldg. World's Fair.

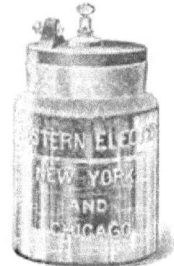
Weco Carbon Battery.

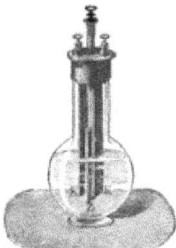
Crenet Battery.

Gravity Battery.

Smee Battery.

Carbon Battery.

Western Electric Company,

CHICAGO. NEW YORK.

ELECTRICITY BUILDING—EXHIBITORS AND THEIR LOCATION.

GALLERY.

MAIN FLOOR.

FAVORABLE
SILVER LEGISLATION
GOOD AS GOLD
MONEY WILL BUY MORE

would undoubtedly be a good thing for certain sections of the country, but we would respectfully remind the trade that our specialties and supplies are

in any section of the country and it is a mistake to withhold orders for material you are going to need in the near future, as your

today than it will thirty days from now.

SUBMITTED FOR YOUR CONSIDERATION BY THE

ELECTRIC APPLIANCE COMPANY,
ELECTRICAL SUPPLIES
242 Madison Street, - - CHICAGO.

THE MATHER ELECTRIC CO.
MANCHESTER, CONN.

Dynamos, Motors, Generators,

Offices, 116 Bedford St., BOSTON.
—AND—
1002 Chamber of Commerce Bldg., CHICAGO.

THE "NOVAK" LAMP.

CLAFLIN & KIMBALL (Inc.)

General Selling Agents.

116 Bedford Street, BOSTON.

1002 Chamber of Commerce Bldg., CHICAGO.

Siemens & Halske Electric Company
of America.
Chicago, Illinois.
Electrical Machinery.

Siemens & Halske,

Berlin.
Charlottenburg.
Vienna.
St. Petersburg.

Enterprise
Electric
Company

307 Dearborn Street.
Chicago . . . .

N. I. R.

GENERAL WESTERN AGENTS

Manufacturers' Agents and Mill Representatives for

Electric Railway,
Telegraph, Telephone and
Electric Light

SUPPLIES OF EVERY DESCRIPTION

Agents for   Cedar Poles,
    Cypress Poles, Oak Pins,
Locust Pins, Cross Arms, Glass
—Feeder Wire,   Insulators,
WIRES, CABLES, TAPE and TUBING

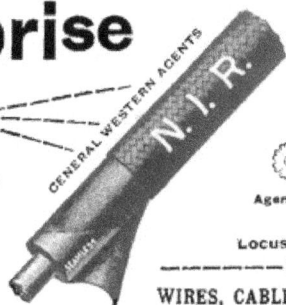

Map of Chicago.

Showing Location of its Electrical and Allied Business Interests, Principal Hotels, Theatres, Depots and Transportation Lines to the World's Fair Grounds. (Index numbers refer to the black squares.)

THE
FERRIS WHEEL

When you visit the World's Fair, you will naturally take a ride on the FERRIS WHEEL and be interested in the ELECTRIC LIGHT INSTALLATION. which is wired throughout with

OKONITE WIRE

FURNISHED BY THE

CENTRAL ELECTRIC COMPANY,

116-118 Franklin Street,
CHICAGO, ILLS.

VISITORS SHOULD NOT FAIL TO SEE THE

First Souvenir Half Dollar

...AT THE EXHIBIT OF...

Remington Typewriters

in the N. E. Corner of the Main Gallery of the Manufactures and Liberal Arts Building.

$10,000 was paid for this coin, making it the most valuable piece of silver in the world.

THE MONTHLY ISSUE FOR AUGUST

ELECTRICAL INDUSTRIES

Should be read by everyone interested in electrical matters.   In its table of contents is the following:

"Incandescent Lighting at the World's Fair."
"The Electric Power Plant of the Chicago City Railway."
"Steam Engine Efficiency– Its Possibilities and Limitations" by Wm. H. Bryan.
"Alternating Arc Lighting for Central Stations" by H. S. Putnam.
"Hard Rubber as an Insulator in Street Railway Work" by W. R. Mason.
"A Brief Review."
Together with illustrations of the recent applications of electricity.
The paper also contains regularly
A Buyer's Directory of Manufacturers and Dealers in Electrical Supplies and Appliances.
A Complete Directory of Electric Light Stations in North America and a Complete Directory of Electric Railways in North America.
These directories are revised each issue to the date of going to press and are to be found in no other electrical journal in the World. Its articles are read carefully and its directories used constantly by all the buyers in the trade. These facts make it without a superior as an advertising medium. Sample copies and rates sent on application.
Subscription price $3 per year. Six months trial $1, if ordered during the next 30 days.

ELECTRICAL INDUSTRIES PUB. CO.,
Monadnock Block, CHICAGO.

CLARK

ELECTRIC

COMPANY, NEW YORK.

192 Broadway and 11 John Street.

MANUFACTURERS OF ARC LIGHTING APPARATUS FOR EVERY PURPOSE A SPECIALTY.
The CLARK ARC LAMPS for use on EVERY CURRENT, have the reputation of being
the best and most durable of any ever made in the United States.

RAWHIDE PINIONS FOR ELECTRIC MOTORS

A SPECIALTY.

RAWHIDE DYNAMO BELTING

Greatest Adhesive Qualities. A Non-Conductor of Electricity. Costs Less. Better than any other Belt.

THE CHICAGO RAWHIDE MANUFACTURING CO.

THE ONLY MANUFACTURERS IN THE COUNTRY

LACE LEATHER ROPE
and OTHER RAWHIDE

GOODS
OF ALL KINDS
BY KRUEGER'S PATENT

Our Belting and Lace Leather is not affected by steam or dampness, is not liable to stretch or slip, and is durable and the most economical Belting made. The Rawhide Rope for Round Belting Transmission is superior to all others.

75 Ohio Street, CHICAGO, ILL.

STANDARD ELECTRIC COMPANY.

GENERAL OFFICES: 625 Home Insurance Building.

WORKS: So. Canal Street,

CHICAGO.

STANDARD SYSTEM

AT THE

WORLD'S FAIR.

MACHINERY HALL, Sec. Q, 2 Standard Arc Dynamos.
Sec. S, 20 " " "
ELECTRICITY BUILDING, Sec. P, Space 2, Arc Lighting Exhibit.

The Standard Lamps Light the Power Plant, Machinery Hall, Agricultural Hall, Shoe and Leather Building, and
Other Buildings and Portions of the Grounds.

See our Double Service All Night Lamp Before Buying an Old Style Two Rod Lamp.

FERRIS WHEEL_____

Wired with mile
upon mile of

Simplex Wire

(We don't claim it all)

GEORGE CUTTER. 851-855 The Rookery, Chicago.

SIMPLEX WIRES

INSURE
HIGH
INSULATION

SIMPLEX

Ever Onward and Upward!

Simplex Electrical Co.
620 Atlantic Ave.,

George Cutter, Chicago.   BOSTON, MASS.

XNTRIC

"That's the Switch"

And we control that movement.

H. T. PAISTE,

10 South 18th St.,
PHILADELPHIA,
PA.

Made 5 amp. S. P.
10 amp. S. P.
5 amp. 3 way.
10 amp. 3 way.

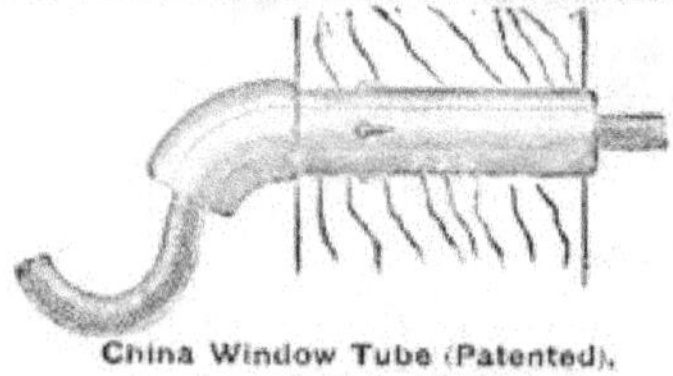

China Window Tube (Patented).

Made only by **PASS & SEYMOUR,**

George Cutter,
CHICAGO.
SYRACUSE, N. Y.

Consolidated Electric Co.

Manufacturers and Dealers in all kinds of

ELECTRICAL . SUPPLIES,

115 Franklin Street,

CHICAGO.

GEORGE PORTER,

Contractor for All Kinds of

ELECTRICAL WORK.

Room 67, 143 La Salle St., CHICAGO.
Crary Block, BOONE, IOWA.

CHAS. A. SCHIEREN & CO.

MANUFACTURERS OF

Genuine Perforated Electric Leather Belting.

46 So. Canal Street,   -   CHICAGO

Section 15, Dpt. F. Clm. 27.   Section D, Space 3
MACHINERY HALL.   ELECTRICITY BUILDING.

CALL AND EXAMINE

Lawton's Call Indicator.

Indispensable for hotels, railroad
offices, school buildings, hospitals, etc.

Section Y, Space 45, Gallery Electricity Building,
WORLD'S FAIR.

CHICAGO, ILL.

WAGNER ELECTRIC FAN MOTORS

For Direct or Alternating Currents.

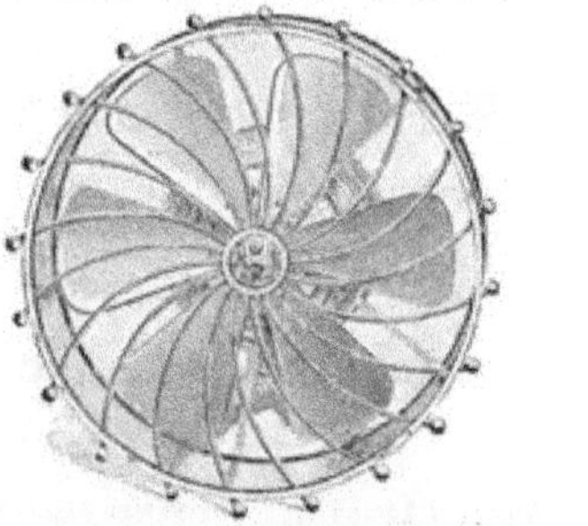

These motors give a steadier hum with less consumption of current than
any other fan motor on the market. They are built 1 8 horse power. No bladed
motor fan. Noiseless. Furnished with or without guards.

IT WILL PAY YOU TO SEE THE WAGNER BEFORE BUYING ELSEWHERE.

TAYLOR, GOODHUE & AMES,
348 Dearborn Street, CHICAGO.

WEEKLY WORLD'S FAIR

ELECTRICAL INDUSTRIES

DEVOTED TO THE ELECTRICAL AND ALLIED INTERESTS OF THE WORLD'S FAIR. ITS VISITORS AND EXHIBITORS.

Vol. I, No. II.   CHICAGO, AUGUST 24, 1893.   FIVE MONTHS $1.00 TEN CENTS A COPY

Arc Lighting Plant of the Standard Electric Company at the World's Fair.

In Machinery Hall, section S, is located one of the largest arc lighting plants of the Exposition supplying current for er. The other lights are distributed in the Agricultural, Anthropological, Forestry, Shoe and Leather buildings and around the Peristyle and in other parts of the grounds.

The 20 dynamos are arranged in two rows and are belted to shafts beneath the floor. The wiring is all concealed be-

ARC LIGHTING PLANT OF THE STANDARD ELECTRIC COMPANY

over 1,100 arc lamps. This plant which was installed by the Standard Electric Company, Chicago, contains 20 dynamos, each having a capacity of 50 2,000 candle power lights while in section Q are one 50 and one 60 light machines, used in the illumination of the power plant prop-

neath the floor, both the leads to the switch board and the wires to the ammeters which are after a special design of the Standard company. These ammeters are raised to standards to a convenient height and are placed sufficiently near to be easily read by the dynamo tender from the dynamo. The metallic cases and standards are nickel plated, presenting a neat appearance. The wires are run from below inside the standards so that there are no exposed wires to be injured or to oppose the free movement of the attendants about the machine.

The dynamos are of the well known Standard type. The frame is an adaptation of the Manchester type. The pole pieces which form the main portions of the frame are of soft cast iron of a character especially designed for this purpose. In the field magnets wrought iron is used. The journal supports are made self adjusting and aligning. The bearings are made long and are fitted with self oiling lining of

ARC LIGHTING PLANT OF THE STANDARD ELECTRIC COMPANY.

an improved pattern. The armature is easily removed by taking out the journal support at either end. This armature has many special features the excellence of which are apparent on inspection. The large open interior permits of a free ventilation of the armature which is so essential to the perfect operation. From this construction both the interior and exterior surface is open to the air.

To remove and replace a coil requires little time or trouble and a coil may be renewed with ease on the armature on account of the armature being free from internal supports. In case of accident to one of the coils it may be disconnected from the commutator and the dynamo run until such a time as repair can be conveniently made. Many visitors have remarked on the smoothness with which these machines run, there being no sparking and heating at the brushes a fact which is quickly noticed by users of many arc machines of other manufacture. All screws, bolts and nuts used about the machine are made from standard commercial taps and dies so that repairs may be quickly made no matter how remote a plant may be from the factory. As a whole the machine commends itself to the visitor as simple, solid and compact.

Three engines furnish the power for this plant. The first engine which is a 24 by 24 by 15 by 24 double tandem compound furnishes power for 12 dynamos. It is a four valve independent exhaust Russell engine. It is fitted with Detroit lubricator and eight level oil cups of the latest pattern. The fly wheel which is 10 feet in diameter and 60 inches wide is belted by a 50 inch George Oberne & Company's belt to a shaft beneath the floor to which each dynamo is independently belted. The next four dynamos are run by a 20 by 20 by 13 single valve Russell engine. The fly wheel which is eight feet in diameter by 30 inches

is belted similar to the above by a 26 inch belt. The last four dynamos are belted tandem to a 200 horse power engine running 275 revolutions per minute. This engine which was built by the Erie City Iron Works is fitted with a Webber automatic governor, Detroit lubricator, sight feed oil cups and other improved devices.

The dynamos in section Q are run by an ideal engine manufactured by A. L. Ide & Son. The power is transmitted by the Ideal system which makes a most compact combination. The dynamos and engine which is 125 horse power requiring no more room than if they were directly connected. This plant though occupying but little space has a capacity of 140 2,000 candle power arc lights and is worthy of more than a passing notice.

In the center of the large plant is erected a model switch board capable of accommodating 24 circuits and 24 dynamos. It is of marble, of the plug pattern and so arranged that the current from any dynamo in the block may be turned on any circuit. The board is extremely simple and compact. The company also furnishes the operating department with a 12 circuit board used as an exchange board. In case the dynamos or engines in any plant should become disabled from any cause, by means of this board current from other plants may be turned on.

The lamps used on the circuits of this plant are all of the Standard company's make. The lamps used in the Machinery Hall power plant, on the patrol circuits, and in the grounds are of the double service weather proof pattern while the others are of the single service pattern. These lamps have been in use a number of years and improvements have been added making the lamps in every way efficient. In the outdoor weather proof lamps the unsightly hood has been abandoned and the lamps present a trim and substantial appearance. All the working parts and connec-

ARC LIGHTING PLANT OF THE STANDARD ELECTRIC COMPANY.

tions are enclosed in a tight weather proof cap which can easily be removed. The lamp may be quickly disconnected from the line without opening the circuit. The carbon feeds smoothly and without any jerking motion, thus giving a clear and steady light.

The hangers used for supporting these lamps were described and illustrated in our July issue but the credit for the designs was given to the department which should have been given to the Standard company, as we have since been informed that the design originated in the factory of this company where the hangers were made.

Every part of this plant has been installed in the most substantial and workmanlike manner and to those interested in central station construction an examination of the plant in detail will be found of interest. Many improvements in construction and methods of operation have been adopted.

World's Congress of Electricians.

After months of preparation the Electrical Congress has assembled. The electricians have been gathering in Chicago for sometime in anticipation of the congress and the attendance at the opening, both in numbers and the promi nence was fully up to expectations. When the congress opened there were on the platform the following gentlemen whose names are familiar to all delegates from their re spective countries to the congress: Dr. H. Von Helmholtz, Dr. E. Voigt, Germany; Prof. E. Hospitalier, Prof. E. Mascart, France; Dr. A. Palaz, M. Thury, Switzerland; Prof. Elihu Thomson, Dr. T. C. Mendenhall, Prof. H. A. Row land, Prof. H. S. Carhart, United States; Prof. Silvanus Thompson, Alexander Siemens, Prof. W. E. Ayrton, W. H. Preece, England; Dr. Johann Sahulka, Austria; Prof. Galileo Ferraris, Italy.

OPENING SESSION.

The Chicago World's Congress of Electricians was called to order in Columbus Hall of the Art Institute by Prof. Elisha Gray, at 3:15 o'clock P.M. on Monday. Prof. Gray read an address welcoming the delegates and the members of the congress, and at the end called for nominations for temporary chairman. Mr. W. H. Preece, a delegate from Great Britain, in a very neat little speech, proposed for temporary chairman Elihu Thomson, of Lynn, Mass., who was elected by acclamation. Prof. Thomson, on taking the chair, made a short speech which was received with ap plause.

On calling for nominations for temporary secretary, Prof. F. B. Cross, of the Massachusetts Institute of Technology, Boston, proposed Prof. F. B. Crocker, of Columbia College, New York, who was elected by acclamation. The chair then appointed as a committee for the nomination of per manent officers Prof. T. C. Mendenhall, Prof. B. F. Thomas, Dr. Louis Duncan, Prof. Silvanus P. Thompson, Prof. E. Hospitalier, Dr. A. Lindeck, and Dr. A. Palaz, who then retired. Prof. Thomson called on Prof. W. E. Ayrton to say a few words on the subject of the World's Fair as seen by a foreigner.

Prof. Ayrton stepped forward, prefacing his speech by saying that the chairman had made a mistake in calling him a foreigner, as no Englishman could be a foreigner in America. He preferred to be called a stranger. He then went on to say that Chicago is a long way from the seas, and he thought that a stranger's impression on electrical matters began long before the arrival at the white city. After describing what impressions a stranger would receive in his journey to Chicago in the various eastern cities he might visit, he said: He is as much astonished with the courtesy with which he is shown every detail in the factories as by the magnitude of the undertakings. His expecta tions are aroused to such a pitch, that he expects that when he comes here to Chicago, to find a display which will cast all previous electrical exhibitions into the shade. Well, take this stranger into the Electricity Building blindfolded, so that his judgment will not be warped by the glories of the outside surroundings, and leave him there, and after a while, if he be candid and if he can master up courage to say a word which might cause pain to those whose kindness has made them dear to him, he will say he is a little disap

pointed that the world has not better answered the invita tion to show the Electricity Building what it could do.

The stranger feels that the real electrical display of America is not in the Electricity Building but in every street where there are trolley wires, in every town and village where there are electric lights—and where is the town in this country where there are not? But if the stranger be thoughtful, he is not disappointed. In Frank fort it was what was inside the Electricity Building that dazzled the mind; at Jackson Park it is what is outside of the Electricity Building that rivets his attention. The committee now returned, and Dr. Mendenhall, as chairman, reported the nominations as follows: For honorary presi dent of the congress his excellency Dr. H. Von Helmholtz, of Germany; as permanent chairman of the congress, Prof. Elisha Gray, of Chicago; as vice presidents, representing the United States, Edw. Weston; Great Britain, W. H. Preece; France, E. Mascart; Germany, Dr. Voigt; Austria, J. Sahulka; Italy, Prof. Galileo Ferraris; Switzerland, Dr. H. F. Weber, who were elected by acclamation. Prof. F. B. Crocker was then nominated as permanent secretary and was elected by acclamation.

In presenting Dr. Von Helmholtz as the honorary presi dent of the congress, the chairman, Prof. Thomson, called attention to his very great achievements and to the fact that at the beginning of his career but a trifle was known of the science of electricity. Dr. Von Helmholtz, on accepting the presidency, was greeted by the audience standing and with the heartiest applause. He spoke shortly of the honor that he considered was conferred on him by the position, and said in his remarks: "The beginning of my career was when the phenomena of electricity were most delicate ex periments which were performed by some physicists in their laboratories. We can move at present great machines of the greatest power. At the time I began the study of electricity, we could not move a little magnetic needle sus pended on a silken cord, the finest that we could find. We could not move the slightest apparatus, but we showed that there were feeble currents. We were obliged to work with the simple elements zinc and copper, without sulphate of copper, elements which altered every moment; which at the first instance had a great electromotive force, then went down and down, so that after some instances there was only a trace of the former force."

In conclusion, he said, "The history of the world and the history of science has grown rapidly during our life time, and it is a great pleasure for us old men, to see how what electricity has reached in its new stages and to admire the newest developments that are collected on this festival occasion here in your great exhibition." Prof. Gray was then introduced by the chair as the permanent chairman of the congress and made a short speech of acceptance, and then introduced in turn Mr. W. H. Preece as vice president representing Great Britain, whose speech of acceptance was brief but heartily applauded. The chairman then introduced Prof. Mascart, who made a short speech. Dr. Mendenhall, on behalf of the program committee, made a few remarks as to the conditions for membership of the congress, etc. The delegates to be distinguished by a white badge, and all persons who have received invitations and those whose name appears on the printed list of reference will receive a red badge; to others they will be passed on the presentation of a card endorsed by the com mittee. Mr. R. W. Pope, W. E. Anderson, and Dr. J. Allen Hornsby were appointed a committee on entertainment and will, from time to time, report such arrangements as were made.

The different sections of the congress assembled in the halls assigned. Section A of pure theory was called to order at 10:30 by its chairman, Prof. H. A. Rowland. The chair appointed a committee to nominate the permanent officers of the section. The following names were recommended by the committee and elected by acclamation: Prof. H. A. Rowland, chairman; Prof. Galileo Ferraris, vice-chairman, and Dr. A. S. Kimball, secretary. After disposing of other details of organization the section adjourned till 10 A. M. Wednesday morning.

Section B was called to order by Prof. Chas. R. Cross, of the Institute of Technology, Boston, at 10:15 A. M. The first business in order was the appointing of a committee to nominate permanent officers for the section, a chairman and a secretary, and at the suggestion of the temporary chairman a third officer was added to the list, which was composed of Prof. Silvanus P. Thompson, London; Prof. Louis Duncan, Baltimore, Md., and Lieut. W. F. C. Hassan, San Francisco. The committee retired and returned in a few minutes recommending that Prof. Cross be continued in the chair, Lieut. Samuel Weber appointed permanent secretary with Prof. Dolbear as the third officer of the section. Prof. Elisha Gray announced that section A would meet in Hall VI., section B in Hall VII. and section C in Hall VIII.

The reading of the regular papers on the program was then taken up and Mr. W. H. Preece, F.R.S., read the first paper on "Signaling through Space by Means of Electro-Magnetic Vibrations." Starting from the discovery by Henry in 1842 of the transmission of signals through the vibration of the magnetic needle in the cellar charged by a Leyden jar in an upper room of his house, Mr. Preece led down to 1884 when it was first reported to him that messages in transit over the underground wires of the London telegraph system had been read over a telephone, the wire of which was 80 feet distant, and incidentally referred to Edison's experiment in 1885 in the transmission of messages from a moving train on the Lehigh Valley Railroad to the wires on the side of the track.

Mr. Preece was last year appointed a member of the Royal Commission to investigate the communication by electricity between light houses, light ships and the shore, and his address was a description in part of the experiments made by this commission on the coast of Bristol Channel. Mr. Preece did not read his paper in full but gave the convention the results of the investigation and the particularly interesting points leading up to them. The result of these experiments were due entirely to the electro-magnetic induction; the self induction and the electro-static induction playing but very slight parts in the general results. The main experiment consisted in the running of a grounded line three-quarters of a mile long on the shore of the channel and the running of a parallel line of the same length on a small island something over three miles distance from the shore, and a third line parallel to the other two and of the same length on a second island some five miles distance from the shore. With a steam launch cables were laid at various distances and experiments made as to the possible reading on such cables through a telephonic connection, of signals sent with an ordinary Morse instrument over the land wire. It was found that when the cable laid on the surface of the water (as it did when being rapidly played out from the steam launch, one end of it being

attached to a buoy, the signals could be very distinctly heard. While, when the cable had sunk under the water to a depth of 6 inches or more the induction ceased. This had Mr. Preece to make certain experiments which established beyond doubt the fact that electro-magnetic waves are reflected from the surface of the water and do not percolate. With 15 amperes of current and a frequency of 192, experiments were made between the land and the three miles island wire, with the result that the signals were read by telephone and the notes were very clear. In experiments with the wire on the five mile island, and with the same frequency and current on the shore wire, the signals were heard but the current was not strong enough to make the Morse code readable. Mr. Preece is of the opinion that had he had five amperes more of current at hand the signals would have been as clear on the five mile island at 20 amperes as they were on the three mile island at 15 amperes. In all of these experiments the readings were made by from three to five different persons to preclude any possibility of the supposed effect being the result of imagination. With wires 30 miles long on each side of the British Channel Mr. Preece is confident that communication by electro-magnetic induction could easily take place between England and France, though he doubts very much if this method will ever come into commercial use for various reasons; principally owing to its being more expensive than the present cable system. The English light houses and light ships, as a result of these experiments, will be connected to the main line by regular cables, though it is highly possible that induction may be used by the light houses and light ships for the transmission of signals to vessels during time of fog. The engineer of the coast of Scotland is at present engaged in the introduction of a system for the protection of vessels approaching the coast by the laying of a cable at the 20 fathom line along the coast and using electro-magnetic induction for the transmission of signals between the shore and the ships through this cable. No practical results have yet been attained in this direction.

In the discussion that followed the reading of Mr. Preece's paper, Mr. T. D. Lockwood, electrician for the American Bell Telephone Company, Boston, delivered a somewhat lengthy expression of opinion and gave a resume of some similar but far less important or interesting experiments that had been made in this country. He also expressed his doubts in regard to the transmission being due entirely to the electro-magnetic induction and inclined to the opinion that water played a very important part. In his final remarks at the close of the discussion, Mr. Preece, in a very few words, showed Mr. Lockwood that he was in error. Owing to the length of Mr. Lockwood's comment Prof. Carhart of Ann Arbor, felt called upon to move that the discussion by any one member of a paper that had been presented be limited to five minutes. The chairman declined to put the motion but suggested that the members so understand it. Dr. Carroll expressed his appreciation of Mr. Preece's paper and recited an incident that came to his attention some 15 years ago where messages had been transmitted over telephone circuit, the wire of which had been broken and the two ends resting some ten feet apart on a large block of stone. The incident at the time was considered so improbable that the Doctor had never mentioned it until Mr. Preece's paper had proven to him the entire possibility of the occurrence being true, Prof. Jamieson desired to inquire what was the greatest distance from the cable in the north of Scotland at which signals could be received by an approaching ship. Mr. Preece replied that the distance would be regulated by the length of the ship, or the length of the cable on the ship.

Mr. Heaviside, the associate of Mr. Preece, who had charge of the experiments described in the latter's paper, was introduced by Mr. Preece and made a short address in which he described an experiment he recently conducted in the pit of a colliery, where he ran a line on two sides of a triangle and made the connection for the third side by means of a ground, it not being possible to run the third side owing to the location. He then constructed a similar triangle on the surface, having all three sides of wire. Signals were transmitted by electro-magnetic induction from the lower circuit to the upper, and received by telephone. The practical result of this experiment can be utilized for communications between imprisoned miners and the outside world, provided cables can be constructed in some way so as not to be broken by the falling rock. Dr. Emery called the attention of the congress to the different principles employed in Mr. Preece's experiments and the American experiments cited by Mr. Lockwood. Messrs. Delaney, Wiegand, Wynn and Lemp asked various minor questions which were replied to by Mr. Preece. Prof. Cross also made a suggestion with reference to the loss of energy in reflection by the electro-magnetic wave.

The second and third papers on the program "Materials for Standards of Resistance and their Construction," by Dr. St. Lindeck and "Variation of P. D. of the Electric Arc Current, size of Carbons and Distance Apart," by Prof. W. E. Ayrton, F. R. S., were not read owing to the absence of the two gentlemen mentioned. The fourth paper on the program was by Dr. Silvanus P. Thompson, entitled "Ocean Telephony." Prof. Thompson finished his paper at 12:45 and the congress adjourned till the following morning.

Section C. was called to order at 10:20 a. m., by the temporary chairman Prof. E. J. Houston. The committee on nominations present the following lists of names: Permanent chairman, Prof. Edwin J. Houston; vice-chairman, George P. Low and secretary Prof. E. P. Roberts; committee for the section George Blodget, Dr. F. H. C. Perine and Townsend Wolcott. The report was accepted and short speeches were made by Prof. Houston and Messrs. Lowe, Roberts, Jackson and others. A very interesting paper was read by Mr. Franz Shulze-Berge, of Brooklyn on "Rotary Mercurial Air Pumps" which was illustrated by comprehensive diagrams. After a brief discussion of the section adjourned and those present attended the session of section B.

WEDNESDAY'S SESSION.

At this session of section A., Prof. A. Macfarlane read a paper on "The Analytical Treatment of Alternating Current" which was discussed by Mr. Chas. P. Steinmetz. The paper of Drs. Bedell and Crehore, "General Discussion of the Current Flow in Two Mutually Related Circuits Containing Capacity," was presented.

Section B was called to order at 10:10 Wednesday by Chairman Cross and after the reading of the minutes of previous meeting by Secy. Weber the discussion of Prof. Silvanus P. Thompson's paper was taken up. Dr. Carroll desired to present the congress a tube made of an alloy of aluminum, silver and copper prepared in his laboratory, and expressed the belief that such an alloy would be better fitted for telephonic purposes than the conductors now used. Prof. Jamieson made the point that whereas the present cables used cost but £40,000 per mile a cable made in accordance with Prof. Thompson's paper with three con-

ductors would cost probably three times as much. He also desired to know how Prof. Thompson would localize faults or flaws.

Mr. Lockwood had to suggest that several pieces of underground cable of similar construction be spliced until a cable be produced with which the practicability of Prof. Thompson's ideas could be tested. Mr. Kennelly called attention to the ideal method devised several years ago by Mr. Oliver Heaviside for absolutely eliminating effects of electro-static induction. Mr. Cuttress, electrician for the McKay-Bennett cable system, described certain experiments made under his direction with a two wire cable connected in straight cord, metallic and multiple circuit, the result of which show but slight variation in the three styles of connection. He also offered to place the cables of his company at the disposal of Prof. Thompson for the conducting of further experiments. Mr. Heaviside expressed his appreciation of Prof. Thompson's paper both from a scientific and commercial point of view and added that his experience since 1877 confirmed the principles laid down in the paper.

Mr. Wilkinson inquired whether the effect of the three wires did not tend to cause further retardation of the signals. Mr. Siemens, of Siemens Bros., London, the cable manufacturers, said that such a cable as Prof. Thompson proposed would be extremely difficult to construct and would cost much more than cables, for telegraphic purposes at least, which would give a similar speed. He called attention to the fact that capacity is not the only enemy of the telegraph but that the insulating materials play an important part. Prof. Cross made the point that while, with the telegraph and the siphon recorder, from the point of extreme legibility to illegibility there is considerable latitude, with the telephone, from audibility to inaudibility there is but a slight step. Prof. Thompson replied quite humorously to the various criticisms made on his paper and expressed himself as believing that he is still on the right side and sincerely hoped that one effect of the reading of his paper will be the giving of his plans a practical trial, which can only be done when the large cable interests cooperate in making such a trial possible.

Dr. St. Lindeck of the Reichsanstalt, Berlin, read an interesting paper on "Materials for Standards of Resistance and their Construction," in which he described the result of the experiments made by Dr. Feussner and himself at the Reichsanstalt during the past four years. He described tests made of platinum silver, german silver, patent nickel and manganese alloy. He found that where the percentage of zinc was less in the alloy there was less variation of resistance. The variation in the first two alloys mentioned proved them to be unfit for standards; while the patent nickel alloy used by Siemens & Halske of Berlin, and composed of 25 parts nickel and 75 parts copper stood the tests remarkably well; an alloy called manganin and composed of 84 parts copper, 12 parts manganese and four parts nickel stood the test for a space of two years with a variation of but a few thousandths of one per cent.

Mr. Edw. Weston of Newark, N. J., discussed Dr. Lindeck's paper at considerable length, the rule as to the time being suspended in this instance. He said that such papers as Dr. Lindeck's tend to increase the confidence of the electrical fraternity in these alloys as standards. In 1884, Mr. Weston began a series of researches of alloys for resistance purposes at his laboratory at Newark, N. J., examining in the neighborhood of 400 of them. About Per

[Continued on page]

ELECTRICAL INDUSTRIES.

PUBLISHED EVERY THURSDAY BY THE

ELECTRICAL INDUSTRIES PUBLISHING COMPANY,

INCORPORATED 1893

MONADNOCK BLOCK, CHICAGO.

TELEPHONE HARRISON 281.

F. L. POWERS, PRES. AND TREAS.   E. E. WOOD, SECRETARY.

F. L. POWERS,   Editor.

H. A. FOSTER,   }
W. A. REMINGTON,   }   Associate Editors.

E. E. WOOD,   Eastern Manager

FLOYD T. SHORT,   Advertising Department

EASTERN OFFICE, WORLD BUILDING, NEW YORK.

World's Fair Headquarters, Y 27 Electricity Building.

SUBSCRIPTION

FIVE MONTHS $2.00
SINGLE COPY   10
Advertising Rates Upon Application

News items, notes or communications of interest to World's Fair
Visitors are earnestly desired for publication in these columns and will
be heartily appreciated. We especially invite all visitors to call upon us
or send address at once upon their arrival in city or at the grounds.
ELECTRICAL INDUSTRIES PUBLISHING CO.,
Monadnock Block, Chicago

For some time Chicago has been the Mecca towards
which the eyes of all true electricians have been looking
and there are now gathered in this city the most prominent men in science and the arts, an array of eminent
names such as this wonderful city of the west has never
previously seen. The number of electricians at the congress is fully up to expectation and the enthusiasm shown
gives evidence of the interest taken in the congress by the
representatives from the different countries. The committees having the preparation for the congress in charge
seem to have been most thorough in their work. The arrangements made have met with the approval of the congress so that it has proceeded directly with the work laid
out for it. The committees have not only provided for the
work of the congress but for many diversions at the Fair
and in the city. Through the kindness of various concessioners and companies at the Fair a most enjoyable program was arranged for yesterday afternoon for visiting
various attractions at the Fair. The banquet this evening
at the Grand Pacific Hotel is the medium through which
the foreign and domestic members will meet in a social
way. The lecture of Nikola Tesla in the assembly hall
of the Agricultural Building at the Fair on Friday evening, at 8 o'clock, will be an interesting feature of the week.

List of the Papers of the Congress.

SECTION A.

"On the Analytical Treatment of Alternating Currents,"
  Prof. A. Macfarlane.
"Complex Quantities and their Application in Electrical
  Engineering," Charles P. Steinmetz.
"General Discussion of the Current Flow in Two Mutually
  Related Circuits Containing Capacity," Dr. Frederick
  Bedell and Dr. Albert C. Crehore.
"Explanation of the Ferranti Phenomenon," Dr. J. Sahulka.
"Measuring the Power of Polyphase Currents," A. Blondel.
"Extended Use of the Name Resistance in Alternate Cur-
  rent Problems," Prof. W. E. Ayrton, F. R. S.

SECTION B.

"Signaling through Space by Means of Electro-Magnetic
  Vibrations," W. H. Preece, F. R. S.
"Materials for Standards of Resistance and their Construc-
  tion," Dr. St. Lindeck.
"Variation of P. D. of the Electric Arc with Current, Size
  of Carbons and Distance Apart," Prof. W. E. Ayrton,
  F. R. S.
"Ocean Telephony," Dr. Silvanus P. Thompson, F. R. S.
"Iron for Transformers from the Magnetic Point of View"
  Prof. J. A. Ewing, F. R. S.
"Note on Photometric Measurement," Prof. B. F. Thomas.
"Some Measurements of the Temperature Variation in the
  Electrical Resistance of a Sample of Copper," A. E.
  Kennelly.
"Various Uses of the Electrostatic Voltmeter," Dr. J.
  Sahulka.
"On a Method of Governing an Electric Motor for Chro-
  nographic Purposes," Prof. A. G. Webster.
"On the Construction of Cables for Subterranean High
  Tension Circuits," Dr. A. Palaz.
"Periodic Variation of the Candle Power of Alternating Arc
  Lights," Prof. B. F. Thomas.
"Transformer Diagram Experimentally Determined," Dr.
  Frederic Bedell.
"London Electrical Engineering Laboratories," Prof. An-
  drew Jamieson.
"On the Source and Effects of Harmonics in Alternating
  Circuits," Prof. H. A. Rowland.
"A Pair of Electrostatic Voltmeters," Prof. H. S. Carhart.
"On the Maximum Efficiency of Arc Lamps with Constant
  Number of Watts," Prof. H. S. Carhart.
"On Direct Current Dynamos of Very High Potential,"
  Prof. F. B. Crocker.
"On an Improved Instrument for Measuring Magnetic Re-
  luctance," A. E. Kennelly.
"The Swinburne-Thompson Unit of Light," Dr. Silvanus
  P. Thompson, F. R. S.

SECTION C.

"Rotary Mercurial Air-Pumps," F. Shulz-Berge.
"A Hundred-Hour Electric Arc Light," L. B. Marks.
"The Conversion of Alternating into Continuous Currents,"
  Dr. C. Pollak.
"The Use of Accumulators in Central Stations," Dr. C.
  Pollak.
"Underground Electric Construction in the United States,"
  Prof. D. C. Jackson.
"A New Incandescent Arc Light," L. B. Marks.

THE CHAMBER OF DELEGATES.

The following topics will be considered by the Chamber
of Delegates:

Adoption of definitions and value of fundamental units
of resistance, current and electro-motive force.

Adoption of definitions and values of magnetic units.

Adoption of definition and value of the unit of self induction.

Definitions and values of light, energy and other units.

The standardization of electric lights.

The consideration of an international system of notation
and conventional symbols and of a more uniform and accurate use of terms and phrases in electrical literature.

A commercial standard of copper resistance.

Together with such other topics as may properly come
before this body.

[Continued from page 4.]

of this number he was able to work satisfactorily and to get from them good and thin wires, and out of this 150 the alloy which suited his purposes best was that described by Dr. Lindeck as manganin. He also expressed the hope that this congress will pass upon these alloys and decide on a unit as a substitute for mercury. Prof. Thompson objected to manganin on the ground that it is rendered entirely useless as a standard when heated above a certain temperature. Dr. Lindeck replied that the use of any standard necessarily involved care, and doubted whether any standard could be established that could not be subject to a sufficient strain to destroy its usefulness.

Mr. A. E. Kennelly read a paper on "Some Measurements of the Temperature Variation in the Electrical Resistance of a Sample of Copper" which he stated was the joint production of Prof. Reginald Fessenden of University of Pennsylvania and himself. The paper described the experiments of Matteson and Dr. Siemens and the varying results obtained by them, together with the experiments of later investigators and scientists to establish the temperature co-efficient of copper. He then described the experiments made by the two authors of the paper at Mr. Edison's laboratory at Newark, N. J., and gave the mathematical result of such researches which coincided very nearly with the results obtained by those investigators, following in the footsteps of Matteson and Siemens. Mr. Preece thanked Mr. Kennelly for his investigations and insured him on behalf of the English electricians that the coefficient established by himself and Prof. Fessenden will be accepted in England without hesitation.

The next paper was a "Note on Photometric Measurement" by Prof. B. F. Thomas. As a result of his investigation Prof. Thomas showed quite conclusively the impossibility of making accurate photometric measurements without taking into consideration the reflection from the rear surface of the chimney used on a standard light. He said that he believes this reflection is sufficient to account for the difference in measurements made by various laboratories of the same light. He suggested that square metallic chimneys with black background and the mica front be used so as to avoid the reflective and lenticular effects. Referring to the tests of the candle power of incandescent lamps Prof. Thompson remarked that the only way to arrive at a correct measurement of the candle power would be to have the lamp revolve at the rate of say 1,000 revolutions per minute, and this can be done only with short or anchored filament lamps. Owing to the lateness of the hour discussion of Prof. Thompson's paper was postponed until 10 a. m., Thursday.

This session of section C was opened by the reading of the paper of Prof. Jackson on "Underground Electrical Construction in the United States." The various systems were described in detail and the discussion which followed was entered into by a number of those present. Mr. A. W. Heaviside described in a general way the conduit work as constructed in England for electric lighting. Prof. Jackson answered the large number of questions in a most satisfactory manner. At the conclusion of the discussion Dr. A. Sahulka read a paper, "The Various Uses of the Electrostatic Voltmeter." Dr. Sahulka was followed by Mr. L. B. Marks with a paper on "A Hundred Hour Electric Arc Lamp." The paper was discussed by Prof. E. L. Nichols and Prof. S. P. Thompson who in his discussion described some of his experiments with the arc lamp. Also by Mr. Geo. P. Lowe, Prof. E. P. Roberts and Dr. N. S. Keith.

The Congress at the Fair.

Members of the congress to the number of 125 assembled Wednesday afternoon at the Van Buren St. Pier to carry out a program arranged by the committee on entertainment. The Whaleback steamer Christopher Columbus through the courtesies of the World's Fair Steamship Company was waiting to convey them to the Exposition grounds. On their arrival at the Casino Pier the party made a trip around the moving sidewalk and stopped to examine the construction. The party then visited the Krupp Pavilion which was placed at their disposal.

Commissioner Carl Richter welcomed the party, and Mr. Lauter, the genial representative of the Krupp company showed the guns and methods of operating them to the party. The Intramural power house was visited and its machinery inspected, then the party made a trip on the Intramural Railway to the north loop and from there went to the parade grounds where seats had been reserved from which they could view the dress parade of the West Point cadets, after which they separated pleased with the trip and expressing their thanks to those through whose kindness they were indebted for the pleasures of the afternoon.

Meeting of the American Institute of Electrical Engineers.

The American Institute of Electrical Engineers met informally at the headquarters in Electricity Building at 8 o'clock Wednesday evening. At 8:30 the party, which numbered about 100, adjourned to the launches which were provided through the courtesy of the General Electric Launch Company. A trip was made around the lagoon, and at the Woman's Building the party disembarked. The Ferris wheel was next visited, and under the care of Mr. Ferris the party made the trip around the wheel. The German village was next visited, where lunch had been provided by Mr. Ralph W. Pope, the secretary, in the absence of the president, Prof. Houston. After the lunch speeches were made by Dr. Keith, Prof. Thomas and others, after which the Institute adjourned.

A very interesting type of engine is that shown by the Dake Engine Manufacturing Company, of Grand Haven, Mich., in Column G, 1-37, Machinery Hall annex. It is exceedingly simple in construction, having but two moving parts outside of the crank itself, and consists of a thin, oblong rectangular box containing the piston. The piston is double, the outside part sliding from end to end of the box; the other part, being inside the first, slides up and down in a direction at right angles to that of the first. Steam is admitted through ports cast in the casing to the center of the shell and exhausts through a circular port surrounding the central admission port. Owing to the double piston and double action, there are no dead points and the engine will start from any part of the stroke. It is said that the engine may be run at any speed up to a thousand revolutions per minute.

THE PHOENIX IRON WORKS COMPANY, 519 The Rookery, has been awarded contract for a complete steam power plant for the city of St. Clair, Mich., comprising one of its 150-horse power tandem compound condensing engines and Manning vertical boiler. Also from Pittsburgh Construction Company two 100-horse power tandem compound engines, for the Ferris wheel lighting plant.

At the Congress.

Among those in attendance at the Electrical Congress are the following:

Dr. Von Helmholtz, Alex. Graham Bell, Prof. Elisha Gray, Prof. Elihu Thomson, Prof. W. E. Ayrton, Dr. T. C. Mendenhall, Washington, D. C.; Dr. Arthur G. Webster, Worcester; Camillo Olivetti, Turin, Italy; Prof. Francisco Grassi, Milan; Prof. Galileo Ferraris, Turin; L. M. Hancock, Chicago; E. Soderholm, Stockholm, Otto Lenitsch, Kagenfurt, Austria; Robert McAlford, New York; Dr. W. F. Geyer, Hoboken, N. J.; B. J. Arnold, Chicago; Dr. Schrader; Dr. Chas. Pollak, Frankfort; Dr. W. Wedding, Berlin; John Cassidy, Honolulu; Dr. Frederick Bedell, Ithaca, N.Y.; Prof. Edw. L. Nichols, Ithaca; Carl Hering, Philadelphia, Pa.; Prof. F. B. Crocker, New York; Geo. P. Lowe, San Francisco.

Lieut. W. F. C. Hasson, San Francisco; E. G. Acheson, Pittsburgh; D. B. Grandy, St. Louis; Austin M. Knight, U. S. Navy; Hermann S. Hering, Baltimore; Chas. F. Kent, Chicago; C. M. Goddard, Boston; E. E. Cabot, Boston; Dr. Johann Sahulka, Vienna, Austria; Kuno Thurmayer, Nuremberg, Germany; E. Braun, Clarendon Hills; Prof. Josef Pechan, Reichenberg, Austria; H. C. Parker, Brooklyn; M. D. Law, Washington, D. C.; Frank T. Layman, Cincinnati; W. S. Jenks, New York; A. Sutton; Rye, N. Y.; A. Wheeler, New York; Gano S. Dunn, New York; S. D. Mott, Passaic, N. J.; Dr. S. S. Wheeler, New York; Wm. J. Danielson, Providence, R. I.; L. P. Hall, New York; Fredk. Beckinzann, West Hoboken; W. A. Preece, London; Prof. Elihu Thomson, Lynn; Prof. Harry A. Rowland, Baltimore; J. E. Cullinane, Dennison, Tex.; F. McCarthy, Chicago; Jos. Wetzler, New York; Max Levy, Galveston, Tex.; Dr. E. Voit, Munich; Geo. P. Squirs, U. S. Army; Luigi Lombardi, Eng.; Drunero Cunco; Walter S. Wiley, South Omaha, Neb.; Chas. W. Livermore, Manchester, N. H.; J. Violle, Paris; Ludwig Weber, Frankfort, Germany; L. Eddy, Darrville, Ky.; Capt. L. Szentorzstevary, St. Petersburg; Prof. Edwin J. Houston, Philadelphia; Camillo Cerrati, Turin, Italy; Tito Galvao, Rio de Janeiro, Brazil; J. W. Meros, Jr., Denver; Ormond Wyman, Ottawa; J. W. Johnson, New York; Cecil P. Poole, Va.; A. T. McKissock, Auburn, Ala.; Harry M. Palmer, Washington, Pa.; J. J. Thoresen, Washington, Pa.; Alfred E. Wiener, Schenectady.

Alex. Siemens, London, Eng.; B Koss, Middletown, Conn.; W. W. Ryder, Chicago; Dr. Chas. E. Emery, New York; Wm. Mayer, Jr., New York; Geo. T. Gibson, Des Moines; Thos. D. Lockwood, Boston; M. E. Rice, Lawrence, Kas.; Wm. Eimer, Jr., Princeton, N. J.; H. W. Frund, Vincennes, Ind.; Hermann Lemp, Lynn; Geo. A. Hamilton, New York; Holbrook Cushman, New York; J. Edward Lissou, St. Petersburg, Russia; A. L. McRae, Rolla, Mo.; Dr. Chas. E. Doremus, New York; C. Courtney, Cincinnati; Dr. T. A. C. Perrine, Palo Alto, Cal.; E. L. Zilinski, U. S. Army; Stephen D. Field, Stockbridge, Mass.; H. B. Fairbanks, Worcester; Capt. A. de Khotinsky, Marlborough, Mass.; Jas. Allen Penta, Philadelphia; Major Capel L. Holden, Woolwich, Eng.; Major A. M. Raguold, R. E. Chatham, Eng.; F. A. Wessel, New York City; Edwin R. Weeks, Kansas City; Francis W. Willcox, Atlanta, Ga.; A. Langstaff Johnston, Richmond, Va.; A. Wickenheiser, Poker, Russia; Otto Frick, Malmoe, Sweden; Douglas Barnett, Brooklyn.

Geo. M. Phelps, New York; Jas. Waring, Manchester, Conn.; Edward A. Colby, Newark; W. B. Cleveland, Chicago; R. Beckinzann, Yonkers; Carl P. Siemens, Berlin; Chas. P. Scott, Pittsburgh; P. B. Delany, South Orange; C. A. Mailloux, New York; E. E. Bernard, Troy; L. B. Marks, New York; L. L. Summers, Chicago; F. W. Jones, New York; Chas. F. Steinmetz, Lynn; H. L. Rodgers, Windsor, Conn.; H. A. Reed, New York; C. H. Meisinc, Newark; F. E. Jackson, Newark; Prof. E. P. Roberts, Chicago; Hammond V. Hayes; T. W. Voeter, Pittsburgh; Harold B. Smith, La Fayette, Ind.; W. E. Goldsborough, Fayetteville, Ark.; J. A. Cabot, Cincinnati; Henry W. Frye, E. W. New York; A. Bays, N. J.; Keijiro Nakamure, Private Secretary to Mr. Nirca; Lewis Searing, Denver, Colo.; J. L. Jayne, U. S. Navy; I. H. Farnum, Boston; H. Bergholtz, Ithaca; Frank C. Perkins, Buffalo; Chas. Cuttriss, New York; D. George Finze, Milan; E. L. French, Pittsfield.

BUSINESS NOTES.

The Brush Electric Company has sold 500 double carbon arc lamps similar to those on exhibition at the Fair to the Edison Light & Power Co. of San Francisco, Cal., and 160 to the Indianapolis Light and Power Co. of Indianapolis, Ind. The larger thirty-thousand light switch board which has attracted so much attention at the Fair is to be shipped to Manila in the Phillipian Islands at the close of the Fair.

The Electric Appliance Company are meeting with considerable success with their new Acme Lamp socket. It has a number of small improvements in the details of construction which are meeting with the approval of the trade and making some very large sales. The recent cool weather has interfered somewhat with the fan motor business, but the Electric Appliance Company reports that it has only a few left of the large stock of fan motors and expect by making some special inducements in price to close them out in a very few days.

DEPARTMENT OF ELECTRICITY.

OFFICES SECTION B, ELECTRICITY BUILDING.

Chief, JOHN P. BARRETT.
Assistant Chief, J. ALLEN HORNSBY.
General Superintendent, J. W. BLAISDELL.
Electrical Engineer, W. W. PRINN.

DEPARTMENT OF MECHANICAL AND ELECTRICAL ENGINEERING.

OFFICES SOUTH OF MACHINERY HALL.

Mechanical Engineer, C. F. FOSTER.
Electrical Engineer, R. H. PRINCE.
First Asst. Mechanical Engineer, JOHN MEADEN.
First Asst. Electrical Engineer, S. G. NEILER.

AMERICAN INSTITUTE OF ELECTRICAL ENGINEERS.

World's Fair Headquarters.
SECTION S, ELECTRICITY BUILDING.

RALPH W. POPE, Secretary.

Open from 9 a.m. to 5 p.m.

CHICAGO WORLD'S CONGRESS OF ELECTRICIANS.

OPENING SESSION MONDAY, AUGUST 21st, 3 P. M.

ADVISORY COUNCIL.

President, DR. ELISHA GRAY, Highland Park, Ill.
Secretary, PROF. H. S. CARHART, Ann Arbor, Mich.

EXECUTIVE COMMITTEE.

Chairman, PROF. ELIHU THOMSON, Lynn, Mass.

COMMITTEE ON INVITATIONS.

Chairman, T. COMMERFORD MARTIN, 203 Broadway, New York.

COMMITTEE ON PROGRAM.

Chairman, PROF. T. C. MENDENHALL, Washington, D. C.

COMMITTEE ON FINANCE.

Chairman, B. E. SUNNY, 175 Adams Street, Chicago.

Amusements.

HOOLEY'S THEATER—Mr. Nat C. Goodwin, in "Mizzoura." 149 Randolph street.

COLUMBIA THEATER—Miss Lillian Russell, in "The Mountebanks." 108 Monroe street.

GRAND OPERA HOUSE—Sol Smith Russell, in "A Poor Relation." 87 Clark street.

AUDITORIUM—Imre Kiralfy's Spectacle "America." Congress street and Wabash avenue.

McVICKER'S THEATER—Denman Thompson, in "The Old Homestead." 82 Madison street.

CHICAGO OPERA HOUSE—American Extravaganza Company, in "Ali Baba, or Morgiana and the Forty Thieves." Washington and Clark streets.

SCHILLER THEATER—Chas. Frohman's Stock Company, in "The Girl I Left Behind Me." Randolph, near Dearborn.

HAVERLY'S CASINO—Haverly's United Minstrels. Wabash avenue, near Jackson street.

TROCADERO—Vaudeville. Michigan avenue near Monroe street.

THE GROTTO—Vaudeville. Michigan avenue near Monroe street.

Buffalo Bill's "Wild West." 63d street. Daily at 3 and 8.30 p.m.

Col. Cody and Mr. Salsbury have added a new feature to their already colossal show. This new feature is the representation of the battle of the Little Big Horn in which Gen. Custer and his band of 310 men were wiped out. Col. Cody was then chief of scouts, and he now has with him a number of scouts that took part in that campaign. Extensive scenery, correctly made from photographs and sketches made on the ground, is used. Thus the scene is realistic. A historical scene, in which many of the original participants take part is something entirely new in the amusement line.

Batteries

D-sque La'Clanche Battery.

Phœnix Dry Battery.

Quad Battery.

Battery Exhibit Electricity Bldg. World's Fair.

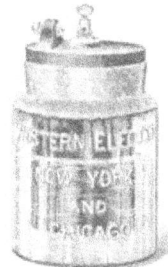
Weco Carbon Battery.

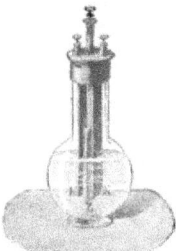
Crenct Battery.

Cravity Battery.

Smee Battery.

Carbon Battery.

Western Electric Company,
CHICAGO.   NEW YORK.

ELECTRICITY BUILDING—EXHIBITORS AND THEIR LOCATION.

GALLERY.

MAIN FLOOR.

A FINAL WHIRL.

We have a few Weston Alternating Current Fan Motors left, which we are going to move. The meaning of this is plain. There is only one way we can do it.

Write for prices. You can get the value of the outfit if you have no other opportunity to use it than during

INDIAN SUMMER.

But there is plenty of hot weather coming in September. Be prepared for it.

ELECTRIC APPLIANCE COMPANY,
ELECTRICAL SUPPLIES
242 Madison Street, - - CHICAGO.

THE MATHER ELECTRIC CO,
MANCHESTER, CONN.

Dynamos, Motors, Generators,

Offices, 116 Bedford St., BOSTON.

AND

1002 Chamber of Commerce Bldg., CHICAGO.

THE "NOVAK" LAMP.

CLAFLIN & KIMBALL (Inc.)

General Selling Agents.

116 Bedford Street, BOSTON.

1002 Chamber of Commerce Bldg., CHICAGO.

Siemens & Halske Electric Company
of America.
Chicago, Illinois.
Electrical Machinery.

Siemens & Halske,

Berlin,
Charlottenburg,
Vienna,
St. Petersburg.

Enterprise Electric Company

GENERAL WESTERN AGENTS

N. W. R.

307 Dearborn Street, Chicago....

Manufacturers' Agents and Mill Representatives for

Electric Railway, Telegraph, Telephone and Electric Light

SUPPLIES OF EVERY DESCRIPTION

Agents for Cedar Poles, Cypress Poles, Oak Pins, Locust Pins, Cross Arms, Glass Feeder Wire. Insulators,

WIRES, CABLES, TAPE and TUBING

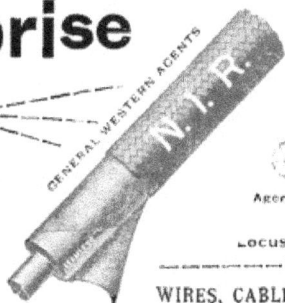

Map of Chicago.

Showing Location of its Electrical and Allied Business Interests, Principal Hotels, Theatres, Depots and Transportation Lines to the World's Fair Grounds. (Index numbers refer to the black squares.)

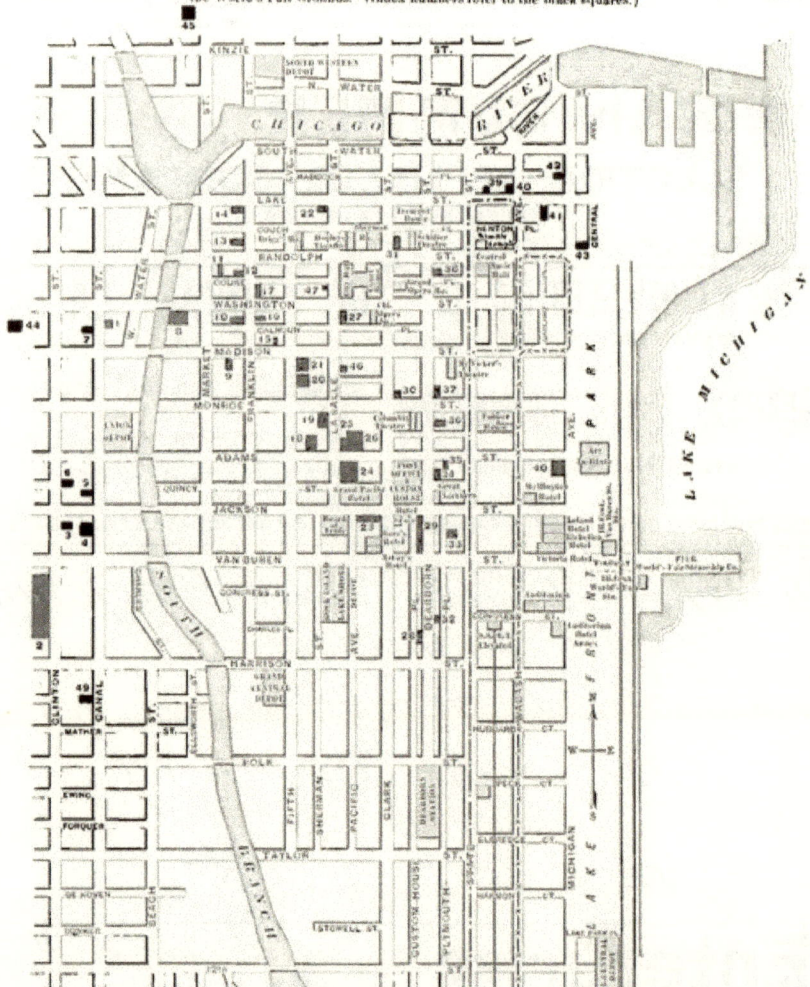

THE
FERRIS WHEEL

When you visit the World's Fair, you will naturally take a ride on the FERRIS WHEEL and be interested in the ELECTRIC LIGHT INSTALLATION, which is wired throughout with

OKONITE WIRE

FURNISHED BY THE

CENTRAL ELECTRIC COMPANY,

116-118 Franklin Street,
CHICAGO, ILLS.

STANDARD ELECTRIC COMPANY.

GENERAL OFFICES: 625 Home Insurance Building.

WORKS: So. Canal Street,

CHICAGO,

STANDARD SYSTEM

AT THE

WORLD'S FAIR.

MACHINERY HALL, Sec. Q, 2 Standard Arc Dynamos.
Sec. S, 20 " " "
ELECTRICITY BUILDING. Sec. P, Space 2, Arc Lighting Exhibit.

The Standard Lamps Light the Power Plant, Machinery Hall, Agricultural Hall, Shoe and Leather Building, and Other Buildings and Portions of the Grounds.

See our Double Service All Night Lamp Before Buying an Old Style Two Rod Lamp.

CLARK ELECTRIC COMPANY, NEW YORK.

192 Broadway and 11 John Street.

MANUFACTURERS OF ARC LIGHTING APPARATUS FOR EVERY PURPOSE A SPECIALTY.
The CLARK ARC LAMPS for use on EVERY CURRENT. have the reputation of being the best and most durable of any ever made in the United States.

RAWHIDE PINIONS FOR ELECTRIC MOTORS
A SPECIALTY.
RAWHIDE DYNAMO BELTING

Greatest Adhesive Qualities.   A Non-Conductor of Electricity.
Equals Leather in Tensile Strength. Will Not Slip.

THE CHICAGO RAWHIDE MANUFACTURING CO.
THE ONLY MANUFACTURERS IN THE COUNTRY.

LACE LEATHER ROPE.
AND OTHER RAWHIDE
GOODS
OF ALL KINDS
BY KRUEGER'S PATENT

This Belting and Lace Leather is not effected by steam dampness, never becomes hard, is at least as durable and tenacious as any Belting made. The Rawhide Rope for Round Belting Transmission is superior to all others.

75 Ohio Street.   CHICAGO, ILL.

THE MONTHLY ISSUE FOR AUGUST

ELECTRICAL INDUSTRIES

Should be read by everyone interested in electrical matters.   In its table of contents is the following:

"Incandescent Lighting at the World's Fair"
"The Electric Power Plant of the Chicago City Railway"
"Steam Engine Efficiency: Its Possibilities and Limitations" by Wm. H. Bryan.
"Alternating Arc Lighting for Central Stations" by H. S. Putnam
"Hard Rubber as an Insulator in Street Railway Work" by W. R. Mason
"A Brief Review"
Together with illustrations of the recent applications of electricity.
The paper also contains regularly
A Buyer's Directory of Manufacturers and Dealers in Electrical Supplies and Appliances.
A Complete Directory of Electric Light Stations in North America and a Complete Directory of Electric Railways in North America.
These directories are revised each issue to the date of going to press and are to be found in no other electrical journal in the World. Its articles are read carefully, and its directories used constantly by all the buyers in the trade. These facts make it without a superior as an advertising medium. Sample copies and rates sent on application.
Subscription price $2 per year. Six months trial $1, if ordered during the next 30 days.

ELECTRICAL INDUSTRIES PUB. CO.,
Monadnock Block, CHICAGO.

FERRIS WHEEL_____

Wired with mile
upon mile of

Simplex Wire

(We don't claim it all)

GEORGE CUTTER, 851 853 The Rookery, Chicago.

SIMPLEX WIRES

SIMPLEX
Ever Onward and Upward!

INSURE
HIGH
INSULATION

Simplex Electrical Co.
620 Atlantic Ave.,

George Cutter, Chicago.   BOSTON, MASS.

XNTRIC

"That's the Switch"

And we control that movement

•

H. T. PAISTE,

10 South 18th St.,
PHILADELPHIA,
PA.

Made 5 amp. S. P.
10 amp. S. P.
5 amp. 3 way.
10 amp. 3 way.

China Window Tube (Patented).

Made only by PASS & SEYMOUR,

George Cutter,
CHICAGO.   SYRACUSE, N. Y.

Consolidated Electric Co.

Manufacturers and Dealers in all kinds of

ELECTRICAL . SUPPLIES,

115 Franklin Street,

CHICAGO.

GEORGE PORTER,

Contractor for All Kinds of

ELECTRICAL WORK.

Room 67, 143 La Salle St., CHICAGO.
Crary Block, BOONE, IOWA.

CHAS. A. SCHIEREN & CO.

MANUFACTURERS OF

Genuine Perforated Electric Leather Belting.

46 So. Canal Street.   -   CHICAGO

Section 15, Dpt. F, Clm. 27.   Section D, Space 3
MACHINERY HALL.   ELECTRICITY BUILDING.

CALL AND EXAMINE

Lawton's Call Indicator.

Indispensable for hotels, railroad
offices, school buildings, hospitals, etc.

Section Y, Space 45, Gallery Electricity Building,
WORLD'S FAIR.

CHICAGO, ILL.

WAGNER ELECTRIC FAN MOTORS

For Direct or Alternating Currents.

These motors give a steady and even with less consumption of current than
any other fan motor in the market. They use full 1-4 horse power. No blood
bleach fan, self-oiling. Furnished with or without guards.

IT WILL PAY YOU TO SEE THE WAGNER BEFORE BUYING ELSEWHERE

TAYLOR, GOODHUE & AMES,

348 Dearborn Street, CHICAGO.

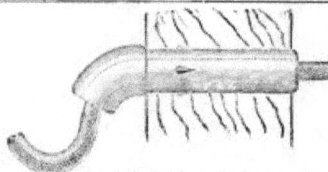

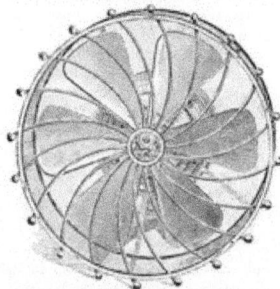

WEEKLY WORLD'S FAIR

ELECTRICAL INDUSTRIES

DEVOTED TO THE ELECTRICAL AND ALLIED INTERESTS OF THE WORLD'S FAIR, ITS VISITORS AND EXHIBITORS.

Vol. I, No. 12.　　　CHICAGO, AUGUST 31, 1893.　　　FIVE MONTHS $1.00 TEN CENTS A COPY

WORLD'S CONGRESS OF ELECTRICIANS.

At the adjournment of the congress on Friday, Electrical Industries secured a photograph of the delegates assembled on the steps of the Art Institute where the sessions of the congress were held. The accompanying engraving was taken from this photograph. Our readers will undoubtedly appreciate the picture, especially those who did not attend the congress and who are not familiar with the appearance of the many eminent scientists present.

Seated in the foreground is the chamber of delegates with Prof. Gray in the center. To the right of Prof. Gray is Dr. Von Helmholtz, next to whom are Mr. W. H. Preece, Prof. Rowland, Prof. W. E. Ayrton, and Alex. Siemens. In the next row, back between Profs. Rowland and Ayrton, is Dr. Silvanus Thompson, to whose left is Prof. Elihu

WORLD'S CONGRESS OF ELECTRICIANS.

Thomson. To the right of Prof. Gray are Prof. É. Mascart, Prof. Galileo Ferraris, Dr. Violle and Prof. É. Hospitalier.

THURSDAY SESSION

On Thursday, section A was called to order at 10 o'clock with Prof. Eddy in the chair. Dr. Bedell furnished the reading of his paper on mutual induction and capacity. Dr. John Sahulka then read a paper on "On the Explanation of the Ferranti Phenomena," after which the section adjourned until Friday.

Section B was called to order by Prof. Cross and the section proceeded with the discussion of Prof. Thomas' paper. The question was discussed by Carl Hering and others. At 10:15 Prof. Henry S. Carhart read a paper on "A Pair of Electrostatic Voltmeters," describing an instrument which he had had in use for some time past. The needles and mirror being suspended by fibre from the top and steadied by a small spiral spring from the bottom. He said that the instrument was very accurate and excellent for laboratory use.

At 11:10 Prof. Webster read a paper on "A Method of Governing an Electric Motor for Chronographic Purposes." The purpose of the governing was to control the revolutions of the chronograph absolutely so that the error in time would be very slight. The means of doing this was through an electro-magnetic arrangement of a tuning fork. The secretary, Lieut. Reber, then read Prof. J. A. Ewing's paper in "Iron for Transformers from the Magnetic Point of View." At 11:30 the section was adjourned until 12 o'clock. On reassembling after adjournment a paper was read by Prof. Andrew Jamieson on "London Electrical Engineering Laboratories." The author said that the title of the paper was somewhat incorrect, and that he had no idea of being called on for such a paper when he arrived. From the way with which the subject was handled it was evident that he is thoroughly familiar with the subject. In his remarks he said: It has often been brought to my notice that the question of a boy's education cannot always be decided by rules and that it must be decided largely from the bent of the boy himself. The Professor said that if the boy evinced a tendency to flightiness of disposition or showed a disposition to slight things and to play with batteries and other small tools, his education certainly should begin with an apprenticeship to some good shop where he would learn the value of work, and would be kept down to it until his flightiness ceased. He should then be sent to some good technical school to complete his education. On the other hand if the lad was studious, steady in his habits and had a good head for mathematics it would be much better for him to be sent immediately to college and get his practical apprenticeship afterwards. He said that in many cases in the city of London schools, the night student who had to work hard all day and studied evenings outstripped his more fortunate competitor who was enabled to attend the colleges during the day. He described the laboratories of the various technical schools, starting first with the city and guilds of London Institute of which Prof. W. E. Ayrton is the head, and which Prof. Jamieson thought was the best type. Others which he described in detail were the Siemens laboratory under the charge of Dr. John Hopkinson, the laboratory conducted by Mr. Kennedy, and entered into quite a detailed description of the Finsbury College conducted by Sylvanus P. Thompson. Others were also described. There was more or less discussion by Dr. Boehm, Prof. Reed of Ann Arbor and Prof. Cross of Boston. Dr Bedell read his paper on "Transformer Diagrams Experimentally Determined."

Section C was called to order at 10:10 A.M. After disposing of the routine business of the section, Prof. F. B. Crocker presented his paper "On Direct Current Dynamos of Very High Potential," in which he described various experimental machines made under his direction having

very high potential. Dr. Keith, at the close of the paper, called attention to the fact that dynamos of high potential had been in use in San Francisco since 1887, and that at the present time there are not more than half a dozen motors in that city which are not operated by high potential current.

At the close of Dr. Keith's remarks the discussion on long distance power transmission was opened with a very able paper by Dr. Louis Duncan of Baltimore. Dr. Duncan was followed by Mr. Scott of the Westinghouse company who described, with the assistance of large diagrams, the Tesla polyphase system as exhibited at the Fair. Lieut. Hassam of San Francisco spoke very humorously and laid stress upon the point that while there are many beautiful theories there has not as yet been a practical application of long distance electrical power transmission within his knowledge which has stood the test of actual use.

Dr. Bell of the General Electric Company spoke decidedly in favor of alternating currents for power transmission because they can be twisted around to better advantage than the direct current, and the single phase motor in its present state of development can be considered nothing more than a poor polyphase motor. There are two systems of polyphase according to Dr. Bell, the independent and the dependent circuit system. Personally Dr. Bell is an advocate of the dependent circuit of the polyphase system for such transmissions of power. Mr. Stillwell of the Westinghouse company made a few remarks, in the course of which he said he hoped that after the large companies had gone to so much labor and expense to perfect a new system for power transmissions, that the consulting engineers would not force them to install the plants at their own expense and guarantee the dividends before they (the engineers) would be convinced of the practicability of the system.

Mr. Frick, of Germany, described the various power transmission plants in his country which have been working from one to two years, giving entire satisfaction, and said he believed no one system the best for all purposes; each has its good points. Prof. Thompson read from notes made by Mr. Thury, of Switzerland, a description of power transmission plants located at various points on the continent, and afterwards made some very interesting remarks, and give it as his personal opinion that the polyphase system is preferable only where no general distribution of power is to take place at the receiving end of the line, but that where distribution is to take place the alternating system is preferable. Prof. Forbes, of London, arrived while Prof. Thomson was speaking, and at the conclusion of the professor's remarks, expressed his belief that while the three phase system is attractive from its beauty the two phase system is much better adapted to practical uses.

Mr. Steinmetz closed the discussion by calling attention to the fact that the continuous current is still more largely used than any other, and expressed the opinion that the single phase system will be the system of the future, that such objections as are now raised to it will be soon overcome, and that we are nearer the ideal condition than is generally supposed.

The next paper on the program was that of Dr. C. Pollak on "The Conversion of Alternating into Continuous Currents," which was appreciatively received but not discussed owing to the hour for adjournment having considerably passed. The machine described in Dr. Pollak's paper can be seen in operation in the German section of the Electricity Building at the Fair.

FRIDAY'S SESSION.

Section A. was called to order at 10:10 a. m. and the Secretary read by title two papers, one by Prof. A. Blondel on " Measuring the Power of Polyphase Currents," the other, by Prof. W. E. Ayrton, F. R. S., on " Extended Use of the Name Resistance in Alternating Current Problems." The session then adjourned to the afternoon session when the report of the chamber of delegates is to be presented.

Section B. After the reading of the minutes of the previous meeting the section listened to a paper read by Prof. W. E. Ayrton, F. R. S., on the " Variation of the P. D. of the Electric Arc, Current, Size of Carbons and Distance Apart. Prof. Ayrton's paper was a very able description of the experiments made under his direction during the past five years for ascertaining an absolutely reliable formula for the potential difference between the carbon points of the electric arc. These experiments have not been fully completed nor the computations made, so Prof. Ayrton was not able to give the congress the corrected formula which will be the final result of his experiments and calculations in this direction.

In the discussion which followed the reading of this paper Prof. Elihu Thomson made a brief address in which he stated that experiments conducted by him in 1881 and since that time coincided in results quite closely with those described by Prof. Ayrton. Prof. Cross, Mr. Henriques, Prof. MacFarlane and Dr. Silvanus P. Thomson participated in the discussion. The next paper was on the " Light and Heat of the Electric Arc" and was read in French by Dr. J. Violle. Dr. Silvanus P. Thompson gave an abstract of the paper in English after the reading of the same in French by the author, in which he said that a translation of a similar article by Dr. Violle could be found in the London Electrical Review for Aug. 4, 1893.

Prof. Elihu Thomson, Prof. Webster and Mr. Henriques participated in the discussion of the paper, some of the gentlemen addressing their remarks to M. Violle in his own language, to which Dr. Violle responded.

Dr. Silvanus P. Thompson presented his paper on " The Swinburne-Thompson Unit of Light" without reading, as the time for adjournment was approaching. The paper was intended to prove that a square millimeter of the surface of the crater of the positive carbon should be taken as the unit of life.

Prof. F. B. Thomas presented a short paper on " Periodic Variation of the Candle Power of Alternating Arc Lights" together with a diagram of curves showing the results of measurements made by him in recent experiments. There was no discussion of this paper. The last paper on the program was that of Prof. H. S. Carhart on " The Maximum Efficiency of Arc Lamps with Constant Number of Watts." The author had intended to simply have this paper read by title but on request gave a brief outline of it without reading it in full. The paper was a suggestion to rate arc lamps by the number of watts expended upon them instead of by the candle power as is the present custom. Prof. Carhart merely offers the proposition as a suggestion, realizing that there are objections as well as points in favor of this idea. Prof. Ayrton expressed his appreciation of Prof. Carhart's paper as it had to do directly with his own line of experiments. The following papers were presented but not read: " On the Source and Effects of Harmonics in Alternating Circuits" by Prof. H. A. Rowland, " Sour l' Arc a Constant de Lumiere" and " Nouvelles Recherches sur l'Arc a Curants Alternatifs" by A. Blondel. This section then adjourned

to meet in Columbus Hall at 3 p. m. to hear the report of the chamber of delegates.

Session C was called to order by Prof. Houston at 10:15 a. m. Dr. Pollak read a few notes on the conversion of alternating into continuous currents for charging storage batteries. Mr. Frick, of Germany, then spoke in reference to the load line of central stations as determined by tests made on days in three different portions of the year. He exhibited diagrams of these load lines and showed how accumulators could be applied for easing up the use of the machinery and that accumulators were in use in all stations in Germany and in nearly all stations on the continent of Europe, and that without exception they were well liked by all who had them in use.

At 10:30 Prof. Geo. Forbes, of London, the consulting electrician for the Cataract Construction Company, of Niagara Falls, reopened the discussion on long distance power transmission that was started yesterday. He said that as 1890 he had stated it as his opinion and had since seen no reason to change it, that the two phase alternating current was the proper thing to use for long distance transmission, the two separate circuits being kept independent. In some shops single phase motors can be used, in others two phase, and where direct current is needed alternating motors of either style could be used to drive direct current dynamos. He called attention to the commutating machines now commonly known as rotary transformers and said that while this was perfectly feasible he thought the expense of the machine and the losses in efficiency due to its construction would prohibit its use. Some simpler form of commutating device might be made which would not consume so much energy.

He expressed as his opinion that not very far in the future we should be in possession of a simple device for this purpose, which would change alternating currents into direct currents and that without the consumption of any large percentage of energy, and instanced as such the very ingenious machine exhibited and explained by Dr. Pollak. He stated that Gramme's first alternating dynamo was two phase with a stationary armature. In reference to the question as to whether he would favor the direct production of the required high potential by the machine itself or would use step-up transformers, the dynamo being run at high pressure, he favored the first method as with the standing armature it would be perfectly feasible to obtain the requisite potential without danger to the machine.

From a financial point of view it was very much cheaper as the transformers for such work would cost nearly if not quite as much as the dynamo itself, and three or more per cent of such loss in such transformation would amount to $3,000 per year, counting the cost per horse power at $20, which, capitalized at 5 per cent. would be $60,000, or more than the entire cost of the dynamo. In regard to the conductors to be used and the location of the same, he very much favored the use of bare wires in a large subway so constructed that the inspector could walk through it. In any overhead construction there would be more or less trouble from sleet and various other difficulties, while the use of conduits for drawing cables in and out would be impossible.

After speaking for an hour Prof. Forbes was followed by Prof. H. A. Rowland, of Johns Hopkins, who argued that the mere loss of three per cent in the efficiency of the transformer could be easily made up by turning on a little more water, that he did not think about any figure at all, and that as for the direct production of the excessively high pressures in the armature itself, he very much favored the use of the step up transformer as liable to be of less trouble for repairs and shut downs. He had considerable to say in regard to the frequency which would be best to use in this general transmission as a frequency that would be proper for motors would not be right for the incandescent lamp as at 35 or less periods per second, the effect on the eye was dreadfully unpleasant, and also could not be used for synchronous motors.

CLOSING SESSION.

The afternoon session was held in Columbus Hall, the chamber of delegates taking their place on the platform at 3.10 P.M. The report of the chamber of delegates was read by Prof. E. L. Nichols, and was as follows: Gentlemen of the International Congress; The chamber of delegates has made a careful investigation in accordance with the program laid down, and has reached certain decisions which it is my duty as secretary to report. The first question which came before the chamber had to do with the adoption of definitions and values of fundamental units of resistance, current and electro-motive force. The following resolutions on this point have been passed by the chamber of delegates:

Resolved, That the several governments represented by the delegates in this international congress of electricians be and they are hereby recommended formally to adopt as legal units of electrical measure the following:

As the unit of resistance the international ohm, which is based upon the ohm equal to 10^9 units of resistance of the initial C. G. S. system of electro magnetic units and is represented by resistance offered to an unvarying electric current of a column of mercury at the temperature of melting ice, 14.4521 grammes in mass, of a constant cross sectional area and of a length of 106.3 centimeters.

As a unit of current the international ampere, which is one-tenth of the unit of current of the C. G. S. system of electro-magnetic units, and which is represented sufficiently well for practical use by the unvarying current, which, when passed through a solution of nitrate of silver in water, in accordance with the accompanying specifications, deposits silver at the rate of 0.001118 of a gramme a second.

As a unit of electro motive force the international volt, which is the electro-motive that, steadily applied to a conductor whose resistance is one international ohm, will produce a current of one international ampere, and which is represented sufficiently well for practical use by 1,000-1,434 of the electro motive force between the poles or electrodes of the voltaic cell known as Clark's cell, at a temperature of 15 degrees centigrade, and prepared in the manner described in the accompanying specifications.

As the unit of quantity the international coulomb, which is the quantity of electricity transferred by a current of one international ampere in one second.

As a unit of capacity, the international farad, which is the capacity of a conductor charged to a potential of one international volt by one international coulomb of electricity.

As the unit of work, the joule, which is 10^7 units of work in the C. G. S. system and which is represented sufficiently well for practical use by the energy expended in one second by an international ampere in an international ohm.

As the unit of power the international watt, which is equal to 10^7 units of power in the C. G. S. system and which is represented sufficiently well for practical use by the work done at the rate of one joule per second.

Concerning the adoption of a definition and value for the unit of induction the chamber of delegates has resolved, upon the motion of Prof. Mascart of France, and seconded by other foreign members, that the unit of induction shall be called the henry, which is the induction in a circuit when the electro motive force induced in this circuit is one international volt while the inducing current varies at the rate of one ampere per second. With regards the question of the selection of names for magnetic units and the units of light and energy, it was resolved that in these cases no specific name should be applied but that these units should be designated as C. G. S. units.

In addition to these topics two others have been considered. The question of standards of light was considered and a committee was appointed to deliberate on this very vexed question of the electrical engineer. I have the pleasure of reading the report of the committee, which report has been received and adopted by the chamber of delegates. The committee says it has had much discussion upon the various forms suggested for standards, and in particular upon the two special forms of lamp known respectively as the amylacetate lamp of Von Hefner Alteneck and the pentale lamp of Vernon-Harcourt. The committee recommends that all nations be invited to make researches in common on well-defined practical standards and on the convenient realization of the absolute unit.

Finally there was before the chamber the question of the modification of electrical notation and nomenclature, and an extended and carefully matured scheme was presented by M. Hospitalier and others. After due consideration and after some amendments and modification of this plan had been made, it was moved that the report be received and printed as an appendix to the regular proceeding and as the report of a sub-committee.

Prof. Gray then stated that he had a few announcements to make relative to to-morrow's work or to-morrow's pleasure and that Mr. Preece also had an announcement to make.

Mr. Preece spoke as follows: "Ladies and Gentlemen: My announcement is an extremely simple one. As president of the Institution of Electrical Engineers in England, it was my wish to invite the members of the Institute of Electrical Engineers in America to meet me at Victoria House in the grounds of the World's Fair, especially as they have done me the great honor of making me an honorary member, there are only two and I am one of them, of their body. I had this great difficulty that the addresses, either of the members of the American Institute or of the Congress, have not been very carefully kept, and therefore I now invite all the members of this Congress, whether members of the American Institute of Electrical Engineers or not, to allow me to have the pleasure of receiving them to-morrow afternoon between the hours of five and seven, in my present British home, Victoria House, World's Fair grounds."

Dr. Von Helmholtz then arose and addressed the congress, saying: "Ladies and Gentlemen: We have come forward here to do a really important work and one which I trust will have good fruit for future time in correcting the incongruities of electrical science and electrical notions, so that all scientific and industrial men can understand each other in the simplest and best way. It was rather a hard piece of work to do in these hot days in continual meetings of the delegates and members of the congress, and he who has had the greatest part of the exertion and work in this connection is our president, Prof. Elisha Gray." I therefore trust you will express to him your thankfulness and give him a vote of thanks." M. Mascart of France then spoke in French in substance as follows:

"My associates join with me in heartily seconding the proposition of Dr. von Helmholtz, to express our thanks to the distinguished President of the Congress of Electricians. We have admired the urbanity, grace and ability with which he has conducted the work, and are grateful for the courtesy with which he has treated us, and for all the hospitality and kindness that we have received in America."

Mr. Preece: "I move that a vote of thanks be accorded to this fine old man, as the newspapers call him, but who is in reality a younger man than myself." The motion was unanimously carried and Prof. Gray responded as follows: "I cannot tell you how gratifying this is to me. I have worked for the last two years in organizing this congress under many difficulties and those difficulties have continued right up to the present moment. These steam engines outside do not even give us a chance to express ourselves. I want to thank you before we leave. I want to thank the members of this congress for the part they have taken and I want to thank you for the good part in which you have taken all the difficulties under which you and I have had to labor. I think on the whole the congress has been very successful and I trust you will go away feeling that this is true, and that you will think of us kindly; and you, gentlemen, who come from foreign shores, when you go home and look back, do not think only of smoke, of noise, and of high buildings, but think of us over here as having warm hearts, as wishing you well, and that we are all praying that you will have smooth seas and a warm-hearted welcome home to your friends and dear ones. Now, gentlemen this closes the work of the International Electrical Congress at Chicago in 1893, and I now declare the same adjourned."

Nikola Tesla's Lecture.

Mr. Tesla's lecture on "Mechanical and Electrical Oscillators" was delivered in Assembly Hall at the southwest corner of the Agricultural Building at the World's Fair grounds on Friday evening as announced in the program. Every member of the congress with one or two exceptions was present and the greatest interest was manifested in the experiments that were shown.

It was twelve minutes after eight when Prof. Elisha Gray introduced Mr. Tesla to the audience in the following words: "I came down here to-night for the purpose of introducing to you a gentleman whose name needs no introduction as it is a household word with all electricians, but I wanted to do him the compliment of coming here as the official head of the World's International Congress of Electricians and to extend to him the compliment of an introduction for them as well as for myself. This compliment loses none of its force when I tell you that I have another official duty to perform to-night and it will be necessary for me to leave immediately, but if I should follow my inclinations I should stay, for no one wants to hear and see these experiments more than myself. I have the honor, ladies and gentlemen, of introducing to you Nikola Tesla. I wish to add one word, inasmuch as I have to leave; I am going to ask Dr. T. C. Mendenhall to do the honors for the rest of the evening."

Immediately the hearty applause subsided Mr. Tesla said: "Ladies and Gentlemen: We are told in a delightful anecdote how, many years ago, when science was still in its infancy, a man to whom the world is largely indebted for the discovery of a great truth was meditating in his garden when the idea came to him, and how on that occasion he was carried away by his enthusiasm. Be that an actual fact or not it is certain that the search for truth through the centuries that have elapsed has exercised an immense power on the imagination of man. It seems as though we are actual witnesses of the researches and experiments made by the discoverers and inventors of ages gone by, and that we are almost in a condition to know how they felt at the moment of sublime inspiration.

"I am sure in this audience there are many who have felt this most exquisite pleasure and whose presence here will forever fix the recollection of this evening upon my mind. To feel this pleasure is accorded to but few; however, I can say myself that I have felt it and am still under the impression of the pleasure which the accomplishment of a few insignificant thoughts that have come to me and the speculation on the possibilities of their future development have caused. In these results I am, inventor like, quite taken up, and I hardly dare hope to be able to develop them as I wish, but I feel that I must explore them; and again I have doubted my ability to present them satisfactorily. Yet I am sure of the practicability of investigations in this manner from the few results I have so far obtained.

"I can best tell you what I have attempted if I comply with the scientific duty and tell you exactly the history of these inventions. It was at a time when I was strenuously endeavoring to solve a question or two which was considered insoluble, first, Was it possible to operate a motor without sliding contact, and second, was it possible to develop constant currents in a certain direction? I found the solution to the one problem and produced dynamos without commutators; and that success emboldened me to go ahead and apply whatever of knowledge and experience I possessed to the solution of the other problem.

"On the occasion of my first visit to this country I stepped into the exposition at Philadelphia and I saw there a very thick copper washer provided with handles that visitors would move within a magnetic field. That day it occurred to me that when the plate was moved slowly in the field there was experienced resistance, and that when the plate was suddenly pressed in there was a rebound as though it struck against something solid. After returning from the exposition the thought occurred to me that if I took a conductor and moved it into the field and then rapidly withdrew it, I could in this way obtain whatever of electro-motive force there was. In the first place I was impressed with the analogy of this device to the induction coil, in which the same process takes place. When we impart current to the primary circuit we put lines of force slowly into the field and when we break the current we take the force swiftly away. I began to think of a mechanism which would be capable of fulfilling these conditions.

"During my work with induction coils and motors which succeeded I became familiar with currents of high frequency, and then I clearly realized the problem before me. I convinced myself that in alternating distributions we must induce currents of more than one phase and that we must have a better organization and a perfected mechanism capable of rendering the current into steady oscillations before we could obtain the desired results. Here then I was confronted with this difficulty. I constructed small machines of the ordinary type which enabled me to investigate, but when I endeavored to construct machines with a greater number of alternations for higher frequency I found that I met insurmountable mechanical difficulties.

"And then the idea came up and I asked, how it would be if I took a very strong field and reciprocated very rapidly a conductor in that field, would I not have a similar machine and one which would not involve loss in the iron? And now I began to consider this matter very seriously and later I considered this question, knowing what has been done in the field of harmonic telegraphy and supposing that instead of the ordinary dynamo we take dynamos constructed like the pendulum of a clock and on a harmonic vibration principle, would we not then obtain alternating currents of a perfectly defined and absolutely constant period, which it is impossible to accomplish with an ordinary machine because it possesses so great inertia? And then again I thought like this, with such powerful machines there is a possibility of transmitting energy through the air and by means of this energy of transmitting messages. This, then, is one of the mainsprings which has driven us into this work."

Here by means of diagrams Mr. Tesla described the construction of the engine which he has invented for producing these absolutely constant oscillations and which is operated by compressed air under 100 pounds pressure and delivered at the rate of 10 cubic feet per minute. One of these engines having a diameter of about two inches and a half and of about that length operated a small motor and has a capacity of nearly one-half horse power. After this Mr. Tesla showed the operation of the larger engine having a piston and plunger weighing 20 pounds which was oscillated at the rate of 78 times a second, and stated that 5,000 or 10,000 oscillations per second could be as readily produced. The next experiment was made by attaching a horse shoe magnet having a wire about two feet in length inserted through it to show the vibration to the plunger of the engine. The engine was then set in motion, a current communicated to the magnet through wires attached to the nodes of the magnet wire. A copper disc revolving on an axis was then introduced by Mr. Tesla between the poles of the magnet while the same was being oscillated by the engine. The result of the experiment was that the copper disc began slowly to revolve, thus disproving the theory that there is no electro-magnetic force at work in the disc and that the force was simply one produced by induction as has been believed universally heretofore.

Mr. Tesla next attempted to show an experiment with what might be called a three coil generator, but the apparatus had been damaged in some way and refused to work. The next experiment was one with a large engine similar to that last described, having attached to the plunger of the oscillator a core of iron which played through four field magnets excited with a current at 5,000 volts from the Westinghouse plant. A current was generated by the apparatus, and was used to run a small registering motor, thus proving that generators can be constructed on the ideas evolved by Mr. Tesla in accordance with these more recent experiments, and having an absolutely constant current. It should be added that one of the claims made by Tesla for these oscillators is that the period will not be varied after the machine is started, whether the pressure be 10 pounds or 100 pounds, and that the electrical currents produced from small units will oscillate just as as the pendulum oscillates, and be unaffected by the circuit load or any imaginable condition. The oscillators are so constructed that they can be adjusted

ELECTRICAL INDUSTRIES.

PUBLISHED EVERY THURSDAY BY THE

ELECTRICAL INDUSTRIES PUBLISHING COMPANY,

INCORPORATED 1893.

MONADNOCK BLOCK, CHICAGO.

TELEPHONE HARRISON 150.

E. L. POWERS, PRES. AND TREAS.   E. E. WOOD, SECRETARY.

E. L. POWERS   -   -   -   Editor.

H. A. FOSTER,
W. A. REMINGTON,   }   - Associate Editors.

E. E. WOOD,   -   - Eastern Manager.

FLOYD T. SHORT,   Advertising Department.

EASTERN OFFICE, WORLD BUILDING, NEW YORK.
World's Fair Headquarters, Y 27 Electricity Building.

SUBSCRIPTION.

FIVE MONTHS,   -   -   - $2.00
SINGLE COPY,   -   -   -   10

Advertising Rates Upon Application.

News items, notes or communications of interest to World's Fair Visitors are earnestly desired for publication in these columns and will be heartily appreciated. We especially invite all visitors to call upon us or send address at once upon their arrival in our city or the grounds.
ELECTRICAL INDUSTRIES PUBLISHING CO.,
Monadnock Block, Chicago

so as to run with practically no noise, and with the entire absence of any pounding effect.

Banquet to the Foreign Delegates.

The event of last week socially was the banquet given by the American electricians to the foreign delegates to the international congress on Thursday evening at the Grand Pacific hotel. In parlor 44 there were gathered the foremost electrical scientists of the world. Dr. Von Helmholtz of Berlin, upon whom all looked as one of the fathers of the science, and Thos. A. Edison, the great American inventor, divided the honors of the evening. The toastmaker, Prof. Elisha Gray, occupied the center chair at the table of honor. At his right sat Dr. Von Helmholtz, Prof. Rowland, Dr. Mascart, Prof. Carhart, Dr. Budde, Prof. Nichols, Dr. Violle, Prof. Houston, Prof. Thompson, Mr. Lockwood, Prof. Dolbear, Prof. Brown-Ayres, Dr. Tourane, Dr. Lummer, Mr. Weinmann and Dr. Sahulka, while on his left sat W. H. Preece, F. R. S., Dr. Mendenhall, Prof. Ferraris, Prof. Thomson, Prof. Ayrton, Prof. Cross, Dr. Shrader, Prof. Crocker, Dr. Palaz, Prof. Thomas, M. Hospitalier, Dr. Voit, M. Chavez, and Messrs. Wetzler, Pope and Phelps. At the table directly opposite Prof. Gray sat Mr. Thos. A. Edison with Messrs. Emery, Wheeler, Insull, and Kennelly. The following is the menu served:

MENU.
Little Neck Clams
Consomme Royal
Fresh Penobscot Salmon, Hollandaise
                    Dressed Cucumbers
Chateau Lanterne
          Fillet of Beef, Larded, Financiere
     Browned Potatoes   New Lima Beans
Mumm's Extra Dry
          Sweetbreads in Cases
               French Peas
               Siberian Punch
     Breast of Young Chicken with Truffles
               Tomato Mayonnaise
Pontet Canet
                              Peaches and Cream
Charlotte Glace   Fancy Cake
Roquefort   Coffee

At 8:30 Prof. Gray arose and presented the first toast, "The International Electrical Congress," and called on

Dr. Von Helmholz to respond, who said, " We Europeans have come over here with the feeling of a good father rejoicing in the success of his children, to which he himself could not attain. Europe is too narrow for the splendid march of electrical progress and America has grandly performed the task set before it. We see in you the result of better conditions and prospects than we have enjoyed, and we rejoice with you in your remarkable advancement. Gentlemen, I drink my glass to the great American Nation." Prof. Gray then proposed " Our Guests, the Official Foreign Delegates," to which Mr. W. H. Preece, Prof. E. Mascart and Prof. Ferraris responded. Prof. W. E. Ayrton of London responded to " The American Electrical Engineers", and Prof. E. J. Houston to "The American Institute of Electrical Engineers." The other speakers were Prof. S. P. Thompson, Dr. Mendenhall, T. D. Lockwood, Prof. H. A. Rowland and Prof. Elihu Thomson.

Exhibit of the Standard Underground Cable Company.

Just west of the exhibit of the Sperry company in the

THE EXHIBIT OF THE STANDARD UNDERGROUND CABLE COMPANY

south western part of Electricity Building on the main floor the Standard Underground Cable Company has placed its exhibit. As a background to the exhibit there has been erected a representation of an elevation of a street showing the pavement on top, the soil beneath and the position and method of laying the Standard Underground system.

The conduit is shown as it actually appears in the ground. Manholes are constructed in the line of different kinds and through the open side the interior is exposed to the view of the visitor. The lead covered cable is there shown with the joints and branches, made according to the methods adopted by the company. The manholes are covered with frames and covers of the latest pattern. All details have been so faithfully carried out that the visitor almost believes himself in a trench in the street. There has also been erected in this space a pole line showing the Standard company's overhead system. Several poles have been placed at convenient distances apart for displaying the system. From these poles are strung a line of duplex cable supported

on hangers from an iron wire. The poles are low enough so that every detail is in plain view of the visitor.

A special feature of the exhibit is the presentation of the various means of terminating the cables in appliances known as terminals, whereby the end of the cable is hermetically sealed and yet presenting an easy and effective method of reaching the ends of the wires for tests or connections. Under this head are shown the distributing boxes by means of which a main cable may be cut into, the wires for various services may be taken out without disturbing the working ability of other lines in the cable. This is practically shown in the taking from the manhole and up the pole a section of underground cable and connected through a terminal to the aerial cable for distribution above ground. On a table near the center of the space are exhibited numerous sections of lead covered cables which the method of making different kinds of joints. Main line joints, branch connections, etc., in electric light and power cables, in telephone, messenger and other cables are shown.

Near by is a pile of cases containing the ozite compound extensively used as an insulation for filling cable joints, cable terminals, splice boxes, converters, etc. Variously distributed about the space are reels, showing the kinds and character of wires and cables manufactured by this company. The line includes wire for every branch of the electrical industry. In pyramid form are arranged reels of wire and cables with the large lead covered messenger and telephone cable at the bottom, above which are placed smaller cables and wires, both braided and waterproof, then the smaller wires, lamp cord and annunciator wires. Several large reels of the lead covered duplex cable specially adapted for the alternating current are shown, also cables for high tension light and power circuits, and numerous samples of the Waring anti-induction cables of the flat and clover leaf styles. A complete set of tools is also shown for removing the lead casing and the insulation from the wires, showing the perfection of this branch of the work and the facilities for performing quickly and easily any alterations.

While the exhibit shows in a general way the range of the business of the company an inspection of the lighting and power distributing system of the Exposition will show the capabilities of the company for complete and extensive installations. Within the grounds of the Exposition 384,000 feet of feeders and mains used in the distribution of the alternating current for the incandescent lighting of the grounds and buildings have been installed and maintained by the Standard Underground Cable Company.

This company has offices both in New York and Chicago, although its main offices and factories are in Pittsburgh, Pa.

The Western Electric Company Sues the General Electric Company.

The exhibitors in the Electricity Building have until recently been friendly and pleasant in their relation with each other and what rivalry there has existed has been unmarred by any ill feeling. Prof. Barrett, the chief of the department, has been most equitable in the management of his department. On July 17th some one removed the signs of the Western Electric Company from its lamp posts in the Electricity Building, a most cowardly act, and one that no one with any manhood or the least appreciation of respectable business methods would degrade himself to do in dealing with a competitor. The result of this is the suit brought by the Western Electric Company against the General Electric Company et al. on Aug. 25. The following is an abstract of the declaration:

STATE OF ILLINOIS, }
County of Cook, } ss.

IN THE SUPERIOR COURT OF SAID COUNTY.

To the September Term, A.D., 1893.

Western Electric Company, a corporation organized and existing under and by virtue of the laws of the State of Illinois, plaintiff, by Williams, Holt & Wheeler, its attorneys, complains of the General Electric Company, a corporation organized and existing under and by virtue of the laws of the State of New York, and E. J. Spencer, R. W. Hofstede-Crull, John Doe, James Rogers, Henry Smith and William Wilson in a plea of trespass on the case.

That heretofore, to-wit, prior to the first day of May, 1893, the World's Columbian Exposition, a corporation of the State of Illinois, by John P. Barrett, chief of its electrical department, solicited the plaintiff to exhibit its electrical devices and products at the Fair or Columbian Exposition, conducted under the auspices of said corporation in the city of Chicago during the summer of the year 1893, and the plaintiff, believing that by so doing, it would not only add to the interest and value of said Exposition, but would secure to itself great advantages by bringing its products and machines into favorable notice among the visitors to said Exposition, consented to make an exhibit, but as a part of such exhibit, the plaintiff was solicited by said Barrett and consented to furnish iron poles or posts, such as are used upon streets for carrying electric lights, which posts to the number of 30 were agreed to be placed at various points on the main floor and in the gallery of the Electricity Building in said Exposition and especially in the public passageways of said building. That it was further agreed between said World's Columbian Exposition and the plaintiff that upon certain of said posts, are lights manufactured by the plaintiff should be placed and upon others of said posts, are lights manufactured by the defendant, General Electric Company, should be placed, the designation of the particular posts to be occupied by the lamps of the respective manufacturers to be made by the said Barrett, chief of the electrical department of said Exposition; and it was further agreed that in consideration that the plaintiff would furnish said posts that the plaintiff should be allowed to mark said posts with the name of the plaintiff, not only upon brass plates screwed upon said posts near the base thereof, but also upon glass signs inserted in the hood or covering at the top of said posts in the manner in which are customarily inserted signs bearing the names of streets in cities. And it was further agreed that the lamps furnished by manufacturers other than the plaintiff and placed upon said posts should

be marked with the names of the respective manufacturers, so that said manufacturers should have due credit for their lamps, and the plaintiff should have due credit for its posts. That the said posts were manufactured and erected by the plaintiff and signs placed thereon in accordance with its agreement, as aforesaid, and that said posts were especially designed and adapted for their purpose, and possessed great advantages over the posts customarily made by other manufacturers, and it was and is of great advantage to the plaintiff to have its name plainly appear on all the said posts and upon the glass signs aforesaid, and that said privilege was and is of great value to the plaintiff, to-wit, of the value of ten thousand dollars ($10,000).

That the defendant, the General Electric Company, prior to the first day of May, 1893, had agreed with the said World's Columbian Exposition that it would also exhibit its products and machines in the said Electrical Building, and among other things, that it would furnish lamps as aforesaid to be placed upon certain of the posts manufactured by the plaintiff, as aforesaid, to-wit, to the number of 27. That the defendant, E. J. Spencer, who, at all the times herein mentioned, was and is the agent of the defendant, the General Electric Company, in charge of its exhibit at said Exposition, conspiring with said General Electric Company to injure the plaintiff and deprive it of its advantage from the exhibition of said glass signs upon its posts aforesaid, repeatedly and continuously endeavored, to-wit, from the first day of May, 1893, to the 17th day of July, 1893, to secure the removal of said glass signs bearing the name of the plaintiff from those posts which had been designated by said Barrett to carry the lamps of said General Electric Company. That at various times between the dates last above mentioned, said General Electric Compay, by said Spencer and its other agents, demanded that said glass signs be removed and replaced by plain glass, and at various times threatened, if its said demands were not complied with, that it would remove its entire exhibit from said Electrical Building, and the said General Electric Company continuously neglected and refused to place its lamps upon the posts designated for that purpose by said Barrett, as aforesaid, unless and until the said signs bearing the name of the plaintiff should be removed from said post, although the General Electric Company did place a small number, to-wit, three or four lamps upon posts adjacent to its own exhibit.

That the said Barrett declined to yield to the improper and unlawful threats and demands of the said General Electric Company and Spencer, made in pursuance of the conspiracy aforesaid, and insisted that the said General Electric Company should comply with its agreement and place its lamps properly marked upon the posts of the plaintiff designated by said Barrett, and that to-wit, on the 17th day of July, 1893, the said defendants, said General Electric Company, E. J. Spencer, R. W. Hofstede, John Doe, Henry Smith, James Rogers and William Wilson, conspired to injure the plaintiff and to take possession of its property and deprive it of its benefit and advantage aforesaid from the exhibition of its posts, and thereupon during the night, between the hours of eleven o'clock P.M. on July 17 and two A.M. on July 18, the said defendants, R. W. Hofstede-Crull, John Doe, Henry Smith, James Rogers and William Wilson, acting in pursuance of said conspiracy and under the direction of said General Electric Company and E. J. Spencer, did enter the said Electrical Building and unlawfully and feloniously remove and take away the glass signs aforesaid bearing the name of the plaintiff, but only from the posts which had been designated as aforesaid to carry the lamps manufactured by the General Electrical Company, and not from the posts designated to carry the lamps of the plaintiff; that the guards employed in and about the said building did not interfere with such removal for the reason that the said defendants, Hofstede-Crull, Doe, Smith, Rogers and Wilson, engaged in such removal were men who were known by said guards to be in the employ of said General Electrical Company, and were therefore supposed to have the right and authority to remove said signs.

By reason whereof and of all the actings and doings of said defendants in the premises the plaintiff has suffered great loss in the value of said signs so feloniously taken and removed, and in the money expended in and about the putting of said signs in place and in the loss of the benefit and advantage of advertising its posts and devices, as aforesaid, to the damage of the plaintiff of ten thousand dollars ($10,000), and therefore the plaintiff brings its suit.

Batteries

D-sque La Clanche Battery.

Phenix Dry Battery.

Quad Battery.

Battery Exhibit Electricity Bldg. World's Fair.

Weco Carbon Battery.

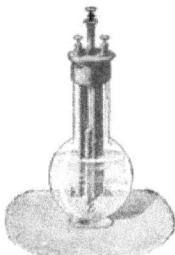

Grenet Battery.

Gravity Battery.

Smee Battery.

Carbon Battery.

Western Electric Company,
CHICAGO.   NEW YORK.

ELECTRICITY BUILDING EXHIBITORS AND THEIR LOCATION.

GALLERY.

MAIN FLOOR.

Exhibitor.	Section.	Exhibitor.	Section.	Exhibitor.	Section.	Exhibitor.	Section.
Abdera		Electrical Review	Y	Jaeger, Chas. L.		Reliance Gauge Co.	T
Ansonia Electric Co.	Y	Electricity	Y	Johns Mfg. Co., H. W.	T	Roessler & Hasslacher Chem. Co.	N
Am. Inst. of Elec. Eng.	S	Electric Gas Co.	B	Jewell Belting Co.	Y	Street Railway Journal	
American Battery Co.	T	Electrical Engineer	Y	Jenney Elec. Motor Co.	L	Strowger Aut. Teleph. Co.	
Axtell, H. M		Electrical World	Y	Knapp Electrical Works		Standard Paint Co.	T
Allg. Elec. Gesellschaft	D	Eddy Electric Motor Co.	B	A. P. Elec. Novelty Co.		Speakody, C. A.	
Bates Mfg. Co.	V	Excelsior Electric Co.	B	Knorpp & Barkley		Star Iron Tower Co.	W
Bryant Electric Co.	S	Electrical Forging Co.		Kennedy Electric Co.	L	Spain	
Ballance & Sprague	R	Equitable Defance Co.		Lawton, H. A.		Schuyler, Chas. A. & Co.	
Briley, W. J.	T	Elektron Mfg. Co.	T	Le Carbone Battery Co.		Schomburg & Seline	
Belknap Motor Co.	K	Electrical Conduit Co.	P	McNeil Yunder Elec. Co.		Soemaw & Baske.	K
Bell Telephone Co.	L	England		Marcus, H. N.		Schuckert & Co.	
Brush Electric Co.		Empire China Works	N	Mesker, Br. G.	S	Short Electric Co.	
Caldwell El. Cloth Cut. Mch. Co.	Y	Franklin Elec. Appliances	S	McIntosh Bt. & Opt. Co.	W	Sperry Elec. Railway Co.	O
Conseil Elec. Storage Co.	H	French Piano Exhibit		Museum of Art, Belmont's		Standard Underg. Cable Co.	
Culter, George	T	Felton & Guillaume		Mather, A. C.	E	Standard Electric Co.	
Conduit Elec. Co.		France	K-P	Mather Electric Co.	M	Samson Battery Co.	S
Chicago Edison Wire Co.		Ft. Wayne Elec. Co.		Newman Clock Co.		Tate Ann. El. Signal Co.	
Copenhagen Fire Alarm Co.		Gamli & Co., N. C.		Nungesser Battery Co.	B	Todd, Applegate Co.	
Central Electric Co.	L	Caldwell Fire Alarm Co.		N. Y. Insulated Wire Co.		Taylor, Goodhue & Ames	
Commercial Cable Co.		General Electric Co.	B-H-N-C-A-J	National Carbon Co.	T	Thomson Elec. Welding Co.	
C. & C. Elec. Motor Co.	S	General Israel's Art Exbt. Co.		Norwich Ins. Wire Co.		Tailandersgit, Kirsting Co.	W
Cleveland Elect. Mfg. Co.	A	Scovley, R. S. & Co.		North Am. Phonograph Co.	N-F	Union Electric Co.	
Chicago Belting Co.		Germany		N. Y. A. L. E. A.		Vetter J. C. & Co.	Y
Pulsing Clock Co.		Righland Wm. & Co.	U	Nat. Engraving of Shoe Co.		Welsh, B. F.	
Department of Electricity	M	Robinson, C. J.		Nat. Engraving of Shoe Co.	S	Weston El. Instrument Co.	
Electrican Pneumatic	Y	Hart & Hegeman Mfg. Co.	S	Owen, D. A.		Washburn & Moen Mfg. Co.	V
Elec. Launch & Nav. Co.		Hope Elec. Appliance Co.	S	Phoenix Glass Co.	N	Western Union Tel. Co.	
Electric Selector	T	Holt, Chas.		Paiste, B. J.		Weare & Burdett Mfg. Co.	
Edgerton, E. M	T	Holmes, S.	W	Pulverbacher Galv. Co.		White, S. S., Dental Mfg. Co.	
Elgin Telephone Co.		Hartman & Braun		Pumpelly, J. S.	T	Western Electrician	
Edison Elec. Mfg. Co.		Hansen & Van Winkle		Post El. Med. Sup. Co.		Wiley Ant. Handel Al. Co.	
Enterprise Electric Co.		Hork, J. S.		Powell, Wm. & Co.		Western Electric Co.	
Eureka Temp. Copper Co.		Holdsmith, E. & Co.	P	Phelps, A. H.		Westinghouse El. & Mfg. Co.	B-H-J
Electric Appliance Co.		Edison Mfg. Co.		Page Belting Co.		Wise & Seawald	
Elec. Sol. & Sig'l Co.		Interna't Aut. Uta El. Tel. Co.		Green & Co.	E	Wing, L. J. & Co.	P
Electric Heat Alarm Co.	Y	India Rubber Comb Co.	S	Hughes, F. A.		Zucker & Levett Chem. Co.	T

LINE SUPPLIES.   CONSTRUCTION MATERIAL.
INCANDESCENT LAMPS.
RUBBER COVERED WIRES.   WEATHER PROOF WIRES.
ELECTRICAL HOUSE GOODS.
TELEPHONE SUPPLIES.   TELEGRAPH SUPPLIES.

ELECTRIC APPLIANCE COMPANY,
CHICAGO. . . .   242 Madison Street,   . . . CHICAGO.

THE MATHER ELECTRIC CO,
MANCHESTER, CONN.

Dynamos, Motors, Generators,

Offices, 116 Bedford St., BOSTON.

—AND—

1002 Chamber of Commerce Bldg., CHICAGO.

THE "NOVAK" LAMP.

CLAFLIN & KIMBALL (Inc.)

General Selling Agents.

116 Bedford Street, BOSTON.

1002 Chamber of Commerce Bldg., CHICAGO.

Siemens & Halske Electric Company

of America.

Chicago, Illinois.

Electrical Machinery.

Siemens & Halske;

Berlin.
Charlottenburg.
Vienna.
St. Petersburg.

Enterprise
Electric
Company

307 Dearborn Street.
Chicago . . . .

GENERAL WESTERN AGENTS

N. I. R.

Manufacturers' Agents and Mill Representatives for

Electric Railway,
Telegraph, Telephone and
Electric Light

SUPPLIES OF EVERY DESCRIPTION

Agents for   Cedar Poles,
Cypress Poles, Oak Pins,
Locust Pins, Cross Arms, Glass
———— Feeder Wire.   Insulators,

WIRES, CABLES, TAPE and TUBING ————

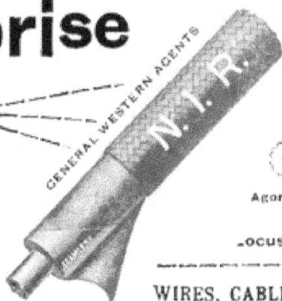

Map of Chicago.

Showing Location of its Electrical and Allied Business Interests, Principal Hotels, Theatres, Depots and Transportation Lines to the World's Fair Grounds. (Index numbers refer to the black squares.)

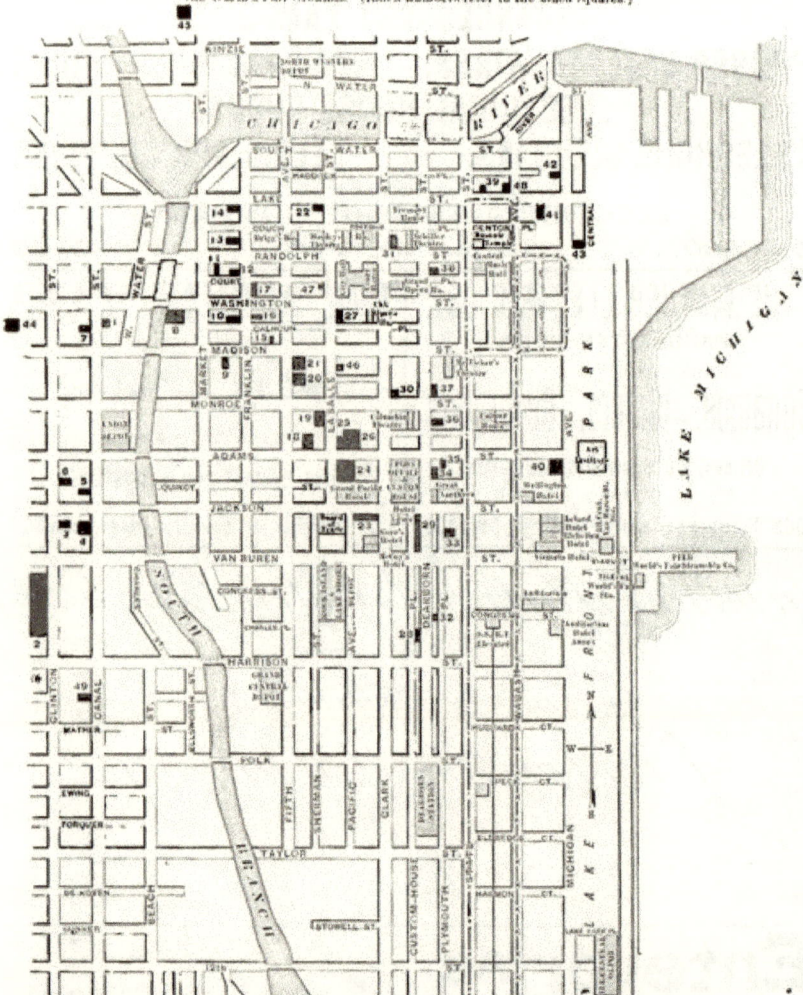

Ansonia Elec Co., Michigan Ave. & Randolph St........31
American Battery Co., 161 Madison St.................35
Bartholomew, Stor & Co., 52 Michigan Ave...........12
Barton & Brown, 118 Monadnock Block................47
Benham, J. B., 367 S. Clinton St....................3
Bell Co., J. G., Phenix Bldg.......................23
Bryant Electric Co.................................
Brush Electric Co., Monadnock Block................
Buckeye Electric Co., 125 Rookery Bldg.............21
Calumet Electric Mfg. & Engineering Co., 174 S.
  Clinton St......................................6
Central Electric Lt. & Pr. Co., 86 Dearborn St.....
Central Electric Co., 178 Franklin St...............31
Central Telephone Co...............................
Chicago Arc Light & Power Co., Washington St. Ply
   South Pl..
Chicago Telephone Co., 3d Washington St............35
Chicago Edison Co..................................36
Chicago Electric Club, 175 Clark St................43
Chicago Electric Motor Co., 328 Canal St...........13
Chicago Insulated Wire Co., 211 The Rookery........
Gallin & Kimball, 1001 Chamber of Commerce Bldg....
Cleveland Electric & Mfg. Co., 129 La Salle St.....98
Cutter, Geo., 161 The Rookery......................51

Curling & Morse, 228 Dearborn St...................34
Commercial Elec Co., The Rookery...................24
Consolidated Electric Co...........................16
C. & C. Electric Motor Co., 3d Madison St...........5
Crocker El. Telephone Co., Stock Exchange Bldg.....4
Detroit Electrical Works, 96 Monadnock Block........
Edward, W. S., Mfg. Co., 3 Lake St.................
Eddy Electric Mfg. Co., 1117 Monadnock Block.......7
Electric Construction Supply Co., Unity Bldg.......38
Electric Appliance Co., 247 Madison St.............
Electrical Instrument Pen Co., Monadnock Block....26
Enterprise Electric Co., 367 Dearborn St...........
Ft. Wayne Elec. Co., 195 Dearborn St...............
Gisle-Gizza & Heat Co., 52 Lake St.................
Gregory, Chas. Z. Co., 419 Jefferson St............
Great Western Mfg. Co., 265 S. Canal St............
General Electric Co., 173 Adams St.................38
General Insul. Arc Lt. Co., 175 Adams St...........
Hood, Wm., 529 LaSalle St..........................
Holleret-Smith Elec. Co., 367 Dearborn St..........27
Illinois Elec. Launch Co., 156-158 S. Canal St....
Kimball Elec. Motor Co., 102 Monadnock Block.......
Keystone Elec. Co., 365 S. Canal St................
Knapp Elec. Works, 56 Franklin St.................13
Kohler Bros., 1117 Monadnock Block................29

McDougall & Cummings, Unity Bldg..................30
McLean & Schmitt, 156 S. Canal St.................
Mather Elec. Co., Chamber of Commerce Bldg........20
New York Insulated Wire Co., 80 Franklin St........17
National Elec. Mfg. Co., Pullman Bldg.............42
Postal Telegraph Co., Phenix Bldg.................
Pullman's Palace Car Co., Pullman Bldg............41
Pumpelly, J. K., 265 S. Canal St.................46
Phoenix Glass Co., Wabash Ave. and Lake St........19
Railway Equipment Co., Pullman Bldg..............
Rockford Elec. & Mfg. Co., 94 La Salle St........17
Standard Elec. Co., 623 Home Ins. Bldg...........
Sawyer-Helsko Elec. Co., 1225 Monadnock Block.....39
Star Elec. Lamp Co., Sub Chamber of Com. Bldg.....
Sterling Co., 946 Pullman Bldg...................40
Schlesener, Chas. A. & Co., 18 S. Canal St........1
Short Elec. Railway Co., Monadnock Block.........28
Taylor, Goodhue & Ames, 316 Dearborn St..........
Todd, Applegate Co., The 340 Dearborn St.........
Wabash Elec. Co., 1852 Monadnock Block...........
Westinghouse Electric & Mfg. Co., Pullman Bldg....45
Western Electric Co., 227 S. Clinton St..........
Wolf-Sanb, E. J., Lake and Franklin Sts..........14
Western Union Telegraph Co., Phenix Bldg.........23

THE
FERRIS WHEEL

When you visit the World's Fair. you will naturally take a ride on the FERRIS WHEEL and be interested in the ELECTRIC LIGHT INSTALLATION. which is wired throughout with

OKONITE WIRE

FURNISHED BY THE

CENTRAL ELECTRIC COMPANY,

116-118 Franklin Street,
CHICAGO, ILLS.

The Standard Underground Cable Co.

ARE THE LARGEST MANUFACTURERS OF

LEAD COVERED CABLES
AND INSULATED WIRES

IN THE UNITED STATES.

And Prepared to offer bids on any or all installations of
Underground or Aerial Systems for Telegraph,

TELEPHONE, ELECTRIC LIGHT AND POWER.

Westinghouse Building,

PITTSBURGH, PA.

Rookery, CHICAGO.   Times Bldg., NEW YORK.

THE MONTHLY ISSUE FOR AUGUST

Should be read by everyone interested in electrical matters.   In its table of contents is the following:

"Incandescent Lighting at the World's Fair."
"The Electric Power Plant of the Chicago City Railway."
"Steam Engine Efficiency—Its Possibilities and Limitations" by Wm. H. Bryan.
"Alternating Arc Lighting for Central Stations" by H. S. Putnam.
"Hard Rubber as an Insulator in Street Railway Work" by W. R. Mason.
"A Brief Review."
Together with illustrations of the recent applications of electricity.
The paper also contains regularly
A Buyer's Directory of Manufacturers and Dealers in Electrical Supplies and Appliances.
A Complete Directory of Electric Light Stations in North America and a Complete Directory of
Electric Railways in North America.
These directories are revised each issue to the date of going to press and are to be found in no
other electrical journal in the World. Its articles are read carefully and its directories used constantly
by all the buyers in the trade. These facts make it without a superior as an advertising medium.
Sample copies and rates sent on application.
Subscription price $3 per year. Six months trial $1, if ordered during the next 30 days.

ELECTRICAL INDUSTRIES PUB. CO.,

Monadnock Block, CHICAGO.

ELECTRICAL INDUSTRIES 15

CLARK
ELECTRIC

COMPANY, NEW YORK.

192 Broadway and 11 John Street.

MANUFACTURERS OF ARC LIGHTING APPARATUS FOR EVERY PURPOSE A SPECIALTY
The CLARK ARC LAMPS for use on EVERY CURRENT, have the reputation of being
the best and most durable of any ever made in the United States.

RAWHIDE PINIONS FOR ELECTRIC MOTORS
A SPECIALTY
RAWHIDE DYNAMO BELTING

THE CHICAGO RAWHIDE MANUFACTURING CO.
THE ONLY MANUFACTURERS IN THE COUNTRY

LACE LEATHER ROPE
and OTHER RAWHIDE
GOODS
OF ALL KINDS
BY KRUEGER'S PATENT

75 Ohio Street, CHICAGO, ILL.

STANDARD ELECTRIC COMPANY.

GENERAL OFFICES: 625 Home Insurance Building.

WORKS: So. Canal Street,

CHICAGO.

STANDARD SYSTEM

AT THE

WORLD'S FAIR.

MACHINERY HALL, Sec. Q, 2 Standard Arc Dynamos.
Sec. S, 20 " " "
ELECTRICITY BUILDING. Sec. P, Space 2, Arc Lighting Exhibit.

The Standard Lamps Light the Power Plant, Machinery Hall, Agricultural Hall, Shoe and Leather Building, and
Other Buildings and Portions of the Grounds.

See our Double Service All Night Lamp Before Buying an Old Style Two Rod Lamp.

Mile after mile of

SIMPLEX WIRE

Supplied to the

FERRIS WHEEL

...

By...George Cutter,

The Rookery, CHICAGO.

SIMPLEX WIRES

SIMPLEX

Ever Onward and Upward!

**INSURE
HIGH
INSULATION**

Simplex Electrical Co.

620 Atlantic Ave.,

George Cutter, Chicago.   BOSTON, MASS.

XNTRIC

"That's the Switch"

And we control that movement.

H. T. PAISTE,

10 South 11th St.,

**PHILADELPHIA,
PA.**

Made 5 amp. S. P.
10 amp. S. P.
5 amp. 3 way.
10 amp. 3 way.

China Window Tube (Patented).

Made only by **PASS & SEYMOUR,**

George Cutter,   SYRACUSE, N. Y

CHICAGO.

Consolidated Electric Co.

Manufacturers and Dealers in all kinds of

ELECTRICAL . SUPPLIES,

115 Franklin Street,

CHICAGO.

GEORGE PORTER,

Contractor for All Kinds of

ELECTRICAL WORK.

Room 67, 143 La Salle St., CHICAGO.

Crary Block, BOONE, IOWA.

CHAS. A. SCHIEREN & CO.

MANUFACTURERS OF

Genuine Perforated Electric Leather Belting.

46 So. Canal Street,   -   **CHICAGO**

Section 15, Dpt. F, Clm. 27.   Section D, Space 3

MACHINERY HALL.   ELECTRICITY BUILDING.

CALL AND EXAMINE

Lawton's Call Indicator,

Indispensable for hotels, railroad offices, school buildings, hospitals, etc.

Section Y, Space 45, Gallery Electricity Building,

WORLD'S FAIR.

CHICAGO, ILL.

WAGNER ELECTRIC FAN MOTORS

For Direct or Alternating Currents.

These motors give a cheaper house with less consumption of current than any other fan motor on the market. They are full 1-8 horse power. Six bladed 12 inch fan. Self-oiling. Furnished with or without guards.

IT WILL PAY YOU TO SEE THE WAGNER BEFORE BUYING ELSEWHERE

TAYLOR, GOODHUE & AMES,

348 Dearborn Street, CHICAGO.

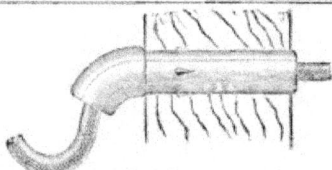

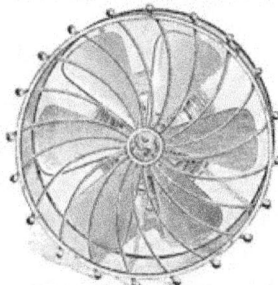

WEEKLY WORLD'S FAIR

ELECTRICAL INDUSTRIES

DEVOTED TO THE ELECTRICAL AND ALLIED INTERESTS OF THE WORLD'S FAIR,
ITS VISITORS AND EXHIBITORS.

Vol. I, No. 13. CHICAGO, SEPTEMBER 7, 1893. FIVE MONTHS $1.00
TEN CENTS A COPY

Exhibit of the Central Electric Company.

Walking through the southwestern part of the gallery of Electricity Building one meets with some mammoth reels nearly six feet in diameter of wire and cable, the well known product of the Okonite Co., of New York. These constitute a portion of the exhibit of the Central Electric Company, of 118

nite trade mark. Something of the variety of electrical conductors which the Okonite company manufacture is shown in this exhibit. Reels of the heavy armored conductors for submarine lines, large, many circuit, lead covered telephone and telegraph cables for underground and conduit lines, braided aerial cables and numerous sizes of the smaller insulated wires are exhibited.

EXHIBIT OF THE CENTRAL ELECTRIC COMPANY.

Franklin St., Chicago. This exhibit as a whole is peculiar in two special features which constitute the principal points of electric light installations, namely the wire used and the system of installing the wire.

The great reels of cable standing almost as high as the visitors' head have attracted more attention since it was used in wiring the great Ferris wheel. These reels are painted white and on the sides appear the well known Oko-

These cables are used by the telephone and telegraph companies in their regular work and were made in accordance with their specifications. The insulation known as Okonite is a special high grade patented rubber compound which is thoroughly seasoned and tested. The Okonite Co. received a gold medal at the Paris exposition in 1889.

Nearly 600 miles of Okonite wire was used in the installation of the police patrol and the fire alarm systems at the

Fair. The feeders for the Intramural Railway and the wires of the lighting system of the Libbey Glass Works are also of this kind of insulated cables and wires. A number of large photographic views of the Okonite factory adorns the walls of the exhibit which show the works and the method of insulating these wires.

One of the most interesting parts of the exhibit of the company is the exhibit of the conduit system of wiring. This system is that of the Interior Conduit Insulation company, of New York, for which the Central Electric Company is general western agent.

On two large panels with projecting tops the conduit and various fittings and appliances are displayed. The conduit is of two kinds, plain and brass armored, the fittings and attachments for each being very much alike except the outward finish. The system is so complete, necessary fittings being made for turning corners and connecting the different lines and branches, that there is no part of the wire in a building but what is enclosed and protected by the conduit,

The conduit is shown in sizes that will admit wire or cable from one fourth inch in diameter to two and one-half inches in diameter. Between these panels is a panel showing samples of carbon of the various kinds handled by the Central company from the works of the Washington Carbon Company, of Pittsburgh.

In contrast is shown the raw unmoulded carbon and the marketable article. Sample carbons for arc lights in a large number of sizes, battery carbons of the shapes best adapted to the different kinds of carbon batteries are shown. Also carbon plates for electrolytic plants, carbon brushes for motors and dynamos and samples showing something of the capacity of the works, are exhibited.

Another of the many articles handled by this company and exhibited in a practical way at the Fair is the Lundel power motor which is used in a number of exhibits. One of these is in constant use in the United States Whip Co's. exhibit in the making of the souvenir pen holders. In the the Linotype type setting exhibit which has received

EXHIBIT OF THE EXCELSIOR ELECTRIC COMPANY.

and the wire in any part or all of it may be withdrawn and new wire inserted.

The stringent rules of the fire underwriters and the extra rates charged for poor methods of wiring have caused this system to be used to a large extent. The system is adapted for any system of wiring and is as easily applied to the three or five wire systems as the two wire. The conduit will all be concealed beneath the plaster or if run open it is furnished in colors to match the interior finish of the building. The joints are all neatly and tastily made and the junction boxes and terminals fitted with covers of any desired finish makes the system present an excellent appearance.

In plastered buildings the terminals are concealed under the fixture canopy and the cut out or junction boxes may appear almost anywhere without marring the appearance of the room, as nothing but the top appears above the plaster over which a handsome cover is placed.

On these panels are shown the porcelain cut out blocks both for the large and small boxes of such size and shape that they just fill the boxes and yet leave room enough for the connections. There is also shown neatly arranged on the panels the necessary coupling and cutting tools, pliers, etc., needed in the installation of this system.

a good deal of attention another of these motors is used. Others are used about the Fair two being put to the novel use of propelling the Japanese launch.

Above the exhibit there have been erected a number of signs the largest of which gives the names and factories of the three large companies for whom the Central Electric Company is the general western agent.

Exhibit of the Excelsior Electric Company.

Just to the left of the east entrance of Electricity Building is located the exhibit of the Excelsior Electric Company. The rows of arc lamps suspended above the railing that surrounds the exhibit immediately attracts the attention of the passing visitor. In this exhibit of lamps are the various styles of arc lamps the company manufactures. Lamps adapted to the different places in which the arc lamp is used, weather proof lamps for the street, lamps for interiors with high ceilings, short lamps for low ceiling with the arc but 15 inches from the top, in this lamp both the upper and lower carbons move.

Lamps for use on steamboats with lenses, focusing lamps for photographic purposes etc., are shown. The latest duplex street lamp is shown which is so arranged that while the first set of carbons is burning the other set is held up. The top of the rod of the first carbon is provided with a button which trips a lever when that carbon is burned out and throws the second set of carbons into use. Special attention has been given in the construction of these lamps to the facilities for cleaning them easily. The switch on the lamp when turned throws the lamp entirely out of circuit, thus the safety of the trimmer is assured.

In the front part of the space is a 30 arc light dynamo with case opened so as to show the armature and bearings, and on counters are shown fan motors of various sizes from one tenth horse power up. On the left is a 10 horse power motor belted to a 120 ampere incandescent 110 volt dynamo at the side of which is placed a skeleton switchboard with ammeter, voltmeter, fuse, switches, etc., very neatly arranged. In the center of the exhibit is a 20 volt 2,100 ampere plating machine around which are placed motors, common tators and parts of motors and dynamos showing the construction of the different parts. Back of this is placed a 100 light 2,000 candle power arc dynamo belted to a shaft beneath the floor. At the sides are placed racks holding a

The Excelsior Company manufacture these outfits in a number of sizes. They have been used successfully in a variety of places and have been found especially efficient for raising water to the upper stories of buildings where the pressure on the street mains was not sufficient.

The exhibit is in charge of Mr. Geo. H. Almon who is looking after the interests of the company at the Fair.

Exhibit of the Stilwell-Bierce & Smith-Vaile Co.

The exhibit of heaters, water wheels and pumps made by the Stilwell Bierce & Smith Vaile Company is located in the annex to Machinery Hall column J, 37 section H, directly opposite the booth of the mechanical trade journal. Since the consolidation of the two companies represented by the above name early in the year the construction of both water power and steam apparatus has been largely undertaken by the new firm.

A prominent position is given to different sizes and types of the victor turbine. A 45 inch horizontal double victor wheel with register gates, a good representation of the

EXHIBIT OF THE STILWELL-BIERCE & SMITH-VAILE COMPANY.

dozen armatures of different sizes with their brightly polished shafts.

On stands near the walls are shown a number of power motors of one, three and five horse power adapted for arc and incandescent light and power circuits, also plating dynamos of different capacities. On the wall to the right are placed connecting boards, rheostats, switches, meters, starting boxes, etc. The name of the company in large white letters covers the west wall. The motors of this company have been used to drive all kinds of machinery. The company building motors for different voltages has been able to supply motors for almost any circuit.

A part of the exhibit that attracts the attention of the passing visitor is the motor and pump shown in operation. Two tanks are placed one above the other. From the lower tank a line of wrought iron pipe leads to the upper in which is placed a small rotary pump. This pump which can be placed in any line of water pipe as easily as making a connection, is operated by a one half horse power motor. This outfit that occupies a space not more than three feet square is said to be able to raise 500 gallons of water an hour 10 feet and at a cost of about 10 cents per hour for current.

horizontal type of modern turbine. Another five inch horizontal wheel for high pressures and using a gate valve are shown. In the regular type with vertical shaft there are shown a 27½ inch with cylinder gate, a 30 inch with register gate and a six inch with the same gate. The wheels are all well finished, simple in construction and have a reputation for very high efficiency. The Stilwell live steam purifier is exhibited with one end open to show the method of construction and to explain the working of the same. It will be remembered that this purifier is a cylinder or chamber placed a few feet higher than the water level in the boiler and to which live steam is admitted at boiler pressure. The feed water enters at the top and being fed very slowly strikes the top one of a series of shallow pans which tip slightly in opposite directions; the water traversing these in a thin sheet is thoroughly heated to the boiler pressure and any impurities contained in the same are deposited either in the pans or are filtered out in the coke filter pans at the bottom. Water enters the boiler from this heater and purifies by gravity and at the same temperature as that in the boiler. The principle is one of common sense and should meet with the most extended success as it will easily save its cost in a very short period.

A sample of the Stillwell vertical improved exhaust heater, purifier and filter contained is shown. This is somewhat similar in principle to the one above described and has an additional convenience in an automatic water inlet valve which controls the supply of feed water and prevents too much entering at a time. The new Stillwell close heater is also exhibited. This device has brass tubes in a U form expanded into a header that is bolted to one end of a cylinder. A partition in the cylinder between the turn of the tubes prevents the feed water going straight through and another partition in the casting bolted into the base directs exhaust steam into one end of the tubes and out of the other. This heater is laid on its side and the support forms a mud drum from which is taken all the deposits from impure water.

In the pump part of the exhibit are samples of various sizes and styles of the Smith-Vaile duplex pumps. A large compound pump with duplex tandem steam cylinders, a large Underwriters pump with duplex cylinders a capacity of 750 gals. per minute with 16 by 9 by 12 cylinders are shown. Also a receiver pump for the return end of steam heating system is shown; this pump has an automatic float controller with 6 by 4 by 6 cylinders.

Other pumps, one a duplex 7 by 4½ by 10, another duplex 5 by 2 by 4 and another a deep well pump 9½ by 30 steam cylinder, are shown. All the products of this company are of the best design and show a degree of finish and workmanship equal to any.

If they will now add a first class water wheel regulator to their other products, one that will do for electric light and power work, the Stillwell-Bierce & Smith-Vaile Co. will be able to supply an equipment that will meet all the wants of the motive power end of any large plant.

A Chicago Branch of American Institute of E. E.

On Saturday evening of last week a meeting of the Chicago members of the American Institute of Electrical Engineers was held and the advisability of establishing a branch at Chicago was discussed. Mr. Pope addressed the meeting on the Institute and its work. Prof. W. N. Stine, of the Armour Institute offered the free use of the lecture room and laboratory for the use of the branch. A number of others spoke, among whom were Dr. Keith, of San Francisco, and Mr. W. H. Preece, of London. A committee consisting of Messrs. B. J. Arnold, H. A. Foster and R. H. Pierce was appointed to canvass the members and ascertain the support such a branch would receive.

California Mid-Winter Exposition.

The management of the California Mid-Winter Exposition announces that the success of the enterprise is assured. A large number of exhibits have already been promised, and the plans for the buildings decided upon. Special arrangements have been made with the railroads for the transportation of exhibits. Just what part electricity will take in the exposition we are as yet uninformed, but it is safe to say that it will do its share. Whether it will pay electrical manufacturers to make extensive exhibits is a matter they alone can judge, but in any event a large amount of light and power will be required to operate the exhibition successfully.

The Mason Battery Company, New York, has placed a line of batteries on exhibition in section 8, just east of the Paiste switch display.

ELECTRICAL INDUSTRIES.

PUBLISHED EVERY THURSDAY BY THE

ELECTRICAL INDUSTRIES PUBLISHING COMPANY,
INCORPORATED 1889.

MONADNOCK BLOCK, CHICAGO.
TELEPHONE HARRISON 250.

E. L. POWERS, PRES. AND TREAS.   E. E. WOOD, SECRETARY.

E. L. POWERS,   EDITOR.
H. A. FOSTER,   }
W. A. REMINGTON,   }   ASSOCIATE EDITORS.
E. E. WOOD,   EASTERN MANAGER.
FLOYD T. SHORT,   ADVERTISING DEPARTMENT.

EASTERN OFFICE, WORLD BUILDING, NEW YORK.
World's Fair Headquarters, Y 27 Electricity Building.

SUBSCRIPTION:
FIVE MONTHS $1.00
SINGLE COPY 10
Advertising Rates Upon Application.

News items, notes or communications of interest to World's Fair visitors are earnestly desired for publication in these columns and will be heartily appreciated. We especially invite all visitors to call upon us or send address at once upon their arrival in city or at the grounds.
ELECTRICAL INDUSTRIES PUBLISHING CO.,
Monadnock Block, Chicago

To many persons the electric fountains at the Fair have been something of a disappointment. The fountains are but a part of the great basin and consequently are not placed so as to form a center of attraction, but to form one part of a great whole. In order that they might harmonize with their surroundings they were placed much below the level of the plaza about the Administration Building; consequently to persons in the rear of a crowd the fountains look small in this "city of grand proportions." They are also contrasted with the Lincoln Park fountain, which on its elevation is a most beautiful sight, appearing as it does against the dark background of the trees. But the wheat-sheafs and other water effects, which form a special feature of the Columbian fountains, are not as perfect. The best view is secured from the basin. To persons in a boat a scene unsurpassed in brilliancy is presented, but the fountains are not the central object and some of their brilliancy is lost in the bright light of their surroundings, but the colors appear clear and beautiful. There will undoubtedly be a great improvement in their operation since they have now passed under the management of the Exposition. The same hand will now manage both the water and light supply and the full power of the fountains will be displayed.

The reception given by the Department of Electricity through its genial chief, Prof. J. P. Barrett, to members of the Electrical Congress on Saturday evening of congress week was an enjoyable ending to a day at the Fair. The members of the congress had spent the day at the Fair viewing more especially the electrical features of the Exposition, and they promptly availed themselves of the opportunity of discussing that part of the Fair. The refreshments served were excellent and of a kind conducive to a friendly interchange of opinions on the various features of the Fair or rather were not, as one member remarked as he arose with both hands and his mouth full to make a speech.

Dr. J. A. Hornsby, assistant chief of the department, was attentive to the wants and comfort of the guests and his remarks on the electrical features of the Exposition were pertinent and interesting. Prof. Elihu Thomson, Dr. Emery, Prof. Houston, Mr. DeCamp, Prof. S. P. Thompson, Prof. G. R. Barker, Mr. Weston and Dr. Keith each responded to the toasts proposed by Prof. Barrett who occupied the chair of honor in the center of the party.

The Department of Electricity has always been very generous in its treatment of visitors and this special entertainment which it provided was greatly appreciated by all who were present.

WORLD'S FAIR NOTES.

The exhibit of the Ansonia Electric Company was closed a portion of last week in consequence of the assignment of the company.

Bartholomew, Stow & Co. are installing an exhibit of Nutting arc lamps at the south end of the gallery, near the phonograph exhibit.

The popular electrical lectures that were being given in the scenic theatre of the Western Electric Co. have been discontinued for the present.

The Belknap Motor Company has added to its exhibit a direct connected dynamo and vertical type engine of small size. At present the dynamo runs the engine.

Official photographer Arnold obtained some very good negatives of many of the prominent delegates to the World's Electrical Congress while they were at the Fair.

The control of the electrical fountains is now in the hands of the Exposition. Displays of the intricate water effects are now made during the day and the fountains are illuminated every evening.

Since Sept. 1st the Elektron Mfg. Co. has carried an average of over fifteen hundred persons daily in its display elevator in the northwest corner of Electricity Building. The accommodation that is thus afforded is highly appreciated by both the public and the gallery exhibitors.

The Jury of Awards has been progressing very rapidly with its work. The work of testing the lamps and instruments has been delayed somewhat but the laboratories are now fully equipped and the tests will be soon under way. It will probably be very near the close of the Exposition before the tests will be completed.

An interesting use of the phonograph may be seen in a number of the educational exhibits in the gallery of the Manufactures and Liberal Arts Building. Records have been secured of the singing by the children in the various schools, and in this way the progress made in the art of reading music is very easily shown.

The popularity of the gallery of the Electrical Building shows the success of the efforts made by the Gallery Exhibitors Club. Neat signs announcing the attractions in the gallery have been placed in conspicuous places, a liberal number of seats have been provided and an artistic little souvenir is given away to visitors. The increased attendance must be a source of satisfaction to the committee who have had the matter in charge.

The four large Westinghouse vertical engines, directly connected to Westinghouse dynamos placed close together in the power plant in Machinery Hall form a very compact plant as well as one of great power. Each engine is of 1,000 horse-power capacity and each dynamo has a capacity of 10,000 16-candle power lights. In this block there is thus a plant of 40,000 lights. The engines are double acting steeple-compound engines and run 200 revolutions per minute. The low pressure cylinder is placed above the

high pressure, and the cylinders are respectively thirty-seven and twenty-two and one-half inches in diameter and have a stroke of twenty-two inches.

Many of the official delegates attending the World's Electrical Congress remained in the city to enjoy the varied features of the Fair. Dr. H. von Helmholtz is spending two weeks in the Yellowstone National Park. Prof. Alex. Siemens and Prof. M. Mascart have already returned home. Prof. Ayrton sails this week on the Germanic, and Prof. Galileo Ferraris will leave Sunday on the Paris. Prof. W. H. Preece will spend some time at Niagara before returning to England.

An attractive addition to the decoration of Electricity Building is being placed this week by the Department of Electricity, between the exhibits of the Brush Co. and the Ft. Wayne Co. An artistic arrangement of flags made of many different colored incandescent lamps will be shown. The lighting of the lamps will be automatically controlled so as to give the flags a waving effect. The artistic signs and electrical effects that have been added during the past month by the various exhibitors has much improved the appearance of the building.

The writer of fairy tales has told of the doors that were opened by unseen hands on the approach of the hero of the story and noiselessly closed after he had passed through. That doors of this kind should become a commercial article, was beyond our comprehension when we read these tales, but among the other wonderful and mysterious things to the uninitiated in Electricity Building are a number of doors that open as a person approaches them and close after him. These doors are operated by the Hicks Troy electric door operator, and a visit to the exhibit in the west gallery will be of interest.

In the German section of the Manufactures Building Heinrich Seitz exhibits two unique fixtures for electric lighting. They are both three quarter size figures in bronze. One represents the figure of an old peasant holding an old-fashioned square lantern at arm's length and shading his eyes with the other hand to get a clearer view into the surrounding darkness. The other is the figure of a country man who is evidently having his first experience with the physical effects of an electrical battery. Clasped in either hand are the electrodes, with sockets for lamps in the upper ends, while the expression of the face and pose of the body plainly indicate that the current is on.

The three Climax boilers in the boiler room annex to Machinery Hall have attracted a great deal of attention on account of their odd shape which is very much different from that of the ordinary water tube boiler. These boilers are two of 500 and one of 1,000-horse power capacity, and 15 and 15½ feet in diameter. The latter standing 37 feet high. This boiler is said to be the largest boiler in the world and is guaranteed to evaporate 30,000 pounds every hour. The total heating surface is 10,000 square feet, the surface of 1,000 tubes 12 feet long. The construction of these boilers is not very complex. It consists of a vertical cylinder of sufficient strength to withstand the internal steam pressure. From its outer surface these tubes come out and then bend back and again enter this vertical cylinder. This is surrounded by a shell lined with material that will prevent the radiation of the heat. The furnace is placed beneath the boiler and as will be readily seen the hot air and gases pass over a large amount of surface before reaching the stack. These boilers were built by the Chadwick Steam Boiler Works, Brooklyn, N. Y.

PERSONAL.

Mr. F. S. Bunting, of the Ft. Wayne Electric Co., is now at the Fair.

Mr. H. A. Lawton returns from an extended trip East, the latter part of this week.

Mr. W. H. McKinloch has associated himself with the Enterprise Electric Co., Chicago.

Mr. Caryl P. Haskins, of the General Electric Co., Boston, is in the city visiting the Fair.

Mr. E. E. Knowles, engineer for the Schuyler Electric Co., Middletown, Conn., is among the arrivals this week.

Mr. A. L. Daniels of the Miamisburg Electric Co., Miamisburg, Ohio, is spending a few days at the Fair this week.

Mr. W. E. Dresser, superintendent of telegraphs from Costa Rica, has been spending some time at the Exposition.

Mr. F. W. Fairfield, city electrician from Nashville, Tenn., has been in the city visiting the Exposition during the past week.

Mr. Coleman Sellers, the well known engineer of Philadelphia, registered at the Windermere last week, and was an interested visitor of the electrical exhibits.

Mr. Miguel F. Horta, an engineer from the Republic of Uruguay, South America, is now at the Exposition, making a special study of electrical railway exhibits.

Mr. D. I. Carson, of New York city, secretary and general superintendent of the Southern Bell Telephone & Telegraph Co., is spending some time at the Exposition this week.

Mr. Henry D. Wilkinson, engineer of the British World's Fair commission, accompanied by Frederic A. Hamilton, M. Inst. E. E. London, were visitors at the headquarters of the American Institute of Electrical Engineers last week.

The New Lamp of the Pennsylvania Electrical Engineering Company.

In the accompanying cut is illustrated the new non-infringing lamp now being manufactured by the Pennsylvania

NEW LAMP OF THE PENNSYLVANIA ELECTRICAL ENGINEERING CO.

Electrical Engineering Company, Penn Mutual Bld., Philadelphia, Pa.

The receiver is not an all glass receiver and the leading in wires do not pass through the glass consequently the lamp does not come under the claims of the Edison lamp

patent. A number of patents have been secured on numerous details in its construction. It is decidedly original in design and contrary to the usual method, the leading in wires are of iron instead of the more expensive metal platinum.

The lamp has been named by the company the "Maggie Murphy" lamp, and it is able to supply them in any quantity to suit the purchaser.

DEPARTMENT OF ELECTRICITY.

OFFICES: SECTION R, ELECTRICITY BUILDING.

Chief, JOHN P. BARRETT,
Assistant Chief, J. ALLEN HORNSBY,
General Superintendent, J. W. BRADSELL,
Electrical Engineer, W. W. PRIDE.

AMERICAN INSTITUTE OF ELECTRICAL ENGINEERS.

World's Fair Headquarters,
SECTION S, ELECTRICITY BUILDING.

RALPH W. POPE, Secretary.

Open from 9 a. m. to 5 p. m.

Amusements.

HOOLEY'S THEATER—Mr. E. S. Willard in "Wealth." 149 Randolph street.

COLUMBIA THEATER—Daniel Frohman's Lyceum Theater Co. in "The Charity Ball." 108 Monroe street.

GRAND OPERA HOUSE—Sol Smith Russell, in "A Poor Relation." 87 Clark street.

AUDITORIUM—Imre Kiralfy's Spectacle "America." Congress street and Wabash avenue.

McVICKER'S THEATER—Denman Thomson, in "The Old Homestead." 82 Madison street.

CHICAGO OPERA HOUSE—American Extravaganza Company, in "Ali Baba, or Morgiana and the Forty Thieves." Washington and Clark streets.

SCHILLER THEATER—Chas. Frohman's Stock Company, in "The Girl I Left Behind Me." Randolph, near Dearborn.

HAVERLY'S CASINO—Haverly's United Minstrels. Wabash avenue, near Jackson street.

TROCADERO—Vaudeville. Michigan avenue near Monroe street.

THE GIARDEN—Vaudeville. Michigan avenue near Monroe street.

Buffalo Bill's "Wild West." 63d street. Daily at 3 and 8.30 p.m.

Daniel Frohman's Lyceum Theatre Company appears at the Columbia this week in "The Charity Ball." This play has been seen here a number of times and has always been well received. This is announced as the opening of the regular season although there has been no intermission during the summer.

Mr. E. S. Willard begins this week another engagement at Hooley's. A new play is presented entitled "Wealth," by Henry Arthur Jones. "A Fool's Paradise," "Judah" and "The Middleman" will follow in each of which Mr. Willard is familiar to Chicago people.

Mr. Sol Smith Russell appears next week in "A Peaceful Valley," a play in which Mr. Russell has been very successful. For its presentation new scenery has been prepared and the third act has been rewritten by the author so that it will be like seeing a new play with all the beauties of the old.

The sign "standing room only" is displayed in the lobby of the Auditorium every evening before the curtain rises so great has been the demand for seats. Extra afternoon presentations of "America" have been instituted. This arrangement will allow thousands to see this spectacle which has come to be considered a part of the visit to the World's Fair, who have heretofore been turned away.

"Ali Baba" continues to draw crowded houses at the Chicago Opera house. On October 1st "Sinbad or Maid of Balsora" will take the place of "Ali Baba". This piece is being rehearsed, the scenery is ready and the costumes designed by Howell Russell of London are being made. New music, new people and lots of fresh dialogue will make the piece lively and amusing.

OUR IMPROVED SYSTEM

... OF ...

Automatic Fire Alarm,

covered by patents recently issued, is the
embodiment of all factors contributing to the

GREATEST SAFETY,

and the MOST RELIABLE

PROTECTION FROM FIRE.

Western Electric Company,

CHICAGO and NEW YORK.

ELECTRICITY BUILDING— EXHIBITORS AND THEIR LOCATION.

GALLERY.

MAIN FLOOR.

Exhibitor.	Section.	Exhibitor.	Section.	Exhibitor.	Section.	Exhibitor.	Section.

LINE SUPPLIES.   **CONSTRUCTION MATERIAL.**

INCANDESCENT LAMPS.

RUBBER COVERED WIRES.   **WEATHER PROOF WIRES.**

ELECTRICAL HOUSE GOODS.

TELEPHONE SUPPLIES.   **TELEGRAPH SUPPLIES.**

ELECTRIC APPLIANCE COMPANY,

CHICAGO. . . .   242 Madison Street,   . . . CHICAGO.

THE MATHER ELECTRIC CO.

MANCHESTER, CONN.

Dynamos, Motors, Generators,

Offices, 116 Bedford St., BOSTON.

—AND—

1002 Chamber of Commerce Bldg., CHICAGO.

THE "NOVAK" LAMP.

CLAFLIN & KIMBALL (Inc.)

General Selling Agents.

116 Bedford Street, BOSTON.

1002 Chamber of Commerce Bldg., CHICAGO.

Siemens & Halske Electric Company

of America.

Chicago, Illinois.

Electrical Machinery.

Siemens & Halske.

Berlin.
Charlottenburg.
Vienna.
St. Petersburg.

Enterprise Electric Company

307 Dearborn Street,
Chicago . . . .

GENERAL WESTERN AGENTS

N. I. R.

Manufacturers' Agents and Mill Representatives for

Electric Railway,
Telegraph, Telephone and
Electric Light

SUPPLIES OF EVERY DESCRIPTION

Agents for   Cedar Poles,
Cypress Poles, Oak Pins,
Locust Pins, Cross Arms, Glass
——————— Feeder Wire,   Insulators

WIRES, CABLES, TAPE and TUBING ——————

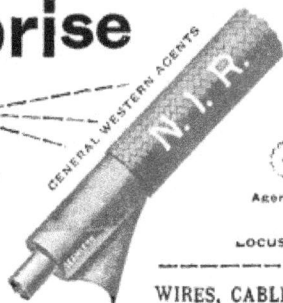

Map of Chicago.

Showing Location of its Electrical and Allied Business Interests, Principal Hotels, Theatres, Depots and Transportation Lines to the World's Fair Grounds. (Index numbers refer to the black squares.)

THE
FERRIS WHEEL

When you visit the World's Fair. you will naturally take a ride on the FERRIS WHEEL and be interested in the ELECTRIC LIGHT INSTALLATION. which is wired throughout with

OKONITE WIRE

FURNISHED BY THE

CENTRAL ELECTRIC COMPANY,

116-118 Franklin Street,

CHICAGO, ILLS.

"PENNSYLVANIA"

THE MAGGIE MURPHY LAMP

IT IS MY DELIGHT ON THE DARKEST NIGHT,
TO ENABLE YOU TO SEE.
I DON'T INFRINGE WHILE OTHERS CRINGE
BEFORE MONOPOLY
I AM A COMPLETE INNOVATION
NO CONDUCTORS PASSING THROUGH GLASS
FOR I'M YOUR "MAGGIE MURPHY",
AN ENTIRELY ORIGINAL LASS.

PENNA ELECTRIC ENGINEERING Co.
PENN MUTUAL BLDG. PHILADELPHIA, PA.

PRICE, 35 CENTS, IN LOTS OF 200.

Dynamos, Power Generators, Motors, Supplies.   Railway and Lighting Construction our Specialty.

ELECTRICAL INDUSTRIES.

WM. BARAGWANATH & SON

BOILER VALVE

MANUFACTURERS OF

**Feed Water
Heaters,
Purifiers,
Power Pumps,
Syphon**
and **Surface
Condensers,
Boiler
Cleaners,**
Etc., Etc.

The Water Jacket Condenser.

The Steam Jacket Feed Water Heater.

The Triple-Acting Power Boiler Feed Pump.

55 West Division St.

CHICAGO, ILL.

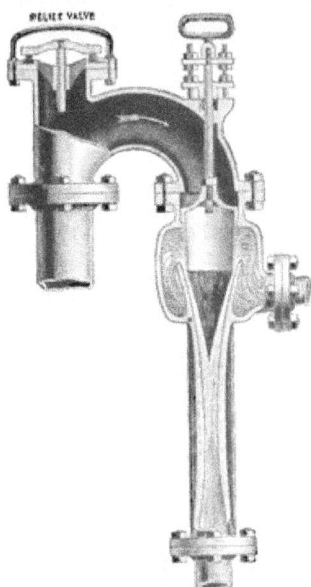

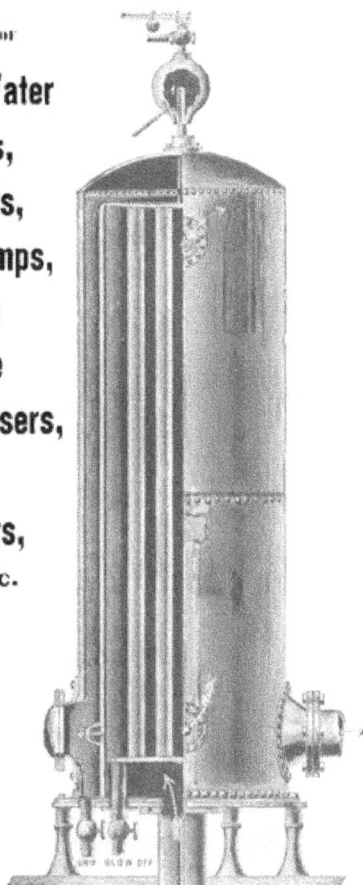

THE MONTHLY ISSUE FOR SEPTEMBER

ELECTRICAL INDUSTRIES

Should be read by everyone interested in electrical matters. In its table of contents is the following:

"World's Congress of Electricians."
"Nikola Tesla's Lecture."
"Love Underground Electric System in Washington." By M. D. Law.
"Gas Engines as Applied to Electric Light Work." By Geo. A. Farwell.
"The Underground System at the World's Fair."
"Electric Railway Plant of the Chicago North Shore Railway Company."
"Obituary—Louis W. Burnham."
"A Brief Review."
"Financial."
Together with illustrations of the recent applications of electricity.

The paper also contains regularly

A Buyer's Directory of Manufacturers and Dealers in Electrical Supplies and Appliances.
A Complete Directory of Electric Light Stations in North America and a Complete Directory of Electric Railways in North America.

These directories are revised each issue to the date of going to press and are to be found in no other electrical journal in the World. Its articles are read carefully and its directories used constantly by all the buyers in the trade. These facts make it without a superior as an advertising medium. Sample copies and rates sent on application.

Subscription price $3 per year. Six months trial $1, if ordered during September.

ELECTRICAL INDUSTRIES PUB. CO.,
Monadnock Block, CHICAGO.

NO ONE interested in Electricity at the World's Fair can afford to be without

WEEKLY WORLD'S FAIR
ELECTRICAL INDUSTRIES.

••••••

IT is the only Paper Published devoted exclusively to the electrical features of the Fair and containing a complete directory of the Exhibitors in Electricity Building.

IT Publishes more electrical news of what is actually going on at the great Exposition and IT is therefore read and used for reference by more visitors than any other Electrical Journal.

Send for free sample copy.

••••••

ELECTRICAL INDUSTRIES,
Monadnock Block, CHICAGO.

Or World's Fair Headquarters, Section Y, 27, Electricity Building.

CLARK COMPANY, NEW YORK.

ELECTRIC

192 Broadway and 11 John Street.

MANUFACTURERS OF ARC LIGHTING APPARATUS FOR EVERY PURPOSE A SPECIALTY
The CLARK ARC LAMPS for use on EVERY CURRENT, have the reputation of being
the best and most durable of any ever made in the United States

RAWHIDE PINIONS FOR ELECTRIC MOTORS
A SPECIALTY.
RAWHIDE DYNAMO BELTING

THE CHICAGO RAWHIDE MANUFACTURING CO.
THE ONLY MANUFACTURERS IN THE COUNTRY

LACE LEATHER ROPE
AND OTHER RAWHIDE

GOODS
OF ALL KINDS
BY KRUEGER'S PATENT

75 Ohio Street,   CHICAGO, ILL.

STANDARD ELECTRIC COMPANY.

GENERAL OFFICES: 625 Home Insurance Building.

WORKS: So. Canal Street,

CHICAGO.

STANDARD SYSTEM

AT THE

WORLD'S FAIR.

MACHINERY HALL, Sec. Q, 2 Standard Arc Dynamos.
Sec. S, 20   ''   ''   ''
ELECTRICITY BUILDING, Sec. P, Space 2, Arc Lighting Exhibit.

The Standard Lamps Light the Power Plant, Machinery Hall, Agricultural Hall, Shoe and Leather Building, and
Other Buildings and Portions of the Grounds.

See our Double Service All Night Lamp Before Buying an Old Style Two Rod Lamp.

Mile after mile of
SIMPLEX WIRE
Supplied to the
FERRIS WHEEL
...
By...George Cutter,
The Rookery, CHICAGO.

SIMPLEX WIRES

INSURE
HIGH
INSULATION

Simplex Electrical Co.
620 Atlantic Ave.,

George Cutter, Chicago.   BOSTON, MASS.

XNTRIC

"That's the Switch"

And we control that movement.

H. T. PAISTE,

10 South 11th St.,
**PHILADELPHIA,
PA.**

Made 5 amp. S. P.
      10 amp. S. P.
      5 amp. 3 way.
      10 amp. 3 way.

China Window Tube (Patented).
Made only by **PASS & SEYMOUR,**
George Cutter,   SYRACUSE, N. Y.
CHICAGO.

Consolidated Electric Co.

Manufacturers and Dealers in all kinds of

ELECTRICAL . SUPPLIES,

115 Franklin Street,

CHICAGO.

GEORGE PORTER,

Contractor for All Kinds of

ELECTRICAL WORK.

Room 67, 143 La Salle St., CHICAGO.
Crary Block, BOONE, IOWA.

KOHLER BROTHERS,

WESTERN BRANCH OF

The Eddy Electric Mfg. Co.

ELECTRIC MOTORS
DYNAMOS FOR LIGHTING
RAILWAY and POWER GENERATORS

1417-1418 MONADNOCK BUILDING,
TELEPHONE 5090.   CHICAGO.

CALL AND EXAMINE

Lawton's Call Indicator.

Indispensable for hotels, railroad
offices, school buildings, hospitals, etc.

Section Y, Space 45, Gallery Electricity Building,
WORLD'S FAIR.

CHICAGO, ILL.

WAGNER ELECTRIC FAN MOTORS

For Direct or Alternating Currents.

These motors give a stronger breeze with less consumption of current than
any other fan motor on the market. They are built 1-8 horse power. Six bladed
18 inch fan. No heating. Furnished with or without guards.

IT WILL PAY YOU TO SEE THE WAGNER BEFORE BUYING ELSEWHERE.

TAYLOR, GOODHUE & AMES,

348 Dearborn Street, CHICAGO.

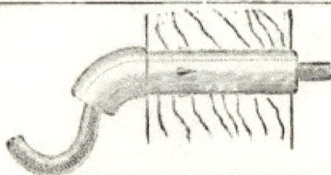

WEEKLY WORLD'S FAIR

ELECTRICAL INDUSTRIES

DEVOTED TO THE ELECTRICAL AND ALLIED INTERESTS OF THE WORLD'S FAIR, ITS VISITORS AND EXHIBITORS.

Vol. I, No. 14.   CHICAGO, SEPTEMBER 14, 1893.   FIVE MONTHS $1.00 / TEN CENTS A COPY

Exhibit of the Electric Appliance Company.

The Electric Appliance Company has one of the most interesting and attractive exhibits made by any of the supply houses. The space is in the southwestern part of gallery facing on gallery front, and is very conspicuous from almost any part of the floor from the fact that large painted signs are used and in addition a Packard lamp sign which runs across the front of the space flashing and spelling out the words letter by letter in red lamps makes a very attractive and prominent display.

A very original result has been obtained by covering the entire space with a canopy of Paranite lamp cord, which has the effect of making the exhibit look very compact and at the same time forms a good back ground for signs and sample boards. At a short distance the cord has the appearance of a Japanese open work curtain, and is certainly unique, being the only thing of the kind in the building. This canopy alone contains about thirty thousand (30,000) yards of Paranite lamp cord made up of upwards of a million and a half feet of number thirty copper wire.

The Electric Appliance Company has made up its exhibit principally of specialties, not showing a general line of supplies. The principal specialties represented are the Packard lamps, Paranite wires and cables, O. K. Waterproof wires, Moston fan motors, Whitney instruments, Elkhart transformers, C E M jack knife switches, New England switches and swinging ball lightning arresters.

As before stated, the principal part of the lamp exhibit consists of the flashing sign made up of the new C E Packard lamps. The sign contains about 500 lamps of various colors. The sign spells out the words Packard lamps a letter at a time and then flashes up both words and also a large lamp and scroll above the sign made up of about 100 small lamps arranged

to show the shape of a regular incandescent lamp. In
addition to those in the sign there is shown a large
assortment of lamps of all candle-powers and voltages and
a number of special lamps made of various kinds of fancy
colored glass making a very original display.

The exhibit of Weston alternating motors is very complete showing the new dental and sewing machine motors
and single and duplex power motors in operation doing
actual work, and also a large number of Weston fan motors
both stationary and revolving which are distributed about
the exhibit in a way to keep up a stiff breeze in all parts of
the space. One of the one-eighth horse power motors is
used to drive the commutator which operates the lamp
sign.

The exhibit of Paranite wires and cables in addition to
the Paranite cord canopy before referred to, is very complete and extensive. The center of the platform is occupied

An interesting sample table showing all sizes and styles
of New England switches and the switch parts from the
cover to the porcelain base adds to the attractiveness of the
display. C. E. M. jack-knife switches are shown in all fini-
shes, shapes and capacities and the swinging ball lightning
arrester is prominently displayed. O. K. weatherproof
wire is also conspicuously displayed by a number of handsome sample reels.

Another part of the exhibit worthy of mention is the
large display board on which is exhibited a line of Von
Cleff & Co.'s all steel pliers, connectors and other tools
used in electrical work.

The exhibit in detail is very interesting and the amount
of machinery, lamps, etc. in operation makes it attractive
and what is more desirable the whole arrangement of the
display is harmonious and symmetrical and the Electrical
Appliance Company is to be congratulated upon having

THE EXHIBIT OF THE STANDARD PAINT COMPANY.

by a large cone about five feet in diameter at the base and
eight feet high covered with Paranite wires running from
heavy submarine cable at the bottom to number eighteen wire
at the top. In addition to this the four corners of the space
are marked with pyramids of fancy reels and several display tables are covered with handsome coils of Paranite
wires and cords.

The Elkhart transformer exhibit consists of the transformers in actual operation supplying the current used for
illuminating and power purposes in the exhibit. The converters are so arranged that their close regulation, high
efficiency and cool running can be readily shown, and being in actual operation it makes a very practical converter
exhibit.

A very fine sample board of Whitney instruments is
shown, also a show case of the same with ammeters and
volt meters connected up to show their operation on direct
and alternating currents.

one of the neatest displays in Electricity Building. Mr.
F. S. Cassoway, assisted by C. C. Hilles, is in charge of
the exhibit and takes pains to explain to visitors the various articles shown.

Exhibit of the Standard Paint Co.

In the manufacture of electrical goods and their installation the matter of insulation plays a most important part.
For this purpose special compounds, paints, varnishes, etc.,
are manufactured and of the manufacturers of this class of
goods none are better known than the Standing Paint
Company.

In the east gallery toward the south end of Electricity
Building this company has an exhibit made up entirely of
electrical insulations. In the center of the space there has
been erected a large stained glass panel in the center of
which appears the well known trade mark of the company.

On each side are draped flags of different nations making an attractive background to the display of the goods.

On the base at the bottom of the panel are placed piles of tape used in insulating joints and connections and the smaller products of its factory. Nearly all the larger manufacturers of dynamos, motors and insulated electric wires use some form of the P. & B. insulations. In the exhibit are shown samples of different kinds of wires in which the P. & B. compounds are used in the insulations. Small electric light wires, lead covered cables painted with the P. & B. compound, braided cables with the braid saturated with P. & B. compound are shown illustrating some of the uses of the P. & B. compound.

Packages of armature varnish for the protection of the armature and field coils and enamels for commutators which are extensively used by electric street railways are shown.

The Exhibit of the Chicago Rawhide Manufacturing Company.

To persons interested in belting, the transmission of power and the various industries in which leather is used, a visit to the exhibit of the Chicago Rawhide Manufacturing Company, of 75-77 E. Ohio St., will be found of great value. It is located in Machinery Hall, section 15, Column J, 28 and 29. The exhibit is composed entirely of the products from the extensive works of the company, treated by the rawhide process.

For many purposes rawhide leather has been found far superior to leather tanned in the ordinary way. It is said that the process is simpler, requires less time and leaves the leather more in its natural state. No lime or acids are used and when finished the leather retains the natural

THE EXHIBIT OF THE CHICAGO RAWHIDE MANUFACTURING COMPANY.

The P. & B. motor cloth for protecting the motors from dust, mud, and water thrown by the wheels, now being used to a large extent by street railways is also shown. There is also displayed samples of the P. & B. wooden underground conduit. The method of putting up these goods for the market is shown in the great variety of sizes and shapes of the packages exhibited, from the small half gill can that forms the apex of the pyramids to the barrels that occupy a corner of the space.

The large works of this company are located at Bound Brook, N. J., from which point the goods are distributed through its various agencies over the United States. The main office of the company is at No. 2 Liberty St., New York, and branch offices at 871 The Rookery, Chicago, and at 116 Battery St., San Francisco, Cal. Mr. P. H. Hover and Mr. Saml. Cochrane are in charge of the exhibit.

strength of the hide. It is especially adapted for belting, ropes, for rope transmission, lace leather, harness leather and many other purposes.

Arranged on tables about the space, as shown in the accompanying illustration, are displayed the different rawhide products manufactured by the company. On the table at the right are placed a number of sizes of rawhide rope from one-quarter of an inch in diameter to an inch and one half, also samples of twist belting from one thirty second of an inch in diameter to three eighths of an inch are shown.

Another interesting part of the exhibit is the display of rawhide packing. For packing, for pumping and hydraulic machinery, it is said to be unsurpassed, and for this purpose the strands are braided into either round or square forms and in size from one quarter inch to as large as is desired

For the past 12 years the company has been making rawhide pinions for various purposes, and with the adoption of electric traction on street railways, it was the first to manufacture rawhide pinions for electric railway motors. With their extended experience as pioneer manufacturers of rawhide goods they have ever since turned out only the best that can be produced. Sample pinions are shown, together with the blanks from which they are cut. The company also manufactures a trolley rope that is rapidly coming into use, as it fills the place better and is much more durable than the cotton rope.

Rolls of flat belting, both single and double, from three-quarters of an inch to 24 inches wide are exhibited, in the manufacture of which great pains are taken to make the best that the hide will produce. Special care is taken in the selection of the hides, and in the stretching, trimming and all other important points, the best of materials only being used. The company makes a specialty of dynamo belting, for which they have an extensive trade both in this country and in Europe, and which is rapidly growing. These belts for driving dynamos being a non-conductor of electricity, users do not experience the heavy static discharge usually found, a point which is recognized as of great importance.

Lace leather is shown in sides and cut in all widths. Immense quantities of this leather is used for this purpose, for which it is specially adapted, being thin and pliable, and yet very strong and durable. These qualities make it valuable for ropes and bridles, for which purpose it is used in the west by the cowboys. The company also furnish large quantities in sides and straps for harness. Samples of these straps and harness leather are also shown.

For a practical exhibit the company has a 3¼-inch belt and an 18-inch belt in daily operation, which connect the Willans engines to the line of shafting in the British section. They run very smoothly. A sample of the rawhide rope can be seen in operation in the neighboring exhibit of the Webster Manufacturing company. It is used to show a particular system of rope transmission for shafting and dynamos, and for any work where it is difficult to apply flat belting. Banners about the exhibit call attention to the company and its exhibit.

Schaffer & Budenberg, of New York, make a very attractive display of engine and boiler appliances on the main aisle of Machinery Hall, just east of the large water tank. The center of the space contains a large glass case in which is exhibited a full line of pressure gauges, ranging through all capacities and fitted for every class of service. A superior display of speed indicators, recording gauges, tachometers, and pyrometers is made. The Thompson indicator and Prof. Carpenter's calorimeter, instruments made only by this company, are also features to be noticed. Upon raised platforms at the sides of the space are shown samples of the "Acme" steam trap, Holtz's reducing valve, steam jet pumps and both exhaust and live steam injectors. Upon the rear walls, which are covered with dark felt, are hung photographs of the company's factory at Brooklyn, N. Y., and a large frame, containing an artistically arranged collection of medals, presented to the company for displays made at various international exhibitions since 1850. An ornamental column, supporting a combination of brewers' gauges, surmounted by a golden eagle marks the corner of the space. A handsome Brussels carpet and a fancy railing do much to complete a very tastefully arranged exhibit.

ELECTRICAL INDUSTRIES.

PUBLISHED EVERY THURSDAY BY THE

ELECTRICAL INDUSTRIES PUBLISHING COMPANY,

INCORPORATED 1892.

MONADNOCK BLOCK, CHICAGO.

TELEPHONE HARRISON 380.

E. L. POWERS, Pres. and Treas.   E. E. WOOD, Secretary.

E. L. POWERS,	Editor.
H. A. FOSTER,	
W. A. REMINGTON,	Associate Editors.
E. E. WOOD,	Eastern Manager.
FLOYD T. SHORT,	Advertising Department.

EASTERN OFFICE, WORLD BUILDING, NEW YORK.

World's Fair Headquarters, Y 27 Electricity Building.

SUBSCRIPTIONS.

FIVE MONTHS $1.00
SINGLE COPY   10

Advertising Rates Upon Application.

News items, notes or communications of interest to World's Fair visitors are earnestly desired for publication in these columns and will be heartily appreciated. We especially invite all visitors to call upon us or send address at once upon their arrival in city or at the grounds.
ELECTRICAL INDUSTRIES Co.,
Monadnock Block, Chicago

The present system of distributing the bands about the grounds of the Exposition and in the galleries of the buildings is a marked improvement. The popular pieces played seem to be better appreciated, and certainly draw as large if not larger crowds.

The present attendance at the Fair is nearly double that of six weeks ago. The railroads are being taxed to nearly their full capacity in the transportation of the crowds, and accommodations on the regular trains are being engaged a long time in advance. The special excursions on state days have brought large numbers that have greatly increased the attendance on those days.

The Jury of Awards has nearly completed its work. Dr. Rowland, Dr. Mendenhall, Dr. Duncan, Prof. Barker and Prof. Dolbear have already returned to their respective homes and duties; many of the other members are away for a few days. A resolution was passed by the Jury recently making five members a quorum for the transaction of business. Tests of lamps and dynamos are still in progress. A report is now being prepared for publication describing the progress made in different branches of the electric art. The awards to competing companies made by the Jury will be announced during the coming week before the Jury disbands.

We copy from the September number of a London contemporary the following paragraph which we are surprised to see appear after such descriptions and illustrations as have appeared in the various journals. We are sorry that the writer has not seen the electrical exhibits in the Exposition, and that he so foolishly accepted without question the opinion of the alleged expert. But the last paragraph would imply that our contemporary was narrow-minded and lacked the cosmopolitan spirit that has characterized the foreign representatives to the Fair.

"Further reports from the Columbian Exposition go to show that, from a manufacturer's stand-point, the electrical section is practically a failure. The spectacular effects with illuminated columns and screens have, of course, been very fine, but, we are informed, on the authority of an expert that there are very few exhibits of new electrical apparatus and that many of the types of machines exhibited are of patterns which would be considered obsolete here."

"When the scheme of the World's Fair was first mooted we suggested that the British manufacturer need expect to derive but little advantage from sending his wares to the Exhibition and in view of the high tariff charges the only result could have been that the American manufacturers would have had a good opportunity of availing themselves of English research and invention as exemplified by the types of apparatus exhibited."

WORLD'S FAIR NOTES.

The operating force of the Electricity Building, consisting of nearly fifty employes, had a group photograph taken last Saturday.

The Electricity Building now boasts of having the only Irish flag on the grounds, outside of Midway. It is one of the electric flags described in our last issue, and is being in honor of the proposed visit of the Lord Mayor of Dublin.

After having given away over 8,000 souvenir horseshoes, the Electrical Forging Company has secured a concession and now sells them at 25 cents each. The number of spectators continually watching the forging process demonstrates the popularity of "working" exhibits.

The west restaurant of the Wellington Catering Company in Electricity Building, which has been closed for some weeks past owing to insufficient patronage, was reopened on Tuesday. This is certainly an indication of the popularity of the electrical displays.

Mr. Tesla is expected to return to Chicago the later part of this week and will then conduct in person some of the high frequency experiments which have made him so famous. The Westinghouse dark room will be used for the displays, but no definite time has yet been fixed.

The Hertz reflectors, which have been promised some time, are now shown in operation in the exhibit of Queen & Co. The reflectors are parabolic in form and when an oscillatory discharge of a high potential current takes place at the focus of one, the effect is reproduced at the focus of the other.

The Westinghouse company has placed a glass floor in one of the cars exhibited in the railway department so that the operation of the motor can be easily seen from above. The Columbus egg, which has proven such an attraction to the Westinghouse space, is now shown the first fifteen minutes of every hour from 5 until 10 p. m.

Visitors to the Electricity Building have found a new attraction in the Hoggson time stamp which has recently been placed in the exhibit of the E. S. Greeley Co. The stamp shows the date and time of day and the machine is electrically connected with a clock so as to change the time indicated every minute. On account of its many commercial uses the device is attracting considerable attention.

A new exhibit in the Electricity Building is the Meyer's Ballot Machine, which has a space near the exhibit of English telegraphs. The machine consists of a metal booth with interlocking entrance and exit doors. On one side of the booth are placed the various party ballots in perpendicular rows and opposite each name a plug, which, when pushed in, registers the vote on an automatic counter and at the same time locks the plugs on the other tickets,

thus preventing a voter from voting for two men for the same office or voting two ballots. When the polls are closed the back of the machine is removed and the vote can be immediately taken from the different registers and announced. No electricity is used except for lighting purposes.

The University of Illinois make a very creditable display from the department of electrical engineering on the main floor of the state building. A direct current dynamo, mounted on a cradle dynamometer and driven by a motor, furnishes current to several arc lamps and for the lighting of a number of incandescent lamps, arranged to form the college monogram placed above the exhibit. A fully equipped switchboard, a water rheostat, a number of sets of apparatus for typical experiments in electrical measurements and various styles of storage batteries, some of which were made by students, are also exhibited. A very good idea of the work done at the school may be gained by examining the photographs of students engaged at work in the rooms of the electrical laboratory.

In the southeastern part of the gallery of Electricity Building, The Consolidated Electric Storage Company has an exhibit of storage batteries. It also gives a practical demonstration of the methods of wiring for them when used in connection with an electric lighting plant. A new battery is shown designated as 15-1 which is designed especially for electric lighting where heavy currents for an uninterrupted period of several hours are required. This battery has a capacity of 350 ampere hours. The great economy of storage batteries in isolated plants has brought many of them into use. In isolated plants, especially in plants for lighting summer residences the storage battery has been found a valuable adjunct. It relieves the machines of excessive loads and furnishes the necessary lights during the hours when but few lights are required.

The scenic theater of the Western Electric Co. is playing to crowded houses and hundreds are turned away disappointed at not obtaining admission. Several new features have been recently introduced. In the day time the scene is now enlivened by a military procession consisting of a band and several regiments of militia and cavalry. During the rain storm, peasants carrying umbrellas pass across the bridge, while toward dusk a load of hay may be seen hurrying toward the the castle. The booth which was formerly occupied by the theater has been remodeled and is now handsomely fitted up as an office. Separating the two rooms is a screen, consisting of a large stained glass window illuminated by incandescent lamps. The subject of the picture is "The Fairy Queen" and the window is said to be the finest stained glass at the Fair. Upon the walls of the rooms, transparencies of some of the large buildings of Chicago, which are fitted with the Western Electric Company's system of incandescent lighting are shown. In the lighting of the booth some new ideas have been introduced. The walls are delicately tinted and have a figured border, covered with a fine gauze, concealing the incandescent lamps entirely from view, but allowing the light to be evenly disseminated through the rooms, thus producing a softened effect. Two miniature illuminated fountains, changing their colors constantly, add to the general attractiveness. These fountains are a novelty being introduced by the Western Electric Co. and are sure to become popular as ornaments for parlors, hotel offices, etc. A model of a new alarm system, which will be of general interest, is now being installed by the company and other additions to the exhibit are promised

PERSONAL.

Mr. R. T. McDonald, of Ft. Wayne, is among the arrivals this week.

Mr. M. C. Canfield, of the Brush Electric Co., Cleveland, O., is at the Fair.

Mr. D. S. Snyder, of Point Pleasant, W. Va., visited the Fair recently.

Mr. T. F. Harris, electrician, Newark, N. J., is visiting the Fair this week.

Mr. H. B. Church, of the electrical department, is taking a ten days' vacation.

Mr. David M. Keith and family, of Denver, Colo., registered at the Everette recently.

Mr. Matt. M. Merritt, of the Boston Inc. Lamp Co., paid the Fair a flying visit last week.

Mr. F. B. Wells, representing the Western Electric Co. at Antwerp, Belgium, is visiting the Fair.

Mr. S. Ekstrom, Supt. of the alternating department of the General Electric Co., is now in the city.

Mr. J. G. Biddle, chief of the electrical department of Queen & Co., spent several days at the Exposition last week.

Mr. Jas. P. Provost, with the E. D. Nuttall Company, Allegheny Pa., is spending a few days at Jackson Park.

Judge Taylor, of the Ft Wayne Co. has been an interested visitor in the Electricity Building during the past week.

Mr. R. S. Williams, of the mining and power department of the Thomson-Houston factory, is spending some time at the Fair.

Mr. Otto Lemisch, electrical engineer, Klagenfurt, Austria, called at the World's Fair office of ELECTRICAL INDUSTRIES Tuesday.

Mr. F. J. Smith, manager of the electrical department of E. S. Greeley & Co., is now at the Fair to attend the operator's convention.

Mr. B. B. Ward, the inventor of the Ward Arc Lamp, accompanied by his wife, has been viewing the sights at the Exposition the past week.

Mr. F. F. Kinney supt. of the Rankin Electric Light & Power Company of Tarkio, Mo., has been a visitor at the Exposition during the past week.

Mr. J. H. Fedeler, who has been with the operating department since the opening of the Exposition, leaves this week to take up studies at Harvard.

Mr. A. W. Cook, secretary of the Susquehanna Electric Light, Heat & Power Company, Susquehanna, Pa., paid his respects to ELECTRICAL INDUSTRIES the fore part of this week.

Mr. Geo. W. Brown, mgr. of the Belknap Motor Company, returned to his home in Portland, Me., on Wednesday of this week. Mr. Winters is now in charge of the company's exhibit.

Mr. W. S. Townsend, electrician for the North End Street Railway Company, Worcester, Mass., is visiting the Fair this week and called at the office of ELECTRICAL INDUSTRIES.

Mr. Ralph W. Pope, secretary of the American Institute of Electrical Engineers, left on the 10th for New York to complete arrangements for the meeting of the Institute on the 20th. He will return about the 25th.

A Correction.

In our last issue appeared a statement that Mr. Wm. H. McKinlock had associated himself with the Enterprise Electric Company. We have since been informed that the statement is incorrect. Mr. McKinlock is located at offices 319 and 320 Manhattan building, where he will be pleased to see his old friends.

BUSINESS NOTES.

The Electric Appliance Company is showing the trade a new iron box bell that has a number of small improvements over the old form of bell in an automatic set screw on the adjustment; an armature spring that is attached to the soft iron armature without rivets; and a frame made entirely of soft stamped iron. They state that the bell is a winner. It is known as the "Acme" iron box bell.

Chief Barrett, of the department of Electricity at the World's Fair, has found the operation of the Reliance Columns on the World's Fair boilers so satisfactory that he has decided to equip the various boilers in the electric and street railway plants in his charge with Reliance columns manufactured by the Reliance Gauge Co., Cleveland, Ohio, as rapidly as possible, believing that these appliances are sources of economy, as well as a protection to life and property. He has already equipped the boilers in the plant of the Cicero & Proviso Street Railway Co. with them.

DEPARTMENT OF ELECTRICITY.

Offices: SECTION H, ELECTRICITY BUILDING.

Chief, JOHN P. BARRETT,
Assistant Chief, J. ALLEN HORNSBY.
General Superintendent, J. W. BLAISDELL.
Assistant, WILLIS HAWLEY.
Electrical Engineer, W. W. PRINN.

AMERICAN INSTITUTE OF ELECTRICAL ENGINEERS.

World's Fair Headquarters
SECTION S. ELECTRICITY BUILDING.

RALPH W. POPE, Secretary.

Open from 9 a.m. to 5 p.m.

Amusements.

HOOLEY'S THEATRE—Mr. E. S. Willard, Mon., Tues., Wed., in "The Professor's Love Story." Thurs. in "Judah." Fri. and Sat. in "Wealth." 149 Randolph street.

COLUMBIA THEATER—Daniel Frohman's Lyceum Theater Co. in "The Charity Ball." 108 Monroe street.

GRAND OPERA HOUSE—Sol Smith Russell, in "A Poor Relation." 87 Clark street.

AUDITORIUM—Imre Kiralfy's Spectacle "America." Congress street and Wabash avenue.

McVICKER'S THEATER—Denman Thompson, in "The Old Homestead." 82 Madison street.

CHICAGO OPERA HOUSE—American Extravaganza Company, in "Ali Baba, or Morgiana and the Forty Thieves." Washington and Clark streets.

SCHILLER THEATER—Chas. Frohman's Stock Company, in "The Girl I Left Behind Me." Randolph, near Dearborn.

HAVERLY'S CASINO—Haverly's United Minstrels. Wabash avenue, near Jackson street.

TROCADERO—Vaudeville. Michigan avenue near Monroe street.

THE GROTTO—Vaudeville. Michigan avenue near Monroe street.

BUFFALO BILL'S "Wild West." 63d street. Daily at 3 and 8.30 p.m.

Everyone who comes to visit Chicago comes with a desire to see "America" at the Auditorium. Though now in the fifth month of its season its drawing powers are unimpaired. The attractive force of the splendid spectacle is as irresistible as a maelstrom.

Trocadero is having crowded houses. To lovers of vaudeville the Trocadero offers a special list of good attractions. This week Mrs. Alice Shaw, the wondrous whistler, and Jules Levy, the cornetist, both of whom are well known, are delighting the guests of the house. Sandow is introducing new evidences of his strength which are surprising and wonderful.

The crowds at Buffalo Bill's Wild West have been increasing. The new scenes, especially Custer's Last Ride, has awakened new interest in these performances. Visitors to the Fair consider this show as one of the most interesting sights of the World's season, and set aside an afternoon or evening for seeing it.

"Ali Baba" began last Sunday night the sixteenth week of its World's Fair run, but this engagement cannot be regarded as unusual or due entirely to the presence of the great Exposition, for last summer the same spectacle played an engagement of twenty-three weeks to business that was every bit as large as it is doing now. With the exception of a new song, recently introduced by Eddie Foy, entitled "They All Take After Me," which has made a big hit, there is nothing new in "Ali Baba." The bright and sparkling character of the entertainment, however, the abundance of good music the attractive scenes, the rich costumes and the general blithe and merry nature of the piece serve to keep it ever fresh and make it interesting, no matter how many times it is seen.

ELECTRICAL INDUSTRIES

OUR IMPROVED SYSTEM

...OF...

Automatic Fire Alarm,

covered by patents recently issued, is the
embodiment of all factors contributing to the

GREATEST SAFETY,

and the MOST RELIABLE

PROTECTION FROM FIRE.

Western Electric Company,

CHICAGO and NEW YORK.

ELECTRICITY BUILDING—EXHIBITORS AND THEIR LOCATION.

GALLERY.

MAIN FLOOR.

WE ARE STILL HARD AT WORK ON

OUR NEW CATALOGUE

which we hope to present to our friends in a very few weeks, in the meantime do not fail to make good use of our last edition which we trust you always have convenient for reference.

If you are without a copy send us

YOUR OPEN ORDERS

and we will guarantee satisfactory prices and first-class material, and that your order will have

INTELLIGENT-EXECUTION.

ELECTRIC APPLIANCE COMPANY,

Electrical Supplies.   242 Madison Street, CHICAGO.

THE MONTHLY ISSUE FOR SEPTEMBER

ELECTRICAL INDUSTRIES

Should be read by everyone interested in electrical matters. In its table of contents is the following:
"World's Congress of Electricians."
"Nikola Tesla's Lecture."
"Love Underground Electric System in Washington." By M. D. Law.
"Gas Engines as Applied to Electric Light Work." By Geo. A. Farwell.
"The Underground System at the World's Fair."
"Electric Railway Plant of the Chicago North Shore Railway Company."
Together with illustrations of the recent applications of electricity.
The paper also contains regularly
A Buyer's Directory of Manufacturers and Dealers in Electrical Supplies and Appliances.
A Complete Directory of Electric Light Stations in North America and a Complete Directory of Electric Railways in North America.
These directories are revised each issue to the date of going to press and are to be found in no other electrical journal in the World. Its articles are read carefully and its directories used constantly by all the buyers in the trade. These facts make it without a superior as an advertising medium. Sample copies and rates sent on application.
Subscription price $3 per year. Six months trial $1, if ordered during September.

ELECTRICAL INDUSTRIES PUB. CO.,
Monadnock Block, CHICAGO.

WM. BARAGWANATH & SON,

LIST OF HEATERS

TO BE SEEN IN OPERATION AT THE WORLD'S FAIR.

Two 500 H. P. East End Boiler Galery doing 1800 H. P. work.
One 300 H. P. heater and receiving tank, Wellington Catering Co's., plant.
One 150 H. P. heater at Hygeia plant.
One 200 H. P. Libby Glass Works.

55 WEST DIVISION STREET,
CHICAGO.

THE
FERRIS WHEEL

When you visit the World's Fair, you will naturally take a ride on the FERRIS WHEEL and be interested in the ELECTRIC LIGHT INSTALLATION, which is wired throughout with

OKONITE WIRE

FURNISHED BY THE

CENTRAL ELECTRIC COMPANY,
116-118 Franklin Street,
CHICAGO, ILLS.

CLARK ELECTRIC COMPANY, NEW YORK.

192 Broadway and 11 John Street.

MANUFACTURERS OF ARC LIGHTING APPARATUS FOR EVERY PURPOSE A SPECIALTY.
The CLARK ARC LAMPS for use on EVERY CURRENT, have the reputation of being the best and most durable of any ever made in the United States.

RAWHIDE PINIONS FOR ELECTRIC MOTORS
A SPECIALTY.
RAWHIDE DYNAMO BELTING

THE CHICAGO RAWHIDE MANUFACTURING CO.
THE ONLY MANUFACTURERS IN THE COUNTRY

LACE LEATHER ROPE
AND OTHER RAWHIDE
GOODS
OF ALL KINDS
BY KRUEGER'S PATENT

75 Ohio Street,   CHICAGO, ILL.

STANDARD ELECTRIC COMPANY.

GENERAL OFFICES: 625 Home Insurance Building.

WORKS: So. Canal Street,

CHICAGO.

STANDARD SYSTEM

AT THE

WORLD'S FAIR.

MACHINERY HALL, Sec. Q, 2 Standard Arc Dynamos.

Sec. S, 20   "   "   "

ELECTRICITY BUILDING, Sec. P, Space 2, Arc Lighting Exhibit.

The Standard Lamps Light the Power Plant, Machinery Hall, Agricultural Hall, Shoe and Leather Building, and Other Buildings and Portions of the Grounds.

See our Double Service All Night Lamp Before Buying an Old Style Two Fed Lamp

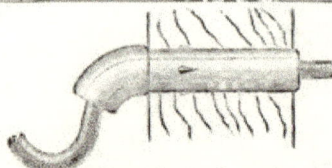

Mile after mile of

SIMPLEX WIRE

Supplied to the

FERRIS WHEEL

...

By...George Cutter,

The Rookery, CHICAGO.

SIMPLEX WIRES

SIMPLEX

Ever Onward and Upward!

**INSURE
HICH
INSULATION**

Simplex Electrical Co.

620 Atlantic Ave.,

George Cutter, Chicago.   BOSTON, MASS.

XNTRIC

"That's the Switch"

And we control that movement.

H. T. PAISTE,

10 South 15th St.,

**PHILADELPHIA,
PA.**

Made 5 amp. S. P.
10 amp. S. P,
5 amp. 3 way.
10 amp. 3 way.

China Window Tube (Patented).

Made only by **PASS & SEYMOUR**,

George Cutter,
CHICAGO.

SYRACUSE, N. Y.

Enterprise
Electric
Company

N. I. R.

GENERAL WESTERN AGENTS

307 Dearborn Street.
Chicago....

Manufacturers' Agents and Mill Representatives for

Electric Railway,
Telegraph, Telephone and
Electric Light

SUPPLIES OF EVERY DESCRIPTION

Agents for   Cedar Poles,
Cypress Poles, Oak Pins,
Locust Pins, Cross Arms, Glass
Feeder Wire,   Insulators,

WIRES, CABLES, TAPE and TUBING

BEAR IN MIND

that the regular monthly issue of ELECTRICAL INDUSTRIES contains the most complete and correct directories published of the electric light central stations and the electric railways in North America.

World's Fair Headquarters Y 27 Electricity Building.

CITY OFFICES, Monadnock Block.

Consolidated Electric Co.

Manufacturers and Dealers in all kinds of

ELECTRICAL . SUPPLIES,

115 Franklin Street,

CHICAGO.

WAGNER ELECTRIC FAN MOTORS

For Direct or Alternating Currents.

These motors give a cooling not breeze with less consumption of current than any other fan motor on the market. They are built 1-8 horse power. Six bladed 12-inch fan. So cooling. Furnished with or without guards.

IT WILL PAY YOU TO SEE THE WAGNER BEFORE BUYING ELSEWHERE

TAYLOR, COODHUE & AMES,

348 Dearborn Street, CHICACO.

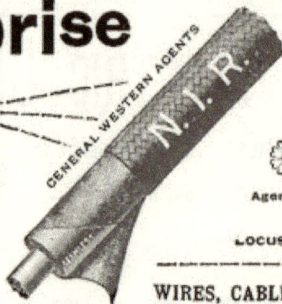

WEEKLY WORLD'S FAIR

ELECTRICAL INDUSTRIES

DEVOTED TO THE ELECTRICAL AND ALLIED INTERESTS OF THE WORLD'S FAIR, ITS VISITORS AND EXHIBITORS.

Vol. I, No. 15.   **CHICAGO, SEPTEMBER 21, 1893.**   FIVE MONTHS $1.00
TEN CENTS A COPY

Exhibit of the Jenney Electric Motor Company.

An exhibit that impresses one with its appearance of comfort and cheerfulness, as well as by the attractiveness of the display, is that of the Jenney Electric Motor Company, of Indianapolis, Ind. Located as it is just at the left of the south entrance to Electricity Building, few visitors pass through the building without noticing it.

The office erected by the company is neatly designed, and, finished without an imitation of hammered copper, it forms an interesting part of the exhibit. The outward appearance is more than equalled by its interior, which has an elaborate wainscoting, a handsome ceiling decorated by raised figures, finished in white and lighted by incandescent lamps. Fans [...] tors keep it cool even on the warmest days. Desk, chairs and an abundance of writing material offer accommodations to the guests of the company.

Above the entrance sparkles in different colors the name Jenney, an effect produced by incandescent lamps shining through the letters formed of cut glass. At each side are the dates 1878 and 1893, indicating the period of the company's successful existence. The floor of the exhibit is an imitation mosaic, and is further ornamented by numerous rugs. The exhibit proper consists of electric light and power generators, direct current transformers, automatic motors and starters, and the switch board which is fitted with the latest improved apparatus for the control of the current and the regulation of the different machines.

The different machines are shown in practical operation, the necessary current being supplied from a 60, 500 volt compound wound generator placed in Machinery Hall. The current from this generator operates a 45 horse power motor directly connected to a 30 kilowatt 110 volt dynamo and a 25 horse power motor operating a 20 kilowatt 110 volt dynamo. These machines are wired to the switch board, from which point the current is distributed as de-

EXHIBIT OF THE JENNEY ELECTRIC MOTOR COMPANY.

sired. Current for the lights and motors in the exhibit, five horse power to a neighboring exhibit, and current for the beacon light on top of Electricity Building are furnished from this switchboard.

Among the dynamos exhibited are a 40 k. w. 500 volt generator, a 30, 20 and 3 k. w. 110 volt dynamos, and a 300 ampere five volt plating machine. The motors shown are of 35 and 25 horse power 500 volt motors and 12, 6, 1½ and ½ horse power 110 volt motors. These machines show excellent mechanical skill in their construction and the smoothness with which they run, and their efficiency shows how well they are constructed electrically. The company has given special attention to the construction of directly connected motors and generators and the machines exhibited are the result of extended experiments. The six horse-power motor runs a 24 ampere dynamo and to the shaft of the half horse power motor is attached a flexible shaft fitted with a buffer, drill or similar tool to show its advantages in machine shop work.

The switchboard forms an attractive part of the exhibit and among the more prominent objects on it are the automatic motor starters. They are ingeniously constructed and are a valuable protection against a sudden rush of current. Attached to the face of the box is an electro magnet, the coils of which are in series with the fields of the motor. Until the main line switch is closed, or the magnet is magnetized, it is impossible to cut out the resistance or to start the motor, as the armature of the magnet is dropped in front of the resistance arm. The moment the switch is closed the armature is drawn up, which then allows the resistance to be cut out and the motor started. After the

field controllers are operated from the front of the board by the small hand wheels, the rheostats being back of the board.

The fact that there is always some one at the exhibit is appreciated by visitors, who often wish to know something

EXHIBIT OF THE LECLANCHE BATTERY COMPANY.

more of the exhibits than can be learned from a glance at the array of machines. Mr. Farnsworth, who has charge of the exhibit, has made it a point either to be at the exhibit

EXHIBIT OF THE JENNEY ELECTRIC MOTOR COMPANY.

resistance is out of the motor running at full speed the arm is held in place by a small lever. In case the current is suddenly or accidentally broken the armature of the magnet drops as the motor slows down allowing the arm to return to its starting position, thus protecting the motor when the current is again turned on.

The switchboard, which is of white marble, is supplied with the necessary switches, ammeters, volt meters, etc., for the regulation of the current. There are two sets of bus bars, one for each dynamo. The light circuits are run from the smaller bus bars and are cut in and out by the small switches or are all thrown out by the main switch. The

or to have an assistant present during the time that the building is open. This method has undoubtedly been beneficial to the interests of his company.

In strong contrast with the large drills used for mining operations is the small dental drill added, this week, to the exhibit of the General Electric Company. The drill is operated by a small motor, the speed of which is easily governed by a switch controlled by the foot, but it is not claimed that this improvement over the foot power formerly used, will make any the less interesting the drilling process preceding the filling of a tooth.

Exhibit of the Leclanche Battery Co.

Near the center of the east gallery of Electricity Building is located the exhibit of the Leclanche Battery Company of New York. Although the goods exhibited consist entirely of batteries, the method of arrangement, the design of the arch that spans the entrance, and the appropriateness of the details, as well as the taste displayed in the colors of the decorations elicit the admiration of the visitors.

At the top of the arch miniature incandescent lamps spell out the word Gonda, the well known trade mark of these batteries adopted some 20 years ago by this company. Prominent on the face of the arch appears the name of the company, the names of the different cells manufactured by the company and near the bottom of the columns the words, strength, endurance, qualities that these batteries are said to possess in a marked degree.

Forming three sides of a square of which the arch forms the front, are the glass cases that contain the exhibit of

is nearly nine inches high and four and one half inches square, and differs greatly in construction from the other cells. The negative electrode consists of a carbon having six vertical wings, over which is stretched a bag so as to form pockets between the wings for the depolarizing compounds. From its upper end a carbon rod projects through the cover of the jar, by which it is suspended. This cover screws down on the neck of the jar, and the cell is thus rendered water tight. The zinc of the cell is circular in form, and nearly surrounds the carbon.

One of the noticeable objects about the exhibit is the shield in white and gold bearing this inscription: "These are the batteries that wind the clocks that furnish the time for the great Exposition."

Exhibit of the Chicago Electric Wire Company.

In the east gallery of Electricity Building, the Chicago Electric Wire Company has a very attractive exhibit as well as very complete in representing the line of goods manu

EXHIBIT OF THE CHICAGO ELECTRIC WIRE COMPANY.

batteries. The glass sides of the cases enable the batteries to be seen from both within and without the space. The batteries shown are designed for open circuit work, and in all the improved battery connection is used. The washers used are made of a non oxidizing metal that is always free from rust and corrosion, and no local action can set up between the connection and the carbon. Among the batteries shown is the Gonda porous cup cell, which is the original form of the Leclanche cell. It is about seven inches high and four and one half inches square. The Aso cell shown is an improved form of battery in which a porous cup is used. The special form of cup used in this battery has a flange that fits over the top of the glass jar, making a closed cell. Thus all dust is excluded and evaporation is prevented. This cell is especially adapted for physicians' use.

Among the other styles of batteries exhibited are the Gonda cell, well known to the trade, and the cylinder cell that possesses a large amount of surface and low internal resistance. The Vok cell is the latest battery introduced by this company. It is constructed for work where a quantity of current and continuous service is required. It

factured by this company. The exhibit has been arranged after a very pretty and artistic design.

In the center of the space is the booth that serves as an office of the company, reels of wire of various sizes making the columns for the support of the canopy which forms the roof of the booth. The canopy is constructed of insulating tubing resembling bamboo, from the edge of which is hung in rope like fashion a beaded curtain of different colors.

On the peak of this booth is placed a circular sign bearing the name of the company. Within the booth are desk and chairs while the colored curtains enclose the space making it a comfortable retreat. Around the space are heavy solid oak tables on which are displayed the smaller samples of wire and insulations. On the front of the tables carved in heavy black letters appear again the well known name of the company.

Between the posts that mark the corners of the exhibit are suspended heavy armored cables of the style manufactured for submarine use. In the exhibit are shown insulated wires and cables of almost every kind used in the various branches of electrical industry, insulated according to the processes used by the company. This exhibit was one of the first arranged in Electricity Building, and it early attracted the attention of visitors. Numerous additions have been made to the exhibit from time to time to freshen its appearance.

Mr. A. A. Cobb, representative of the company is in charge of the exhibit, to whom very much credit is due for the excellence of the display.

Wonder has been caused in the past, by the seven great search lights which nightly pierce the darkness with their powerful light, but Fair visitors of the future should prepare themselves for a greater surprise. Arrangements are being completed for the placing of an immense projector on the roof of Manufactures and Liberal Arts Building with which it is intended to throw advertisements on the clouds. The various attractions along Midway are advised to make early applications for space.

The gondolier who follows the business of his forefathers on the grand canals of Venice is liable to be driven to some other form of employment. The practical test of the electric launch on the lagoons of the Exposition grounds has shown their many advantages. According to the Engineering News, from which we take the following paragraph they have made a favorable impression on the commissioners from sunny Italy. "Electric launches may supersede gondolas on the canals of Venice, as one result of the Columbian Exposition. At least one of the Chicago launches has been sent to Venice by a company which is said to include several members of the Royal Italian Commission. The Electric Launch & Navigation Co. owning the boats at the Exposition, say this company has an option until Oct. 15 on 30 of the 50 launches of the company. Fast steam launches of a considerable size now run regularly on the Grand Canal in Venice and are well patronized, as the fare is small and the speed great compared with the gondolas. The smoke is said to be objected to, and these larger boats cannot traverse the crooked smaller canals with their low and frequent bridges. It is expected that the electric launches will find favor owing to their small size, lowness and noiseless operation, without smoke."

The Action Gesellschaft fur Chemische Industrie.

The extensive use of various chemicals in the different branches of the electrical industry, especially sal ammoniac or muriate of ammonia and blue vitrol, or sulphate of copper, makes the exhibit of the above company of interest to the electrical trade. This exhibit is located in the chemical division of the German section in Manufactures and Liberal Arts Building.

The exhibit has been arranged with the aim of not only showing the varied products of the extensive works of the company, but also of presenting an arrangement that will be pleasing to the eye. Paintings form a handsome background to the piles of cylinders that contain the exhibits. The painting on the left shows the works of the company at Reinau, while the one on the right shows the works at Rauen.

The company is one of the largest exporters to the United States, and it is said that nearly three-quarters of the imports of this character can be traced to the Mannheim company. The exhibit at the Fair was installed under the direction of Mr. C. F. Holland, one of the managers of the company.

ELECTRICAL INDUSTRIES.

PUBLISHED EVERY THURSDAY BY THE

ELECTRICAL INDUSTRIES PUBLISHING COMPANY,
INCORPORATED 1893.

MONADNOCK BLOCK, CHICAGO.
TELEPHONE HARRISON 395.

E. L. POWERS, PRES. AND TREAS.   E. E. WOOD, SECRETARY.

E. L. POWERS . . . . EDITOR.
B. A. FOSTER } . . . . ASSOCIATE EDITORS.
W. A. REMINGTON }
E. E. WOOD . . . . EASTERN MANAGER.
FLOYD T. SHORT . . . . ADVERTISING DEPARTMENT.

EASTERN OFFICE, WORLD BUILDING, NEW YORK.
World's Fair Headquarters, Y 27 Electricity Building.

SUBSCRIPTION.

FIVE MONTHS, ... $1.00
SINGLE COPY, ... 10
Advertising Rates Upon Application.

News items, notes of communications of interest to World's Fair Visitors are earnestly desired for publication in these columns and will be heartily appreciated. We especially invite all visitors to call upon us or send address at once upon their arrival in city or at the grounds.
ELECTRICAL INDUSTRIES PUBLISHING CO.,
Monadnock Block, Chicago.

Owing to the report of the Jury of Awards not being complete in certain details the report of the jury will not be made public until after another meeting of the jury, which will be called in the near future.

Through the efforts of the Department of Electricity a beacon light has been placed on Electricity Building, at the south end. It is of the type used in the French navy and was placed in position by Barbier et Cie. It has three lenses, red, white and green, which revolve in the frame about the light. A small motor placed just below the light turns the lenses, the shaft of the motor being connected to the frame-shaft through a number of gears. A 64-candle-power incandescent lamp, which will soon be replaced by a 150-candle power gives the light. The wiring for the light was done by the Exposition and the current is furnished by The Jenney Electric Motor Company from its exhibit at the south end main floor of Electricity Building.

The committees appointed by the Mayor of Chicago and President Higinbotham, of the Exposition, have adopted a plan for the distribution of the funds secured by popular contributions for the relief of the injured and those dependent on the killed in the Cold Storage fire at the Exposition grounds. It has required a considerable time to get at the facts in each case and to plan for an equitable distribution of the funds. In general the plan proposes that a certain sum be paid the adults in cash; that a certain sum be set aside for the widows and all the money set aside for the children be placed with one or more trust companies to hold in trust, until they become of age and in case of widows until death, or perhaps re-marriage, when it is to revert to the children. The funds will allow each widow $2,000 in cash and the interest on a second $2,000, and the children receive what will be equivalent to $2,000 on their coming of age.

WORLD'S FAIR NOTES.

The large Columbus sign at the south end of Electricity Building has been greatly changed by the new colossal lamps recently put in. The work of changing the lamps was watched with interest.

Contracts have been let for the two principal buildings of the California Mid-winter Exposition, the Manufactures and Liberal Arts and Mechanical Arts Buildings, and work will begin immediately. The contract price of the two buildings is said to be $172,000.

The first meeting of the present season of the American Institute of Electrical Engineers was held last evening at 12 West 31st street, New York. President Houston delivered his inaugural address, taking for a subject the Chicago International Electrical Congress.

It is now the custom to turn off the lights on the cornice of Electricity Building, Administration Building and Machinery Hall, on the side of the buildings toward the electric fountains while the fountains are playing. By this arrangement the lights and colors of the fountains appear to better advantage.

Chief Allison of the Manufactures and Liberal Arts Building, recently secured from Mr. Geo. B. Clark, of the executive committee of the Gallery Exhibitors Club, an outline of the plan followed for increasing the attendance in the gallery of the Electricity Building, and expects to adopt similar measures for his department.

The good natured rivalry which has always existed between the employes of the electrical engineering department and those of the mechanical department has resulted in arrangements for a foot ball game to be played in the stock pavilion in the near future. The elevens have been chosen, and, judging from the amount of practice being indulged in, a hard struggle may be expected.

"The oldest telegram in the world" is the title of an exhibit in the space occupied by the University of the City of New York in the gallery of the Manufactures and Liberal Arts Building. The telegram was sent and recorded by Prof. Morse in the chapel of the University on Jan. 24th, 1838. The tape containing the original record in the Morse alphabet and the translation in Prof. Morse's handwriting is shown in a glass case. The telegram, which was dictated by Prof. Cummings, reads: "Attention—The Universe by kingdoms—Right wheel."

The improvised tug of war between the new General Electric Company's electric locomotive and the B. & O. switch engine on last Saturday afternoon did not result in a great victory for the electric locomotive. The event had been well advertised and a large crowd watched the affair with interest. The wire fence along the terminal sheds kept the crowd back while the more eager climbed upon the exhibition cars stationed on the neighboring tracks much to the detriment of their summer clothes and the elegant finish of the cars.

The illumination of the Texas state building on the evening of the day celebrated by the Lone Star State was accomplished by using the portable electric lighting plant which has been exhibited by the Danube Motor Co. in the south end of Transportation Building. The apparatus consists of a twelve horse power oil engine and a ten arc light dynamo, mounted compactly on a carriage, so that they can be quickly moved and easily handled. While intended primarily for use in military operations, the apparatus is adapted for any service where the rapid installation of a temporary plant is called for. Chief Pierce, of the Electrical Engineering Department, speaks in high terms of the accommodation afforded by the plant, and hopes to make use of it on future occasions.

The crowds of visitors these days come to see something and those places where are to be found the most startling displays are the places where the largest crowds assemble. By simply fusing small wires connected across the terminals of a small plating dynamo the Jenney Electric Motor Company attracted crowds of visitors during the evenings of last week. It is the live exhibits that attract the visitors. The General company has a small water wheel operated by a small motor which stirs up the water in a water tight box located near the center of the building. The water is so agitated that the little propellor is concealed from view and it is a mystery to many passing by what is creating the disturbance.

A good display of small motors may be found by examining the exhibit of the Western Electric Co. Various designs adapted to special service are shown. Near the north entrance to the Egyptian temple four motors of the multi-polar type are shown ranging from two to fifteen horse power. These motors are especially designed for low speed and high efficiency and show most excellent results. There are also two forty-horse power motors directly connected to the two generators which supply the current for the scenic theater and the column of light. The loads on these motors are constantly changing, but the shunt and compound windings are so arranged to oppose each other and make the motors perfectly self regulating.

Street Railway Association of the State of New York.

The eleventh annual meeting of the Street Railway Association of the State of New York is held this week at Rochester, N. Y. An interesting program has been prepared, which includes a number of papers by prominent railway men. Mr. Geo. W. McNulty, engineer of the Broadway & Seventh Av. railroad, presents a paper on "The Recent improvements in Cable Traction," and Mr. Thos. J. McTighe, electrical engineer of the Atlantic Ave. Ry., Brooklyn, reads a paper on "The Return Circuit of Electric Railways." A large attendance is expected and excellent accommodations have been secured at Powers' Hotel.

The popular route to Milwaukee is the Chicago Milwaukee and St. Paul Railway. Its double track between Chicago and Milwaukee has been completed, and with its block signal system, absolutely preventing accidents from collisions, trains may be run at the highest rate of speed obtainable. The equipment of the line is perfect.

Trains leave the Union Passenger Station, Canal and Adams streets, as per following schedule:

8:30 a. m. daily.
11:30 a. m. daily, except Sunday.
1:00 p. m. daily, except Sunday.
5:00 p. m. daily.
6:00 p. m. daily.
8:00 p. m. daily.
10:30 p. m. daily.

City ticket office, No. 207 Clark street.

PERSONAL.

Mr. W. E. Jackson, Jr., of Augusta, Ga., was among the arrivals during the past week.

Mr. A. Newman returned the forepart of the week from a somewhat extended trip east.

Mr. Wilder, superintendent of the electric plant at Kalamazoo, was a visitor at the Fair last week.

Mr. F. L. Southack, the well known electrical engineer of Boston, Mass., is now at the Fair.

Mr. A. J. Wright, president of the Reliance Gauge Co., of Cleveland, Ohio, is visiting the Fair with his family.

Mr. C. A. Paul, superintendent of the Bridgeport, Conn. Electric Light Co., spent some time at the Fair recently.

Mr. Jerome Penn, of the Electric Light & Power Co., of Washington Court House, Ohio, is in the city attending the Fair.

Mr. Jas. E. Cole, electrical inspector from Boston, Mass., was among the visitors to the electrical features of the Fair this week.

Mr. D. L. Davis, superintendent of the Electric Light & Power Co., of Salem, Ohio, was among the visitors of Electricity Building last week.

Mr. Morgan Brooks, general manager of the Electrical Engineering and Supply Co., of St. Paul, Minn., is in the city visiting the Exposition.

Mr. Herbert M. Price, who is interested in the installation of a power plant at Mt. Morency Falls, seven miles from Quebec, paid the Fair a short visit last week.

Prof. W. J. Herdman, a member of the Jury of Awards, Department of Electricity, was recently elected president of the American Electro-Therapeutic Association.

Mr. H. H. Eustis, of Boston, president and electrician for the Eastern Electric Cable Co., registered at the headquarters of the American Institute of Electrical Engineers this week.

Herr Prüssman and Herr Otto Arnold, of the main factory of Schaffer & Budenberg, at Magdeburg-Buckau, Germany, accompanied by Mr. L. Portong, the American manager of the company from New York, were interested visitors at the Fair recently, and expressed themselves as highly pleased with the exhibits.

BUSINESS NOTES.

The ANSONIA ELECTRIC COMPANY has been given permission by the court to continue business. The business is carried on under the direction of the assignee, Mr. J. B. Waller. The same line of specialties and standard supplies will be carried as heretofore.

The CHICAGO ELECTRIC WIRE COMPANY is among the few manufacturing companies who have sufficient orders to keep their full force at work. The recent contract received from the U. S. government has kept the New York factory going night and day.

The eight boilers of the Citizen's Street Railway Co., at Detroit, have recently been equipped with No. 3 Reliance safety water columns, manufactured by the Reliance Gauge Co., Cleveland, Ohio. They were specified by Mr. Harry Knowlton, of the Hartford Steam Boiler Inspection & Insurance Co.

The Electric Appliance Company reports a number of nice orders for the new Hammond porcelain cleat, which they have been introducing to the western trade. Its peculiar merit consists in the fact that it is only one piece of porcelain, but at the same time holds the wire away from the wall or ceiling like an insulating knob instead of clamping it against the surface. The price is also very reasonable.

DEPARTMENT OF MECHANICAL AND ELECTRICAL ENGINEERING.

OFFICES SOUTH OF MACHINERY HALL.

Mechanical Engineer, C. F. Foster.
Electrical Engineer, R. H. Pierce.
First Asst. Mechanical Engineer, John Meados.
First Asst. Electrical Engineer, S. G. Neiler.

DEPARTMENT OF ELECTRICITY.

OFFICE: SECTION B, ELECTRICITY BUILDING.

Chief, John P. Barrett.
Assistant Chief, J. Allen Hornsby.
General Superintendent, J. W. Brassill.
Electrical Engineer, W. W. Prinn.

AMERICAN INSTITUTE OF ELECTRICAL ENGINEERS.

World's Fair Headquarters
SECTION S ELECTRICITY BUILDING.

Ralph W. Pope, Secretary

Open from 9 a. m. to 5 p.m.

Amusements.

HOOLEY'S THEATER—Mr. E. S. Willard, Thurs. and Sat., in "The Professor's Love Story," Fri. in "A Fool's Paradise." 149 Randolph street.

COLUMBIA THEATER—Daniel Frohman's Lyceum Theater Co. in "The Charity Ball." 108 Monroe street.

GRAND OPERA HOUSE — Sol Smith Russell, in "A Peaceful Valley." 87 Clark street.

AUDITORIUM—Imre Kiralfy's Spectacle "America." Congress street and Wabash avenue.

McVICKER'S THEATER—Denman Thompson, in "The Old Homestead." 82 Madison street.

CHICAGO OPERA HOUSE—American Extravaganza Company, in "Ali Baba, or Morgiana and the Forty Thieves." Washington and Clark streets.

SCHILLER THEATER—Miss Rose and Charles Coghlan, in "Diplomacy." Randolph, near Dearborn.

HAVERLY'S CASINO Haverly's United Minstrels. Wabash avenue, near Jackson street.

TROCADERO Vaudeville. Michigan avenue near Monroe street.

THE GROTTO Vaudeville. Michigan avenue near Monroe street.

Buffalo Bill's "Wild West." 63d street. Daily at 3 and 8.30 p.m.

THE STODDARD LECTURES—Central Music Hall, commencing Oct. 2d. Seven courses exactly alike six evenings and matinee each week. Week of October 2, Eastern Japan; week of October 9, Western Japan; week of October 16, China; week of October 23, Farther India; week of October 30, Nearer India.

On Monday evening, October 2, Mr. Stoddard will commence his fourteenth season of lectures in Chicago. The lectures this season are on Japan, China and India, the material for which was gained in Mr. Stoddard's tour around the world. These lectures will be elaborately illustrated.

The managers of "America" did a wise thing when they decided on an extra Monday matinee. The great Auditorium has been crowded on these occasions by people who could not secure seats for the regular performances. The spectacle has been freshened up and improved by a new and superior force of chorus singers. Another agreeable change is the engagement of Annie Cameron, a well-known light opera contralto. She takes the part of Bigotry which Miss Russell has had since the engagement opened. The Schaffer family continue to perform their acrobatic feats to delighted and astounded audiences.

Last Sunday night "Ali Baba" began its fortieth week in Chicago, and on Wednesday evening, September 20, at the Chicago Opera House, the 290th performance of the extravaganza will be given. Business continues as large as ever. All the best seats for the evening performances are sold away in advance, and the standing room only sign is now put out in the afternoons. The wisdom of the lavish expenditure always made upon the American Extravaganza Company's productions is now shown in the splendid preservation of the scenery, costumes and properties of "Ali Baba." The rich satins, velvets and brocades show not a sign of wear, and the beautiful pictures are as bright and charming as when they first came from the painter's brush.

THE
FERRIS WHEEL

When you visit the World's Fair. you will naturally take a ride on the FERRIS WHEEL and be interested in the ELECTRIC LIGHT INSTALLATION. which is wired throughout with

OKONITE WIRE

FURNISHED BY THE

CENTRAL ELECTRIC COMPANY,

116-118 Franklin Street,

CHICAGO, ILLS.

ELECTRICITY BUILDING—EXHIBITORS AND THEIR LOCATION.

GALLERY.

MAIN FLOOR.

ELECTRIC APPLIANCE COMPANY,

242 Madison Street, CHICAGO.

.... HIGH GRADE ....

ELECTRICAL SUPPLIES

of every description at

LOWEST PRICES

consistent with

FIRST CLASS MATERIAL.

Always ready to open orders and requests for quotations to the

ELECTRIC APPLIANCE COMPANY,

242 Madison Street, CHICAGO.

ELECTRICAL INDUSTRIES

Should be read by everyone interested in electrical matters.

The paper contains regularly

A Buyer's Directory of Manufacturers and Dealers in Electrical Supplies and Appliances.

A Complete Directory of Electric Light Stations in North America and a Complete Directory of Electric Railways in North America.

These directories are revised each issue to the date of going to press and are to be found in no other electrical journal in the World. Its articles are read carefully and its directories used constantly by all the buyers in the trade. These facts make it without a superior as an advertising medium. Sample copies and rates sent on application.

Subscription price $3 per year. Six months trial $1 if ordered during September.

ELECTRICAL INDUSTRIES PUBLISHING COMPANY, MONADNOCK BLOCK, CHICAGO.

JENNEY
HIGH GRADE ELECTRICAL APPARATUS

WE MAKE A SPECIALTY

.. OF ..

Dynamos
For **Electric Lighting,
Power Circuits,
Transmisssion of
Power, Electro
Plating, Etc.**

.. AND ..

MOTORS FOR ALL PURPOSES.

Write for catalogue.

CHICAGO OFFICE, 932 Monadnock Block.

TYPE OF MOTOR FROM 1 TO 9 H. P.

JENNY ELECTRIC MOTOR CO., . Indianapolis, Ind.

OUR IMPROVED SYSTEM

...OF...

Automatic Fire Alarm,

covered by patents recently issued, is the
embodiment of all factors contributing to the

GREATEST SAFETY,

and the MOST RELIABLE

PROTECTION FROM FIRE.

Western Electric Company,

CHICAGO and NEW YORK.

CLARK ELECTRIC COMPANY, NEW YORK.

192 Broadway and 11 John Street.

MANUFACTURERS OF ARC LIGHTING APPARATUS. FOR EVERY PURPOSE A SPECIALTY.
The CLARK ARC LAMPS for use on EVERY CURRENT, have the reputation of being
the best and most durable of any ever made in the United States.

RAWHIDE PINIONS FOR ELECTRIC MOTORS
A SPECIALTY.
RAWHIDE DYNAMO BELTING

THE CHICAGO RAWHIDE MANUFACTURING CO.
THE ONLY MANUFACTURERS IN THE COUNTRY

LACE LEATHER ROPE
AND OTHER RAWHIDE
GOODS
OF ALL KINDS
BY KRUEGER'S PATENT

75 Ohio Street, CHICAGO, ILL.

STANDARD ELECTRIC COMPANY.

GENERAL OFFICES: 625 Home Insurance Building.

WORKS: So. Canal Street,

CHICAGO.

STANDARD SYSTEM

AT THE

WORLD'S FAIR.

MACHINERY HALL, Sec. Q, 2 Standard Arc Dynamos.
Sec. S, 20 " " "
ELECTRICITY BUILDING, Sec. P, Space 2, Arc Lighting Exhibit.

The Standard Lamps Light the Power Plant, Machinery Hall, Agricultural Hall, Shoe and Leather Building, and
Other Buildings and Portions of the Grounds.

See our Double Service All Night Lamp Before Buying an Old Style Two Rod Lamp.

Mile after mile of
SIMPLEX WIRE
Supplied to the
FERRIS WHEEL
...
By...George Cutter,
The Rookery, CHICAGO.

SIMPLEX WIRES

INSURE
HICH
INSULATION

SIMPLEX

Ever Onward and Upward

Simplex Electrical Co.
620 Atlantic Ave.,

George Cutter, Chicago.   BOSTON, MASS.

XNTRIC

"That's the Switch"

And we control that movement.

H. T. PAISTE,

10 South 11th St.,
PHILADELPHIA,
PA.

Made   5 amp. S. P.
        10 amp. S. P.
        5 amp. 3 way.
        10 amp. 3 way.

China Window Tube (Patented).

Made only by PASS & SEYMOUR,
George Cutter,   SYRACUSE, N. Y.
CHICAGO.

Enterprise Electric Company

307 Dearborn Street.
Chicago . . . .

GENERAL WESTERN AGENTS

N. I. R.

Manufacturers' Agents and Mill Representatives for

Electric Railway,
Telegraph, Telephone and
Electric Light

SUPPLIES OF EVERY DESCRIPTION

Agents for   Cedar Poles,
              Cypress Poles, Oak Pins,
Locust Pins, Cross Arms, Glass
——— Feeder Wire,   Insulators,
WIRES, CABLES, TAPE and TUBING

BEAR IN MIND

that the regular monthly issue of ELECTRICAL INDUSTRIES contains the most complete and correct directories published of the electric light central stations and the electric railways in North America.

World's Fair Headquarters Y 27 Electricity Building.

CITY OFFICES, Monadnock Block.

Consolidated Electric Co.

Manufacturers and Dealers in all kinds of

ELECTRICAL . SUPPLIES,

115 Franklin Street,

CHICAGO.

WAGNER ELECTRIC FAN MOTORS

For Direct or Alternating Currents.

These motors give a stronger breeze with less consumption of current than any other fan motor on the market. They are full 1-8 horse power, six bladed 12-inch fan. Self-oiling. Furnished with or without guards.

IT WILL PAY YOU TO SEE THE WAGNER BEFORE BUYING ELSEWHERE.

TAYLOR, COODHUE & AMES,

348 Dearborn Street, CHICAGO.

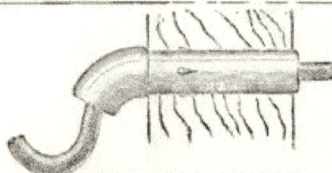

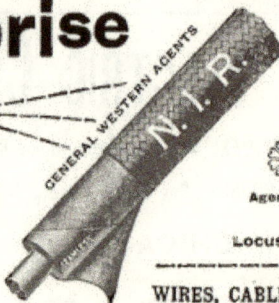

WEEKLY WORLD'S FAIR

ELECTRICAL INDUSTRIES

DEVOTED TO THE ELECTRICAL AND ALLIED INTERESTS OF THE WORLD'S FAIR, ITS VISITORS AND EXHIBITORS.

Vol. I, No. 16. **CHICAGO, SEPTEMBER 28, 1893.** FIVE MONTHS $1.00 TEN CENTS A COPY

Exhibit of the E. S. Greeley & Company.

In the northeastern part of Electricity Building, Section F, E. S. Greeley & Company, of New York, has an extensive exhibit containing a large and comprehensive display of electrical goods. The large floor space has been used to the best advantage by a neat arrangement of show cases and elegant display boards which contain the finely constructed articles and delicate instruments from the factories of this firm.

The space is enclosed by neat railings above which is placed a large sign bearing the name of the company that extends the entire length of the exhibit. One of the more prominent figures in the exhibit is the pyramid of primary batteries surmounted by a statue of liberty 36 inches high holding aloft a 16-candle power lamp. This exhibit of batteries comprises a large number of styles of batteries among which is the Foote Primach battery. This battery is of an improved form and construction; to the center of the copper element is fastened a wooden post coated with asphaltum that supports the zinc element. The zinc is star-shape and has a recess on the under side into which the post fits. This post keeps the two elements always the same distance apart, thus preventing the copper working

FIG. 1.—EXHIBIT OF THE E. S. GREELEY & COMPANY.

up and all short circuits are in this way avoided. A new telephone battery is shown which is a modification of the Fuller cell which is used so extensively in that class of work. Instead of having a bell shaped zinc, it has a pencil zinc which is protected by a rubber tube with an opening at the bottom. The zinc feeds down and the life of the battery is said to be much longer. Its voltage is two and the initial current about three amperes. This exhibit of batteries contains 88 varieties of wet and dry batteries.

An extensive line of telegraph instruments are shown, many of new and special designs. The most popular part of this display is the Victor key which the president touched when he opened the Exposition. By the simple pressure of this key the current from the battery of Exeter cells operated the starting device that opened the valve of the Allis engine and the great Worthington pump. The Victor key is the only telegraph key that has no trunnions, the fulcrum being a knife edge. This manner of construction presents all side motion and consequent poor contact and also eliminates the so-called jamming motion. A short stout spring near the fulcrum holds the lever in place. This

The main circuit runs through an electro magnet which when the circuit is closed holds the armature up to which there is a lever attached. A pin in this lever catches in a slot of the main lever or the pull of the box. The box cannot be operated except when the pin is held in the slot of the pull so that when another box is being pulled the circuit is closed only momentary but not long enough to magnetize the electro-magnet. The non-interfering magnet is short circuited when the outside door is closed, thus preventing loss of current. The mechanism of the gong is so constructed that when a wire is broken it will not continue to ring.

A large double jack telegraph switch board is so mounted as to show the connections. It is also of the Western Union standard type and will accommodate 200 loops and 50 circuits. In the cases along the walls are displayed watchmen's time detectors, burglar alarms, annunciators of a variety of forms for hotels, houses, etc. Mounted on handsome hardwood boards are bells of all descriptions and push buttons in a variety of designs.

The Pratt speed indicator is shown which can be used in

FIG. 2.—EXHIBIT OF E. S. GREELEY & COMPANY.

key is also modified to meet the requirements of cable transmission.

The Victor relays and sounders shown exhibit the same attention to the details of construction. The relay has a plunger by which the adjustment of the spring is quickly determined. Western Union relays and Giant sounders are also mounted with the Victor keys. The Greeley ink writing self starting and stopping register is shown in several modifications, also connected with the Victor keys and district call boxes. As a fire alarm instrument it is made with either one, two, three or four pens. The four pen register is the first of its kind built and it is able to record four signals of various length of duration at one time.

These instruments are connected to a single jack switch board of the standard Western Union pattern. Each jack will accommodate four loops and the board is built for 25 lines. The batteries supplying the current for these instruments are the above mentioned Foote Primach type. There are 30 cells used which also operate the Greeley fire alarm system, consisting of four non-interfering boxes, an indicator, an electro-mechanical gong and two registers.

both directions and does not register until the handle is pressed.

The W. B. G. protectors which are exceedingly sensitive to any excess of current are mounted in various ways to accommodate different circuits. These protectors are said to fuse at their rated capacity to within one-tenth of an ampere. Its value is readily seen and it is used extensively for protecting telephone, fire alarm and other systems.

At the south end of the space is an octagonal electric light pole on which is mounted a patented electric light crane. It is constructed of ¾ inch pipe and of such a form as to be strong and rigid. Being pivoted in the center the crane may be raised or lowered. Weights are placed on the end opposite the lamp to balance it so that but a slight force is required to raise and lower it. When it is lowered the lamp comes within a foot of the pole enabling the trimmer to trim the lamp from the ground. The Wildt door opener is shown in operation.

An extensive line of testing instruments are shown in the cases among which are the Thompson galvanometers, portable testing sets containing six chloride of silver cells, a galvanometer sensitive to a variation of one ohm in one thousand and resistance coils of .001 to 11,000,000 ohms. The Frey portable galvanometer for street railway use is shown in which the galvanometer is suspended by a universal joint so that it is always level. It also contains an improved compensating device. Other instruments are displayed for a variety of uses.

Mr. L. J. Auerbacher is in charge of the exhibit and will be found a valuable assistant in an examination of the display of electrical goods.

Exhibit of the Billings & Spencer Company.

The Billings & Spencer Company has a very complete exhibit of drop forgings such as are used in electrical work in the south gallery of Electricity Building, just back of the Weston Electrical Instrument Co.'s exhibit. On a six-

early Edison machines were made in two parts, the blade rolled out to the required shape and the shank forged by hand. These two pieces were joined by small dowel pins and then brazed, a long process and one which gave by no means satisfactory results. Mr. Billings, who had watched the process offered to make them of one piece. His offer was accepted but their confidence in his ability was not perfect until they had seen the bars produced. From that time until the present the manufacture of these goods has increased and to one manufacturer of electrical dynamos and motors this company has furnished as high as 16 tons a month.

The method of manufacture is so perfect that the fibre or grain of the metal is preserved so that it is everywhere parallel to the axis of the arm. They are used by the principal manufacturers of electric generators and motors.

EXHIBIT OF THE BILLINGS & SPENCER COMPANY.

sided pyramid standing some ten feet high, are arranged samples of forgings in pure lake copper, aluminum, aluminum bronze, iron and steel in the shape of commutator bars of various shapes and sizes, name plates for machinery, eye bolts for dynamos and other machines, wrenches, commutators, etc.

This pyramid is surmounted by a sign painted on a large fac simile of a copper commutator bar which is supported on four ornamented wrought iron brackets, secured to the sides of the pyramid. At the four corners of the exhibit are simple commutators in copper and bronze assembled without insulation to show how perfectly each bar fits into its place. Eye bolts of large size up to two inches, drop forged from iron are shown at the corners of the platform. This company has been long known for its commutator bars and the origin of this process of manufacture is now an interesting piece of history. The commutator bars of the

Besides the exhibit in Electricity Building the company has an extensive exhibit in Machinery Hall Annex, main aisle, in which are found many tools manufactured for the use of linemen and wiremen. Billing's hand vise is extensively used by linemen in telephone, telegraph and electric lighting line construction. Each part is interchangeable and can be duplicated at slight expense. There are also shown gas pliers and wire cutters specially adapted for the use of fixture men, carbon tongs which are largely used by trimmers, being light and strong and a great variety of machinists' tools and drop forgings among which are various bronze forgings used on the Whitehead torpedoes.

A large part of this exhibit shows the capabilities of the company in the line of drop forgings, the forgings being shown just as they were forged, no attempt being made to finish the goods. In the show cases are exhibited tools, etc., finished with the greatest care.

ELECTRICAL INDUSTRIES.

PUBLISHED EVERY THURSDAY BY THE

ELECTRICAL INDUSTRIES PUBLISHING COMPANY,

INCORPORATED 1893.

MONADNOCK BLOCK, CHICAGO.

TELEPHONE HARRISON 310.

E. L. POWERS, Pres. AND TREAS.   E. E. WOOD, Secretary.

E. L. POWERS   Editor.

H. A. FOSTER }
W. A. REMINGTON } · · · Associate Editors.

E. E. WOOD · · · Eastern Manager.

FLOYD T. SHORT.   Advertising Department.

EASTERN OFFICE, WORLD BUILDING, NEW YORK.

World's Fair Headquarters, Y 27 Electricity Building.

SUBSCRIPTION:

FIVE MONTHS, · · · · · · · · · · · · · $2.00
SINGLE COPY, · · · · · · · · · · · · · · 10

Advertising Rates Upon Application.

News items, notes or communications of interest to World's Fair Visitors are earnestly desired for publication in these columns and will be heartily appreciated. We especially invite all visitors to call upon us or send address at once upon their arrival in city or at the grounds.
ELECTRICAL INDUSTRIES PUBLISHING CO.,
Monadnock Block, Chicago.

The List of Awards Department of Electricity.

The following is a list of the awards made at the present time. The tests of incandescent lamps, wires, storage batteries, of some of the transformers and a few others have not been completed. The diplomas will soon be ready and will be placed so that parties interested may inspect them.

UNITED STATES.

Albert & J. M. Anderson, Boston, Mass.
1. Trolleys.
2. Railway insulators.

Brush Electric Co., Cleveland, Ohio.
1. Direct current dynamos for series arc lighting.
2. Direct current dynamos for series arc lighting coupled to engine, 125-2000 c. p. lamps.
3. Alternating current dynamos constant potential, 30.-150 K. W.
4. Arc circuit switch board.
5. Direct current dynamos constant potential 20-100 K. W.
6. Arc lamps all types.

Bryant Electric Co., Bridgeport, Conn.
Snap switches.

W. R. Brixey, New York.
Underground, aerial and submarine telegraph and telephone cables.

J. H. Bunnell & Co., New York.
1. Standard dry batteries.
2. Telegraphic apparatus.

C. C. Electric Motor Co., New York.
1. Direct current motors, constant potential 3-50 H. P.
2. Electric motor, fan and blower combination.

Cutter Mfg. Co., Philadelphia, Pa.
Push switches for electric lights.

Commercial Cable Co., New York.
1. Ocean telegraphic apparatus operating through Muirhead's artificial resistances.
2. Cuttriss improved cable telegraph apparatus.

Crane Electric Co., Chicago, Ill.
Electric passenger elevator complete.

Carpenter Enamel Rheostat Co., Bridgeport, Conn.
Rheostats.

Copenhagen Fire Alarm Co., Chicago.
Automatic fire alarm.

Geo. Cutter, Chicago, Ill.
Lamp supporting pulley.

Electrical Forging Co., Boston, Mass.
Electric heating and welding apparatus.

Electric Heat Alarm Co., Boston, Mass.
Thermostat for automatic fire, hot journal and hot grain alarms.

Electrical Conduit Co., New York.
Underground conduit for electrical wires.

Electric Launch & Navigation Co., New York.
Electric launches.

Eddy Electric Mfg. Co., Windsor, Conn.
Direct current motors constant potential.

Excelsior Electric Co., New York.
1. Arc lamps series circuits.
2. Direct current dynamos for series arc lighting.

Elektron Mfg. Co., Springfield, Mass.
Direct current motors constant potential multipolar, slow speed.
2. Automatic motor starter.

Electric Selector and Signal Co., New York.
Electrical system for locking and unlocking.

Edison Mfg. Co., New York.
Edison, LeLande primary battery.

Eureka Tempered Copper Co., North East, Pa.
Tempered copper for use in electrical construction.

Fort Wayne Electric Co., Fort Wayne, Ind.
1. Direct current "Wood" dynamo for series arc lighting.
2. Alternating current "Wood" dynamo constant potential compound wound 150 K. W.
3. Arc lamps for constant current.

I. P. Fink, New York.
Screen reflectors for incandescent lamps.

General Electric Co., New York.
1. Electric locomotives for factory and switches services.
2. Electric elevated railway system.
3. Long distance power transmission, plant in operation, tri-phase.
4. Arc lamps for direct current series circuits.
5. Search lights and focusing lamps.
6. Transformers 250-125,000 watts.
7. Engine dynamos.
8. Automatic overload switch.
9. Electrically illuminated fountains.
10. Thomson eccentric coil anameters and volt meters for alternating currents.
11. Pumping machinery driven by electric motor.
12. Electrically driven rock working machinery.
13. Mine locomotive.
14. Haskins Astatic armeter.
15. Arc lamps for constant potential circuits direct and alternating (Knowles).
16. Jaw switches, fuses, sockets and branch blocks.
17. Direct current dynamos for series arc lighting.
18. Alternating current dynamos for series arc lighting.
Alternating current dynamos constant potential 30-300 K. W.
19. System of street railway service.
20. Direct current dynamos constant potential (direct connected excepted) and direct current shunt wound motors, constant potential.
21. Edison feeder system for distribution of electricity.
22. Slate switch board for arc light circuits.
23. Ventilating set, portable, Government standard.
24. Historical apparatus.
25. Edison three-wire system for distribution of electrical energy.
26. Exhibit of incandescent lamps all styles, ¼ to 250 c. p.
27. Underground system complete in all details.
28. Hoisting apparatus driven by electric motors.
29. Integrating watt motor.

The E. S. Greeley & Co., New York.
1. Testing instruments.
2. Exter dry battery.
3. Telegraph apparatus.
Gamewell Fire Alarm Telegraph Co., New York.
Automatic fire alarm telegraph system.
General Incandescent Arc Lamp Co., New York.
Arc lamps for constant potential circuits.
Hart & Hageman Mfg. Co., Hartford, Conn.
Snap switches.
Helios Electric Co.
Arc lamp for alternating current.
The Hanson Battery Co., Washington, D. C.
Primary batteries.
Interior Conduit & Insulation Co., New York.
1. System of interior insulating conduits.
2. Snap switches.
Jenny Electric Motor Co., Indianapolis, Ind.
Direct current dynamos and motors constant potential.
H. W. Johns Mfg. Co., New York.
Vulcabeston and molded mica insulating material worked into all kinds of insulations.
LeClanche Battery Co., New York.
LeClanche batteries, especially the "Vole" and "Cylinder" cells.
Mather Electric Co., Manchester, Conn.
Direct current dynamo, constant potential 500 volts.
McIntosh Battery & Optical Co., Chicago, Ill.
Electro medical, dental and surgical apparatus.
National Carbon Co., Cleveland, Ohio.
Carbons for arc lamps.
Nutting Electric Mfg. Co., Chicago.
Nutting arc lamps.
Otis Bros., New York.
1. Electric pump.
2. Electric motor and controlling devices for elevator and hoisting service.
H. T. Paiste, Philadelphia, Pa.
Snap switches.
Police & Signal Co., Chicago.
System of police patrol telegraph.
Phoenix Glass Co., Chicago.
Electric and gas globes and shade cut, etched and colored.
Queen & Co., Philadelphia, Pa.
1. Electrometer (Ryan).
2. Galvanometers.
3. Testing sets and resistances.
4. Portable medical induction apparatus for physician's use.
5. Commercial ammeters and voltmeters.
J. A. Roebling's Sons Co., Trenton, N. J.
Bare copper and trolley wire.
F. A. Ringler & Co., New York.
Half tone photo electro type steel faced.
Standard Electric Co., Chicago.
1. Arc lamps for direct current series circuits.
2. Direct current dynamos for series arc lighting.
Stevenson Hoggson Electric Co., St. Louis, Mo.
Automatic electric time-stamp.
Sperry Electric Railway Co., Cleveland, Ohio.
Electric railway system.
Short Electric Railway Co., Cleveland, Ohio.
Short Electric Railway system.
Self Winding Clock Co., New York.
Special application of an iron-clad solonoid magnet.
Thomson Electric Welding Co., Boston, Mass.
Apparatus for electric welding and forging.
Union Electric Works, Chicago.
Primary battery.
J. C. Vetter & Co., New York.
1. Incandescent current adapter.
2. Dry LeClanche battery.
Western Electric Co., Chicago.
1. Columbian street lamp post.
2. Telegraph apparatus.
3. Telephone cables, Paterson.
4. Annunciators and signaling apparatus.
5. Multiple switchboard for telephone service.

6. Direct current dynamos for series arc lighting.
7. Application of electric lights for the production of scenic effects in theaters and for the decorations of rooms, etc.
8. Arc lamps various styles for series circuits.
9. Arc lamps for constant potential circuits.
10. Direct current dynamos and motors constant potential.
Waite & Bartlett Mfg. Co., New York.
1. Holtz induction machines in air tight case with 6 40 inch revolving plates.
2. Special faradic apparatus for varying the tension and strength of current (Engleman's apparatus).
Washington Carbon Co., Pittsburgh, Pa.
Carbons for arc lamps, batteries and dynamos, and motor brushes.
Wm. Wallace, Ansonia, Conn.
Historical electric light exhibit.
Walworth Mfg. Co.
Poles for trolleys and arc lamps.
S. S. White Dental Mfg. Co., Philadelphia, Pa.
Acid gravity batteries "Partz."
Westinghouse Electric & Mfg. Co., Pittsburg, Pa.
1. Engine dynamos.
2. Transformers, 250-12500 K. W.
3. Direct current dynamos and motors constant potential bipolar and multipolar (except direct connected dynamos).
4. Alternating current dynamos constant potential 750 K. W.
5. Electric street railway system.
6. Alternating current dynamos for series arc lighting.
7. Long distance power transmission plant in operation.
8. Two phase alternating current motors (Tesla).
9. Incandescent system of street lighting.
10. Switches.
11. Complete switchboard for controlling 15 dynamos and 40 currents.
12. Lightening arresters.
Weston Electrical Instrument Co., Newark, N. J.
1. Alternating current instrument including watt meters.
2. Standard resistance and bridges.
3. Electrical measuring instruments for physicians use.
4. Switchboard instruments.
5. Direct current ammeters and voltmeters, standard and portable.
Western Union Telegraph Co., New York.
Instruments used in quadruplex telegraph, latest design.
Zucker & Leavett Chemical Co., New York.
Collection of chemicals and appliances for electro-plating.

GERMAN.

Prof. Aron, Berlin.
Electric meters.
J. Berliner, Hanover.
Universal transmitter, long distance.
Geo. Carette & Co., Nuremburg.
Optical, physical and mechanical instruments and toys.
Dr. Edelmann, Munich.
Electro medical apparatus.
Felton & Guilleaume, Mulhhausen on Rhine.
1. Electric cables with special armor.
2. Electric cables.
Gebrueder, Nuremburg.
Carbons for arc lights.
Hartmann & Braun, Frankfort.
1. Differential arc lamps.
2. Galvanometers.
3. Electrical measuring instruments including instruments of precision.
4. Photometer, large universal.
5. Apparatus for testing iron and steel with respect to magnetic permeability.
6. Reading telescopes, mirrors and scales.
7. Portable measuring apparatus used in laying cables.

W. A. Hirshman, Berlin.
Electro medical apparatus.
Imperial German Postal Telegraph Dept., Berlin.
1. Maps and drawings showing system of Government telegraph lines using armored underground cables.
2. Telegraphic apparatus of the German telegraph service.
3. Historical telegraphic apparatus.
Korting & Matthiesen, Leipzig.
Arc lamps for constant potential circuits.
H. Schomburg & Sons, Berlin.
1. Insulators and insulating material.
2. Dry batteries.
Schuckert & Co., Nuremburg.
1. Search lights with parabolic glass mirrors.
2. Annunciators for vessels.
3. Arc lamps for constant potential circuits.
Siemens & Halske, Berlin.
1. Arc lamps for constant potential circuits.
2. Direct current dynamos, constant potential 750 K. W.
3. Historical apparatus.
Schmidt & Haensch, Berlin.
Lummer Brodhun photometer with accessories.
J. Zacharias, Berlin.
Dry batteries.

TURKEY.

Imperial Ottoman Government.
Telegraphic apparatus.

GREAT BRITAIN.

British Government Postal Telegraph Dept.
1. Modern telegraph apparatus in operation.
2. Historical telegraph apparatus.
Corporation of Birmingham.
Original Woolwich dynamo.
General Electric Co., Ltd., London.
H. L. switches and other incandescent house fittings.
James White, Glasgow.
Electro magnetic balances, Kelvin.

AUSTRIA.

F. Hardtmuth & Co., Vienna.
Carbons for arc lamps, etc.

RUSSIA.

Imperial Russian State Paper Manufactory, St. Petersburg.
Collection of electrotypes.
Imperial Artillery Arsenal, St. Petersburg.
Electrical registering attachment for testing machine.
(Prince Gagarian.)

ITALY.

Prof. Galebo Ferraris, Turin.
Historical alternating current motors.

JAPAN.

Imperial University, Tokio.
1. Seismographs and accessory apparatus.
2. Model of an earthquake.
Department of Engineering Imperial University, Tokio.
Automatic electric current recorder.

BRAZIL.

Directoria Geral dos Telegraphos, Rio de Janeiro.
Telegraphic apparatus.

Arrangements are being made for a street railway day at the Fair. It is proposed to have the day following the close of the convention at Milwaukee street railway day and have a special program prepared for the delegates on their return from Milwaukee.

Irish day at the Exposition Saturday the 30th, bids fair to be one of the most notable of the red letter days. The Department of Electricity on this occasion will not be behind in doing honor to Erin. Not only is Chief Barrett the proud possessor of the only Irish Electric flag on the grounds pre-

sented him by the Brush Electric Co., the current for which is supplied by the Westinghouse company, but he is also the owner of a very handsome silken flag, presented him by admiring friends. The graceful folds of this flag will float over the Electricity Building in honor of Lord Mayor Shanks, of Dublin, and all other true followers of St. Patrick. Frank D. Millet, the artist of the Fair, has objected to the hoisting of this flag and a lively fight has been going on between him and Prof. Barrett, but as usual, the Chief has come off victorious and the flag will be unfurled. The electric flag taken with the other banner will certainly do credit to the department on that day.

PERSONAL.

Mr. J. H. Smith, of Inverness, Ohio, was a visitor at the Fair last week.

Mr. Emile Berliner, of Washington, D. C., is making a visit to the Fair.

Mr. W. E. Shepard, electrical engineer of Lincoln, Neb., is visiting the World's Fair.

Mr. Wm. J. Hammer, of New York, registered at Institute headquarters last week.

Mr. Wm. Grosvenor Ely, Jr., of Norwich, Ct., is taking in the electrical features of the Exposition.

Mr. C. E. Billings, president of the Billings & Spencer Company, of Hartford, Ct., is at the Fair.

Mr. C. E. Clifford, assistant electrical engineer of the Buffalo Street Railway, is in Chicago and the Fair.

Mr. H. F. Parshall, with the Thomson-Houston Company, at Boston, Mass., was among the arrivals last week.

Mr. E. E. Schlosser, manufacturer of electrical appliances from Denver, Colo., is stopping at the Palmer House.

Mr. W. E. Denning, city electrician of Minneapolis, Minn., registered at the space of ELECTRICAL INDUSTRIES recently.

Mr. Robt. Lundell, engineer for the Interior Conduit & Insulation Company, of New York, is registered at the Great Northern.

Mr. G. H. S. Young, manager of the C. & C. Electric Company, at Philadelphia, spent some time at the Fair during the past week.

Mr. Elias Nusbaum, superintendent of the Pennsylvania Electrical Engineering Company, of Philadelphia, has been at the Fair this week.

Mr. Ed. W. Norton, of Newton, Kan., vice-president of the Armature Electric Union, was an interested visitor in the Electrical Building a short time ago.

Mr. J. Hamblet, manager of the time service of the Western Union Telegraph Company, of New York City, returned this week after a two weeks' visit at the Exposition.

Mr. W. J. Gilmour, district manager of the Bell Telephone Company, at Brockville, Ont., is at the Fair, and is examining with special interest the exhibits of storage batteries.

Mr. L. W. Robinson and Mr. W. B. Cosgrove, of the Westinghouse Electric Manufacturing Company, from Pittsburg, while visiting the Fair last week, called upon ELECTRICAL INDUSTRIES.

Among the visitors to the Electricity Building who have registered during the past week at the booth of ELECTRICAL INDUSTRIES, are the following: Mr. H. K. McCay, of the McCay-Howard Engineering Company, of Baltimore, Md.; Mr. Elias Nusbaum, superintendent of the Pennsylvania Electrical Engineering Company, Philadelphia, Pa.; Mr. J. M. McGrath, general sales agent for the Acme Company, of New York; Mr. J. S. Romine, president of the Electric Light & Power Company, of Chadron, Neb., and Mr. A. M. Hanbrich, of Detroit.

Mr. W. H. McKinlock is now busy organizing the Metropolitan Electric Co. Chicago, a corporation of which he will be the head, with ample capital for promoting its business. Mr. McKinlock has had a wide experience in the electric business extending from the time of the introduction of electricity and is therefore one of the best known men in the trade. With a full knowledge of the wants of the trade, the success of his new enterprise is assured from the start. The new company will do a large manufacturing and supply business comprising the most desirable specialties of all kinds in the line of electric apparatus and appliances.

THE
FERRIS WHEEL

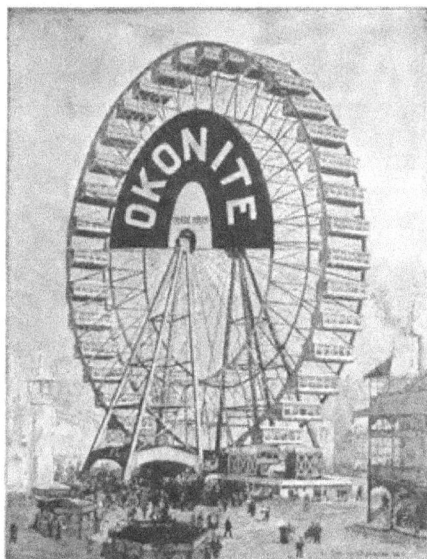

When you visit the World's Fair, you will
naturally take a ride on the FERRIS WHEEL
and be interested in the ELECTRIC LIGHT
INSTALLATION, which is wired throughout
with

OKONITE WIRE

FURNISHED BY THE

CENTRAL ELECTRIC COMPANY,

116-118 Franklin Street,

CHICAGO, ILLS.

ELECTRICITY BUILDING—EXHIBITORS AND THEIR LOCATION.

GALLERY.

MAIN FLOOR.

Exhibitor.	Section
Acetic ...	T
Ansonia Electric Co.	E
Am. Inst. of Elec. Eng.	S
American Battery Co.	T
Axford, H. M.	
Allg. Elec. Gesellschaft	D
Bates Mfg. Co.	N
Bryant Electric Co.	M
Bishops & Spencer	
Brixey, W. T.	E
Belknap Motor Co.	E
Bell Telephone Co.	E
Brush Electric Co.	L
Caldwell El. Cloth Cal. Mch. Co.	Y
Consol. Elec. Storage Co.	K
Cutter, George	
Croston Elec. Co.	T
Chicago Elec. Wire Co.	T
Okonite Co.	T
Consolidated Fire Alarm Co.	V
Central Electric Co.	T
Commercial Cable Co.	T
A. C. Ex. Motor Co.	
Cleveland Elec. Mfg. Co.	A
Chicago Beltage Co.	
Delaney Clock Co.	
Department of Electricity	B
Electrical Industries	
Elec. Launch & Nav. Co.	X
Electric Separator	
Edgerton, E. M.	T
Elgin Telephone Co.	T
Edison Elec. Mfg. Co.	
Interstate Elec. Co.	S
Kawkin Temp. Copper Co.	U
Electric Appliance Co.	T
Elec. S. & Sig'l Co.	T
Electric Heat Alarm Co.	Y
Electrical Review	Y
Electricity	Y
Electric Gas Co.	T
Electrical Engineer	Y
Electrical World	Y
Eddy Electric Motor Co.	B
Excelsior Electric Co.	B
Electrical Forging Co.	D
Reynolds Dynamo Co.	D
Elektron Mfg. Co.	D
Electrical Conduit Co.	P
England	O
Empire China Works	
Franklin Elec. Appliance Co.	
French Power Exhibit	
Felten & Guilleaume	D
France	K P
Ft. Wayne Elec. Co.	M
Gault & Co.	N
Gamewell Fire Alarm Co.	V
General Electric Co., B H N C & J	
Gregg Belt and Car Axe L T Co.	E
Gowley, E. S. & Co.	
Germany	X
Holland Mfg. & Co.	T
Hofmann, C. J.	S
Hart & Hegeman Mfg. Co.	M
Hope Elec. Appliance Co.	S
Holtzer Cabot	W
Holmes, W. L.	W
Hartman & Braun	
Havelock & Von Winkle	L
Hawk, J. M.	
Mouthcuath, P., & Co.	
Illinois Alloy Co.	
Internat. Am. T & A P'y Co.	J
India Rubber Comb Co.	S
Jaeger, Chas. L.	T
Johns Mfg. Co., H. W.	Y
Jewell Belting Co.	
Jenney Elec. Motor Co.	L
Knapp Electrical Works	L
K. A. P. Elec. Novelty Co.	
Knapp & Buckley	
Kennedy Electric Co.	L
Lawton, B. A.	
LeClanche Battery Co.	V
McNeil Tanded Elec. Co.	
Marris, W. N.	
Meeker, Dr. G.	
McIntosh Bat. & Opt. Co.	W
Muncon, C. J., Belting Co.	T
Mather, A. C.	
Mather Electric Co.	M
Newman Clock Co.	V
New Magnetic Watch Co.	T
N. Y. Woodhead Wire Co.	T
National Carbon Co.	T
Norwich Ins. Wire Co.	
North Am. Phonograph Co.	S P
N.Y.A.I.E.A.	Y
Nat. Ind Fire Alarm Co.	V
Nat. Engraving Machine Co.	
Owen, Dr. A.	
Phoenix Glass Co.	
Porter, H. T.	
Pulverfabrik Coöp. Co.	T
Pempely, J. S.	T
Pratt Ed. Med. Sup. Co.	T
Powell, Wm. & Co.	
Phelps, A. H.	
Page Belting Co.	
Queen & Co.	E
Rougher, F. A.	E
McIntosh Gauge Co.	T
Roessler & Hasslacher Chem. Co.	S
Street Railway Journal	
Struecger Ant. Telph. Co.	
Standard Paint Co.	T
Spamhold, C. B.	
Star Iron Tube? Co.	
Spies	V
Schieren, Chas. A., & Co.	D
Schoenherr & Sohn	E
Stearns & Halske	
Scheckerl & Co.	S
Short Electric Co.	
Sperry Elec. Railway Co.	L
Standard Underg. Cable Co.	C
Standard Electric Co.	
Samson Battery Co.	V
Tate Ant. El. Signal Co.	V
Todd, Applegate Co.	
Taylor, Goodhue & Ames	
Thomson Elec. Welding Co.	O
Thomson-Houghton, Elmha Gray	J
Union Electric Co.	
Vetter J. C. & Co.	W
Webb, F.	
Weston El. Instrument Co.	
Washburn & Moen Mfg. Co.	
Western Union Tel. Co.	
Waite & Bartlet Mfg. Co.	
White, S. S., Dental Mfg. Co.	
Western Electrician	
Witt-Ant. Burglar Al. Co.	
Western Electric Co.	A
Westinghouse El. & Mfg. Co.	B H
Wilox & Seabald	
Wing, L. J., & Co.	
Zucker & Levett Chem. Co.	

ELECTRIC APPLIANCE COMPANY,
242 Madison Street, CHICAGO.
.... HIGH GRADE ....

ELECTRICAL SUPPLIES
of every description at

LOWEST PRICES
consistent with

FIRST CLASS MATERIAL.
Always send your open orders and requests for quotations to the

ELECTRIC APPLIANCE COMPANY,
242 Madison Street, CHICAGO.

THE MONTHLY ISSUE FOR SEPTEMBER

ELECTRICAL INDUSTRIES

Should be read by everyone interested in electrical matters.   In its Table of Contents is the following:
"World's Congress of Electricians."
"Nikola Tesla's Lecture."
"Love Underground Electric System in Washington." By M. D. Law.
"Gas Engines as Applied to Electric Light Work." By Geo. A. Farwell.
"The Underground System at the World's Fair."
"Electric Railway Plant of the Chicago North Shore Railway Company."
Together with illustrations of the recent applications of electricity.
The paper contains regularly
A Buyer's Directory of Manufacturers and Dealers in Electrical Supplies and Appliances.
A Complete Directory of Electric Light Stations in North America and a Complete Directory of Electric Railways in North America.
These directories are revised each issue to the date of going to press and are to be found in no other electrical journal in the World. Its articles are read carefully and its directories used constantly by all the buyers in the trade. These facts make it without a superior as an advertising medium. Sample copies and rates sent on application.
Subscription price $3 per year. Six months trial $1, if ordered during September.

ELECTRICAL INDUSTRIES PUB. CO.,
Monadnock Block, CHICAGO.

WM. BARAGWANATH & SON,

LIST OF HEATERS

TO BE SEEN IN OPERATION AT THE WORLD'S FAIR.

Two 500 H. P. East End Boiler Gallery doing 1800 H. P. work.
One 300 H. P. heater and receiving tank, Wellington Catering Co's., plant.
One 150 H. P. heater at Hygeia plant.
One 200 H. P. Libby Glass Works.

55 WEST DIVISION STREET,
CHICAGO.

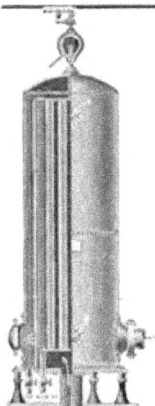

OUR IMPROVED SYSTEM

... OF ...

Automatic Fire Alarm,

covered by patents recently issued, is the
embodiment of all factors contributing to the

GREATEST SAFETY,

and the MOST RELIABLE

PROTECTION FROM FIRE.

Western Electric Company,

CHICAGO and NEW YORK.

CLARK ELECTRIC COMPANY, NEW YORK.

192 Broadway and 11 John Street.

MANUFACTURERS OF ARC LIGHTING APPARATUS FOR EVERY PURPOSE A SPECIALTY. The CLARK ARC LAMPS for use on EVERY CURRENT, have the reputation of being the best and most durable of any ever made in the United States.

RAWHIDE PINIONS FOR ELECTRIC MOTORS
A SPECIALTY.
RAWHIDE DYNAMO BELTING

Greatest Adhesion Qualities. A Non-Conductor of Electricity. Greater Loss Friction than any other Belt.

THE CHICAGO RAWHIDE MANUFACTURING CO.
THE ONLY MANUFACTURERS IN THE COUNTRY

LACE LEATHER ROPE
AND OTHER RAWHIDE

GOODS
OF ALL KINDS
BY KRUEGER'S PATENT

The Lacing and Lace Leather is not affected by change of moisture, is not become hard as the most durable and therefore most practical Belting made. The Raw-Hide Rope for Round Belting Transmission is superior to all others.

75 Ohio Street, CHICAGO, ILL.

STANDARD ELECTRIC COMPANY.

GENERAL OFFICES: 625 Home Insurance Building.

WORKS: So. Canal Street,

CHICAGO.

STANDARD SYSTEM

AT THE

WORLD'S FAIR.

MACHINERY HALL, Sec. Q, 2 Standard Arc Dynamos.
Sec. S, 20 " " "
ELECTRICITY BUILDING, Sec. P, Space 2, Arc Lighting Exhibit.

The Standard Lamps Light the Power Plant, Machinery Hall, Agricultural Hall, Shoe and Leather Building, and Other Buildings and Portions of the Grounds.

See our Double Service All Night Lamp Before Buying an Old Style Two Rod Lamp.

Mile after mile of
SIMPLEX WIRE
Supplied to the
FERRIS WHEEL
...
By...George Cutter,
The Rookery, CHICAGO.

SIMPLEX WIRES

SIMPLEX
Ever Onward and Upward

**INSURE
HIGH
INSULATION**

Simplex Electrical Co.
620 Atlantic Ave.,

George Cutter, Chicago.   BOSTON, MASS.

XNTRIC

"That's the Switch"

And we control that movement.

H. T. PAISTE,
10 South 18th St.,
PHILADELPHIA,
PA.

Made 5 amp., S. P.
      10 amp., S. P.
      5 amp., 3 way.
      10 amp., 3 way.

China Window Tube (Patented).
Made only by **PASS & SEYMOUR,**
George Cutter,   SYRACUSE, N. Y.
CHICAGO.

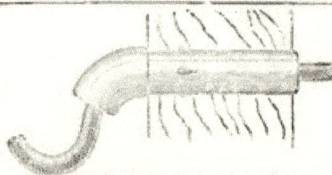

Enterprise
Electric
Company

GENERAL WESTERN AGENTS

N. I. R.

307 Dearborn Street.
Chicago....

Manufacturers' Agents and Mill Representatives for

Electric Railway,
Telegraph, Telephone and
Electric Light

SUPPLIES OF EVERY DESCRIPTION

Agents for   Cedar Poles,
              Cypress Poles, Oak Pins,
Locust Pins, Cross Arms, Glass
        Feeder Wire,   Insulators,

WIRES, CABLES, TAPE and TUBING

BEAR IN MIND

that the regular monthly issue of ELECTRICAL IN-
DUSTRIES contains the most complete and correct
directories published of the electric light central stations
and the electric railways in North America.

World's Fair Headquarters Y 27 Electricity Building.

CITY OFFICES, Monadnock Block.

Consolidated Electric Co.

Manufacturers and Dealers in all kinds of

ELECTRICAL . SUPPLIES,

115 Franklin Street,

CHICAGO.

WAGNER ELECTRIC FAN MOTORS

For Direct or Alternating Currents.

These motors give a strong clear current less consumption of current than
any other fan motor on the market. They are full ⅛ horse power. Six bladed
9 inch fan. Self-oiling. Furnished with or without guards.

IT WILL PAY YOU TO SEE THE WAGNER BEFORE BUYING ELSEWHERE.

TAYLOR, GOODHUE & AMES,
348 Dearborn Street, CHICAGO.

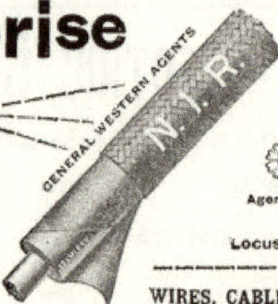

WEEKLY WORLD'S FAIR

ELECTRICAL INDUSTRIES

DEVOTED TO THE ELECTRICAL AND ALLIED INTERESTS OF THE WORLD'S FAIR, ITS VISITORS AND EXHIBITORS.

Vol. I, No. 17.   CHICAGO, OCTOBER 5, 1893.   FIVE MONTHS $1.50 / TEN CENTS A COPY

Exhibit of The Standard Electric Company.

One of the most prominent exhibits of arc lamps and arc apparatus is that made by the Standard Electric Company of Chicago in Electricity Building. The arc lamps hung above the borders of the space include a number of styles and a mast arm such as has been adopted by the city of Cincinnati and which possesses special features adapting it for street use. The lamp is easily drawn into the pole and lowered to a point where it can easily be reached by the trimmer. These devices are shown in such a way as to be easily inspected by the visitor. The Standard lamp is also

EXHIBIT OF THE STANDARD ELECTRIC COMPANY

of arc lamps, adapted for the various places in which arc lamps are used.

In connection with the lamps are shown the safety lamp hanger by which the lamp is disconnected from the circuit as it is lowered to be trimmed thus protecting the lamp and trimmer from any injury from the current while trimming, shown in detail. These lamps although the different styles vary in certain features, which are changed as special uses demand, are all of the independent feed type and are so constructed that under varying loads no difference in light is perceptible. An open circuit is almost impossible, no matter what position the rod may be in. The clutches are

insulated so that no current can pass through the frame of
the lamp.

The feeding device is sensitive and the automatic cut-out
which is one of the most essential features of a lamp, acts
sharp and quickly under all conditions and great care has
also been taken with its insulation. The same lamp may
be changed to a 6, 8 or 10 ampere lamp without changing
the windings or the adjustment exc pt the adjustment of
the arc spring which is the only change required.

Among the many improvements in the system shown in
the exhibit is the double service, elliptical carbon, weatherproof lamp using no hood and but one carbon rod. One
feeding mechanism is thus required to keep in repair in
place of two which with the time saved in trimming and the
decrease in the investment a large saving is made in the
operation of the system.

An ingenious way of showing this lamp in operation has
been devised by the company. In a cabinet placed near
the center of the space one of these lamps is burning and
by looking through a slide in this cabinet the arc of the
lamp is seen projected on a screen greatly magnified. The

ning. The heat developed in these machines never ex-
ceeds a rise of over 60 degrees above the surrounding
atmosphere. The journal boxes are self oiling and self
adjusting, thus insuring at all times the perfect alignment
of the armature shaft. To remove or put in an armature
requires simply the removing of two bolts of the journal
stands from either side of the dynamo. The armature is
of the Gramme ring type which has long been conceded to
be the best adapted for machines of high voltage. It is
made of thin laminated rings of Norway iron compressed
tightly together under hydraulic pressure, mounted upon a
brass spider at one end. Steel rods extend horizontally
therefrom, the width of pole pieces and support the armature. The danger of grounding on the core is obviated
and under this construction the ventilation of the armature is made perfect. The commutator is simple in construction containing 126 sections made of the best grade of
hard drawn copper thoroughly insulated. Power is furnished the exhibit by a 40-horse power iron clad Rockford
motor placed near the center of the space.

The regulations of this dynamo is so perfect that the en-

EXHIBIT OF THE EDDY ELECTRIC MANUFACTURING COMPANY.

other lamps are also shown burning, current being supplied by one of the dynamos in the exhibit.

About the space are distributed arc dynamos of the differ
ent sizes manufactured by this company. One dynamo of
50 2000 candle power lights capacity, a 50 1200-candle
power light, a 40 1600-candle power light, a 30 and a 20
2000-candle power light machines are shown. There is also displayed an armature showing the open commutator
and its appearance when finished and an armature partly
completed showing the armature core and method of winding. In design the Standard arc dynamo presents some
special features in which the form of the well known Manchester type of field is carried out. The most essential
features desired in a dynamo are efficiency, durability, simplicity and ease of operation. In the Standard dynamo
these requirements have been as nearly fulfilled as is possible in the present age of dynamo building. The use of
wrought iron for the cores of the field magnet coils enables
the machine to attain its required amount of magnetism
with a much smaller expenditure of electrical energy than
if cast iron was used. The wire on the field magnets as
well as that on the armature is of unusual size, which
means less energy used in the machine; hence cooler run

tire load or any part of it may be thrown on or off without
subjecting the machine to an injurious strain. The device
is very quick and positive in its action. In the combination of the Standard dynamo and arc lamp an efficiency of
92 per cent is said to be secured. It is certainly an efficient combination and worthy of a careful study. This
exhibit shows more especially the mechanical features of
the various machines and appliances manufactured by the
Standard company while a more practical demonstration is
made by the large plant in Machinery Hall and the hundreds of arc lamps distributed about the grounds and buildings which the company installed.

Exhibit of the Eddy Electric Manufacturing Com-
pany.

The Eddy Electric Manufacturing Company of Windsor,
Conn, has quite an extensive exhibit in the east side of
Electricity Building, Section B. It occupies a space over
70 feet long and contains sample machines of the several
styles manufactured by the company.

A brass railing surrounds the space and desks and chairs
furnish accommodations to visitors as well as an office for

the representatives of the company. The exhibit is confined to motors and generators of which there are a variety shown. A part of the exhibit is located in Machinery Hall, over 1,000-horse power of generators being used for furnishing power about the grounds. These machines were in constant use during the early part of the Fair often carrying a load much above their rated capacity until other machines were in order to take part of the load.

These machines have shown great strength under the continued heavy service. The four 200 K. W. machines composing this plant are of the multipolar type.

The motors in the exhibit are supplied with current and are connected to the different generators so as to show them in operation. A 50-horse power 500-volt motor of the style designated as type C, operates a 110-volt, multipolar dynamo rated at 45 kilowatts. It is compound wound; has self

among which are one 20-horse power 220-volt motor, one seven and one-half 220-volt, a two horse power 500-volt and a one horse power 500-volt motor. Generators of a number of sizes are shown and of different patterns. These machines show care and skill in their construction. The magnetic circuits are short, the windings well made and their durability and efficiency is said to be of a high order.

The rheostats and other appliances manufactured by the company have met all requirements as to safety and durability. The exhibits of the company are in charge of Mr. U. C. Ross, manager and Mr. H. A. Balcom, electrician.

Exhibit of H. T. Paiste Switches.

In the accompanying cut is shown the exhibit of H. T.

EXHIBIT OF THE H. T. PAISTE SWITCHES.

oiling bearings and is a very smooth running machine. Another motor of 15-horse power capacity is supplied with current from the 500-volt circuit and runs at 600 revolutions. It is the latest type manufactured by the company. With two field coils and four poles it suggests the old Wenstrom motor.

This motor operates a 3,000 gallon nickel plater, known as No. 2. It runs 1,200 revolutions and furnishes current at eight volts. For the amusement of visitors a number four copper wire is connected across the terminals of this machine. The wire is soon heated to a bright red and becomes so soft that it will not support itself. The plating machine does not seem to waver while under this test. Other plating machines are shown, numbers 0 and one, which give sufficient range in sizes to supply the wants of any customer.

Various sizes of the regular type of Eddy motor is shown.

Paiste, of 10 S. 18th street, Philadelphia. The ornamental background is finished in pure white touched out with gold. The arrangement of ceiling cut-outs in the center illustrating the "xntric" movement of the switch with the words "xntric that's the switch" is unique.

Within the very neat railing, finished in natural wood, that marks the space, are placed several easels and stands. On the easels, one placed at each end of the space, are panels showing the growth of the "xntric" switch and the evolutions of the Paiste switch. On the first are placed a number of porcelain bases, each succeeding one having one piece more than the preceding one, so that the set shows the growth of the switch, step by step, from the plain porcelain base to the completed switch with its binding posts, its spring, contacts, etc.

On the second are mounted numerous switches showing the gradual development of the Paiste switch, the first losing the oldest form of the Paiste switch and the last the latest and most improved form of the "xutric."

The panels are of polished oak and present a very fine appearance. On another panel are displayed numerous porcelain block cut outs of a great variety of sizes. Visitors interested in switches will find an attendant at the exhibit who will furnish any desired information in regard to the articles shown.

The Cloud Projector.

The cloud projector spoken of in a previous issue was given a public trial last Thursday evening and proved to be all that inventors have claimed for it. The apparatus has been temporarily erected upon the southwest corner of Manufactures and Liberal Arts Building and replaces one of the search lights. It consists of a large drum containing a powerful arc lamp and fitted with a parabolic mirror for collecting and reflecting the light. In the front aperture a combination lens is placed and four iron arms extend some distance from the drum to carry another lens. By the use of these lenses it is intended to reproduce on the clouds pictures or words. When nature fails to provide the clouds, vapor, produced by exploding bombs high in the air, will be used. The method of producing the pictures is the result of experiments which have been carried on by Mr. L. H. Rogers, of the Brush Electric Co. The experiments were first made upon Mount Washington and it is said, that in the tests made there, pictures which could be plainly distinguished fifteen miles away were thrown upon the clouds. The present machine is manufactured by The Independent Electric Co., 39th and Stewart Ave., Chicago. At a private exhibition given recently at the factory President Higinbotham and other Fair officials were so satisfied that the scheme was feasible, that permission was given to conduct the experiments upon the grounds of the Exposition. Mr. Elmer E. Sperry, the well known electrician, is in charge of the apparatus, and with the assistance of Mr. E. E. Stark, has been working to simplify the adjustments necessary for the perfect control of the light. The difficulty that has been experienced has been to properly adjust the arc, mirror and lens in their relative positions, a combination very hard to secure. Mr. Sperry regards the invention as a perfect success and calls it a new discovery in optical engineering. It has been found necessary to move the apparatus to an elevated platform constructed above the walk about the roof of the Manufactures Building. When this change has been completed it is promised to give the visitors at the Fair an exhibition during which will be thrown upon the clouds the pictures of Columbus, Cleveland, President Higinbotham and Director General Davis. On Chicago day the picture of Mayor Harrison will be added to the list. Just how much the public may expect of the new device is not yet certain, but it is said that the development of the invention will lead to the spacing of the heavens and make it possible to rent certain square miles to the proprietors of Hood's soap and Pear's sarsaparilla for the purpose of advertising the merits of their respective articles. Mr. F. L. Rogers is the business manager of the Cloud Projector Company and is now making arrangements for the placing of the machines in several of the larger cities and also at Niagara Falls, where the spray arising from the falls will be used as a screen for the pictures. Several large contracts are said to have been signed, among which is one with a prominent soap manufacturer of this city.

ELECTRICAL INDUSTRIES.

PUBLISHED EVERY THURSDAY BY THE

ELECTRICAL INDUSTRIES PUBLISHING COMPANY,
INCORPORATED 1892.

MONADNOCK BLOCK, CHICAGO.
TELEPHONE HARRISON 530.

E. L. POWERS, Pres. and Treas.   E. R. WOOD, Secretary.

E. L. POWERS,   Editors.
R. A. FOSTER,
W. A. REMINGTON,   Associate Editors.
E. R. WOOD,   Eastern Manager.
FLOYD T. SHORT,   Advertising Department.

EASTERN OFFICE, WORLD BUILDING, NEW YORK.
World's Fair Headquarters, Y 27 Electricity Building.

SUBSCRIPTION:

FIVE MONTHS, $1.00
SINGLE COPY,   .10

Advertising Rates Upon Application.

News items, notes or communications of interest to World's Fair Visitors are earnestly desired for publication in these columns and will be heartily appreciated. We especially invite all visitors to call upon us or send address at once upon their arrival in city or at the grounds.
ELECTRICAL INDUSTRIES PUBLISHING CO.
Monadnock Block, Chicago.

For testing the speed of electric cars and locomotives the use of a five mile straight track and the necessary electricity has been offered. To the manufacturers of electric railway motors an excellent opportunity is thus afforded for comparing the merits of their machines. A similar trial of speed between steam locomotives exhibited at the Fair representing the different countries, England, France, Germany and the United States is being arranged.

In the additional report of the jury of awards published this week, it will be noticed that no award has been made to the Allgemeine Elektricitats Gasellschaft for their excellent exhibit in the electricity building. This is due to the fact that Director General Rathenau is a member of the jury and it was early agreed that no award should be given to a display in which any member of the jury was personally interested. For this same reason the historical exhibit in the offices of the American Institute of Electrical Engineers has received no recognition from the jury.

Chicago Day, Oct. 9, bids fair to be the greatest of all red letter days at the Exposition. No effort is being spared by the committee in charge to make the day so attractive in special features that the crowd to be seen on that day will be the largest ever gathered together in this country. Guesses on the attendance vary all the way from 400,000 to 700,000. Everything, however, depends upon the weather. Given a fine day the attendance will be all that could be desired or expected. Among the business houses in the electrical line there is a disposition to close almost without exception and give employes the opportunity to be present. There will be on this day the same electrical features as on every regular day which taken with the other special attractions will make the day one long to be remembered. Our advice to every patriotic Chicago man in whatever line of business is to be present with his friends.

WORLD'S FAIR NOTES.

The tests of insulated wires will be continued for six months under the direction of Prof. Jackson, consequently it will be some time before these awards will be announced.

With the crowds who come to Electricity Building, both during the day and evening there have come hucksters of every kind, class and description who weary the exhibitor, beguile the visitor and grow fat on the profits of their sales. The long days during the early summer and spring when customers were few are now forgotten as they reckon up their profits at the end of the day.

One of the clearest and most comprehensive souvenir guides of the buildings at the Fair was, until a few days ago, being distributed by the Hale Elevator Co., of Chicago. It contained floor plans of the different buildings and maps of the grounds completely indexed. It was too good to give away, so the catalogue monopoly thought, and the company has been restrained from distributing them on the grounds.

Irish Day at the Fair last week has the distinction of having been the worst day so far as weather is concerned since the Exposition opened. In spite, however, of the unceasing torrent of rain a large number were present to do honor to Erin. The Irish flag was hoisted over Electricity Building by Chief Barrett as announced in last issue. It is true that it was hauled down once or twice by Mr. Millet's men but the Department of Electricity came out ahead as usual, and the flag was allowed to remain.

A collection of electrical apparatus of historical value is shown in a glass case occupying a prominent position in the exhibit made by Princeton college in the gallery of Manufactures and Liberal Arts Building. The most interesting objects in the collection are a glass rod covered with sealing wax used at one time in experiments made by Franklin, a galvanic trough brought to the United States in 1800, by John Maclean and a number of roughly constructed galvanometers and electro-magnets, made by Henry. A series of induction coils are also shown with which Prof. Henry first investigated the induction of electric currents and showed the possibility of producing an electric current of high pressure from one of low pressure. Near the case stands a large electro-magnet made by Henry, which has supported a weight of 3,500 pounds. An electric machine made under the supervision of Benj. Franklin, and used by Priestly also forms part of the exhibit.

The Western Electric Co. have understood from the beginning of the Exposition, the value of moving effects and by placing on their space a number of startling features, have succeeded in making their exhibit one of the most attractive in the Electricity Building. That the public appreciates the efforts made by the company, one has only to notice the crowd gathered around the writing finger, the groups of visitors watching the bands of light chase each other up the column of incandescent lamps, and the great number always standing in line to be admitted to the scenic theatre. Another addition has been made to the exhibit, last week which is sure to attract its share of attention. A device consisting of a large wheel within a wheel, made of different colored incandescent lamps has been placed near the south entrance of the Electricity Building. The two wheels revolving in opposite directions and with the figures upon them changing constantly in figure and form make a novel and beautifully effect and are the first thing to catch the eye of the visitor upon entering the building.

Awards. Department of Electricity.

In addition to the list of awards published in the last issue there have been made the following:

UNITED STATES.

American Battery Co.
   Storage batteries.
Electrical Engineer.
   Historical electric railway model (Davenport's).
Elisha Gray.
   Telautograph.
General Electric Co.
   Incandescent lamps used for decorating rooms and other structures.
A. H. Phelps.
   Electro Pyro gravure process.
Westinghouse Electric & Mfg. Co.
   High tension experimental apparatus.

AUSTRIA.

Schindler & Jenny.
   Electrical cooking apparatus.

GERMANY.

Cosmotern Fabrick "Deutz."
   Dynamo, directly connected to gas engine.
Hartmann & Braun.
   Galvanometers of special form.
Chas. Pollock.
   Storage batteries.
Reiniger, Gebhardt & Schall.
   Electro medical apparatus.

GREAT BRITAIN.

Epstein Accumulator Co.
   Storage batteries.

Awards for Street Railways and Street Railway Appliances. Group 81.

NEW YORK

J. M. Jones' Sons, West Troy.
   Body open electric car; Body closed electric car.
John Stephenson Co., L't, New York.
   "Broadway" cable car; electric motor car.
Peckham Motor Truck & Wheel Co. Kingston.
   Electric motor trucks.
James H. Steelman, Rochester.
   Detective transfer.

MASSACHUSETTS.

Lambeth Cotton Rope Co., New Bedford.
   Lambeth Cotton Rope.
Benas Box Car Co., Springfield.
   Electric motor truck.
Robinson Electric Truck & Supply Co., Boston.
   Electric radial truck.
Coburn Trolley Track Mfg. Co., Holyoke.
   Overhead Carrying Track.
Washburn & Moen, Worcester.
   Cables for street railways.

ILLINOIS.

Columbian Intramural Railway Co., Jackson Park.
   Electric elevated railway.
Pullman Palace Car Co., Chicago.
   Single and double deck street car.
Wm. Wharton Jr., & Co., Chicago.
   Rails, fittings and special work for street railways.
Street Railway Review, Chicago.
   Street Railway Review.
International Register Co., Chicago.
   Conductor's portable register.

McGuire Mfg. Co., Chicago.
Trucks.
Gennett Air Brake Co., Chicago.
Air brake equipment for electric and cable railway street cars.

PENNSYLVANIA.

Johnson Company, Johnstown.
Street railway appliances
Robinson Machine Co., Altoona.
Electric car truck.
Westinghouse Electric & Mfg. Co., Pittsburgh.
Street railway electric car equipments.
E. H. Wilson, Philadelphia.
Open and closed vestibuled street cars.

CALIFORNIA.

A. S. Hallidie, San Francisco.
Passenger and grip car; ropeway and grip; historical collection of cable systems; pulley.
California Wire Works, San Francisco.
Wire ropes and cables.

NEW JERSEY.

Trenton Iron Co., Trenton.
Aerial tramways and rolling stock; interlocked wire ropes and cables.
Rowell-Potter Safety Stop Co., Trenton.
Automatic block and safety stop system.

GERMANY.

Hoerder Mining & Steel Co., Hoerde.
Street railway switches; grooved rails; wheel tires, axles, etc.
Phoenix Actiengesellschaft für Bergbau and Hüttenbetrieb, Laar.
Construction of street R. R. tracks, sections, profiles. etc.
Daimler Motor Co., Cannstatt.
Street car motor.
J. Pohlig, Cologne.
Photographs and plans of cable roads.
George Mary Mining Co., Osnabrück.
Rails.
Felton & Guilleaume, Mülheim.
Ropes for cable roads.

BRAZIL.

Carrio Urbana Co., Rio de Janeiro.
Tramway street horse car.

PERSONAL.

Dr. J. Allen Hornsby, of the Department of Electricity is in St. Louis, Mo.

Prof. H. S. Carhart, president of the Jury of Awards has returned to Ann Arbor.

Mr. Charles M. Wilkins, of the Partrick & Carter Co., of Philadelphia, is at the Fair.

Mr. Thorburn Reid, of Lynn, Mass., was noticed upon the Fair grounds last week.

Mr. D. E. Thompson, of the Mt. Vernon Electric Co., paid the Fair a short visit last week.

Mr. J. W. Crowther, of the Quaker City Electric Co., Philadelphia, is among the visitors.

Mr. H. K. McCay, of the McCay-Howard Engineering Company, Baltimore, is at the Fair.

Mr. J. L. Cochran, of the Edgewater Electric Light Co. was a visitor at the Fair the other day.

Mr. R. H. Moses, of Cleveland, Ohio, spent some time among the electrical exhibits last week.

Mr. Otto Lembich, an electrical engineer from Austria, is a visitor in the city and at the Fair.

Mr. R. W. Pinkerton, of the Elgin City Railway Co., Elgin, Ill., was among the visitors last week.

Mr. A. L. Daniels, of the Miamisburg, Ohio, Electric Co. was a caller on ELECTRICAL INDUSTRIES last week.

Mr. J. W. Leech, of the Keystone Electric Co., Erie, Pa., spent a few days in Chicago and at the Fair recently.

Mr. M. Freeman, superintendent of the Mt. Clemens, Mich., Electric Co. has returned home from a visit to the Fair.

Mr. M. C. Austin, secretary of the Light & Power Co., of Effingham, Ill., was in the city attending the Fair last week.

Mr. M. Kellog, superintendent of the Spencer, Ia., Electric Light Co., was among the recent arrival of visitors to the Fair.

Mr. Sydney Smith, general manager of the Hoosac Falls Electric Light & Power Co., spent part of last week at the Fair.

Mr. M. Nippert, Superintendent of the Lake Geneva, Wis., Electric Light Co. spent some time at the World's Fair recently.

Mr. T. H. Bentley, secretary of the Hastings, Mich., Electric Light and Power Co., is now in the city visiting the Exposition.

Mr. W. C. Gotshall, superintendent of the Citizens' Street Railway Co., Muncie, Ind., spent some time at the Fair last week.

Mr. W. G. Walter, of Pittsburg, Pa., designer of the booth occupied by the Westinghouse Co. has returned to the Fair for a short time.

Mr. Wm. L. Dresser, president of the Electric Light & Power Co., at Newnan, Ga., was an interested visitor last week in Electricity Building.

Amusements.

HOOLEY'S THEATER—M. Coquelin and Mme. Jane Hading, repertoire. 149 Randolph street.

COLUMBIA THEATER—Mr. Henry Irving and Miss Ellen Terry, "The Merchant of Venice." Saturday, "Louis XI." 108 Monroe street.

GRAND OPERA HOUSE—Hoyt's "A Trip to China Town." 87 Clark street.

AUDITORIUM—Imre Kiralfy's Spectacle "America." Congress street and Wabash avenue.

McVICKER'S THEATER—Wm. H. Crane, in "Brother John," 82 Madison street.

CHICAGO OPERA HOUSE—American Extravaganza Company, in "Ali Baba, or Morgiana and the Forty Thieves." Washington and Clark streets.

SCHILLER THEATER—Felix Morris, in "The Old Musician" and "Champagne." Randolph, near Dearborn.

HAVERLY'S CASINO—Haverly's United Minstrels. Wabash avenue, near Jackson street.

TROCADERO—Vaudeville. Michigan avenue near Monroe street.

THE GROTTO—Vaudeville. Michigan avenue near Monroe street.

Buffalo Bill's "Wild West." 63d street. Daily at 3 and 8.30 p.m.

THE STODDARD LECTURES—Central Music Hall, commencing Oct. 2d. Seven courses exactly alike six evenings and matinee each week. Week of October 2, Eastern Japan; week of October 9, Western Japan; week of October 16, China; week of October 23, Farther India; week of October 30, Nearer India.

This season has brought to Chicago a line of attractions never before equaled, and the appreciation of these eminent players by the people and visitors to Chicago, is shown by the crowded houses at the different theatres. At the Columbia, Mr. Henry Irving appears as Shylock and Miss Ellen Terry as Portia in the "Merchant of Venice" this week except Saturday evening. On Saturday and Monday evenings "Louis XI" will be presented and the balance of next week "Becket."

At Hooley's M. Coquelin and Mme. Hading have made a very favorable impression on the first evenings of their month's engagement. Although they speak a foreign language their art cannot be hidden, but is apparent in every movement and expression. The repertoire for a week is: Thursday evening, "Tartuffe" and "Les Preciouses Ridicules;" Friday, "La Dame Aux Camelie;" Saturday matinee, "Mlle. de la Seigliere;" Saturday evening, "Nos Intimes;" Monday and Wednesday evenings, "La Megre Appelvoise," and Tuesday evening. "La Joie Fait Peur."

THE
FERRIS WHEEL

When you visit the World's Fair, you will
naturally take a ride on the FERRIS WHEEL
and be interested in the ELECTRIC LIGHT
INSTALLATION, which is wired throughout
with

OKONITE WIRE

FURNISHED BY THE

CENTRAL ELECTRIC COMPANY,

116-118 Franklin Street,
CHICAGO, ILLS.

ELECTRICITY BUILDING—EXHIBITORS AND THEIR LOCATION.

GALLERY.

MAIN FLOOR.

Exhibitor.	Section.	Exhibitor.	Section.	Exhibitor.	Section.	Exhibitor.	Section.
Austin	T	Electrical Review	Y	Jaeger, Chas. L.	7	Reliance Guage Co.	T
Acousta Electric Co.	K	Electricity	Y	Johns Mfg. Co., H. W.	7	Roesch & Hasebacher Chem. Co.	S
Am. Inst. of Elec. Eng.	S	Electric Gas Co.	Y	Jewell Belting Co.	7	Street Railway Journal	T
American Battery Co.		Electrical Engineer	T	Jenney Elec. Motor Co.	J	Stronger Aut. Telph. Co.	T
Ashoft, H. M.	N	Electrical World	Y	Knapp Electrical Works.	V	Standard Paint Co.	T
Allg. Elec. Gesellschaft	D	Eddy Electric Motor Co	D	K. A. P. Elec. Novelty Co.	V	Sponholz, C. L.	T
Balco Mfg. Co.	D	Excelsior Electric Co.	D	Knapp & Beckley	S	Star Iron Tower Co.	W
Bryant Electric Co.	D	Electrical Forging Co.	D	Kennedy Electric Co.	U	Spain	Y
Billings & Spencer	C	Equitable Dynamo Co.	D	Lawton, H. A.		Schirren, Chas. A. & Co	D
Bracy, W. F.	Z	Electron Mfg. Co.	P	LeChancho Battery Co.	V	Schoenherr & Sohne	C
Belknap Motor Co.	E	Electrical Conduit Co.	D	McNeil-Trader Elec. Co.	U	Siemens & Halske	C
Bell Telephone Co.	G	England		Marcus, W. N.	S	Seluckert & Co.	L
Brush Electric Co.	L	Empire China Works	S	Mecker, Dr. G.	S	Short Electric Co.	L
Caldwell El. Cloth Cut Mch. Co.	S	Franklin Elec. Appliances	S	McIntosh Bat. & Opt. Co.	W	Sperry Elec. Railway Co.	M
Consol. Elec. Storage Co.	N	Frost Piano Exhibit	V	Mason, C., Belting Co.	7	Standard Ludwig, Cable Co.	J
Cutter, George		Felton & Guilleaume	C	Mather, A. C.	E	Standard Electric Co.	N
Onuteo Elec. Co.	7	France	K P-Q	Mather Electric Co.	M	Samson Battery Co.	S
Chicago Elec. Wire Co.	7	Ft. Wayne Elec. Co.	M	Newman Clock Co.	S	Tele Aut. El. Signal Co.	S
Copenhagen Fire Alarm Co.	S	Gault & Co., N. C.	S	New Magnetic Watch Co.	H	Todd, Appricate Co.	B
Central Electric Co.	L	General Fire Alarm Co.	7	N. Y. Insulated Wire Co.	J	Taylor, Goodhue & Ames	D
Commercial Cable Co.	V	General Electric Co.	B-H-N-C-E	National Carbon Co.	J	Thomson Elec. Welding Co.	O
C & C Elec. Motor Co.	J	General Incand't Arc L't Co.	E	Norwich Ins. Wire Co.	J	Telautograph, Elisha Gray	X
Cleveland Elec. & Mfg. Co.	E	Greeley, E. S. & Co.	Y	North Am. Phonograph Co.	S-P	Union Electric Co.	A
Chicago Belting Co.	7	Germany	C	N. Y. J. K. K. A	J	Vetter J. C. & Co.	W
Delancy Clock Co.	T	Halstead, Wm. J. & Co.	V	Nat. Aut. Fire Alarm Co.	S	Webb, H. J	W
Department of Electricity	K	Harleman, C. J.	S	Nat. Enameling Machine Co.	S	Western El. Instrument Co.	M
ELECTRICAL INDUSTRIES	T	Hart & Hegeman Mfg. Co.	D	Owen, Dr. A.	I	Washburn & Moen Mfg. Co.	7
Elec. Launch & Nav. Co.	U	Hotel Elec. Appliance Co.	S	Phoenix Glass Co.	S	Western Union Tel. Co.	V
Electric Separator	S	Hall, Chas. F.	S	Paste, H. T	S	Waste & Bartlett Mfg. Co.	T
Edgerton, E. M.	7	Holmes, N. S.	K	Pulvermacher Galv. Co.	Y	White, S. S., Dental Mfg. Co.	U
Elgin Telephone Co.	T	Hartman & Braun	E	Pumpelly, J. R	S	Western Electrician	Y
Edison Elec. Mfg. Co.	D	Hoosic & Van Winkle	J	Prest El. Med. App. Co.	I	Wilder Aut. Burglar Al. Co.	S
Enterprise Elec. Co.	T	Honk, J. M	V	Powell, Wm. & Co.	U	Western Electric Co.	A
EurekaTemp. Copper Co.	U	Hardmuth, F., & Co.	P	Phelp, A. H	S	Westinghouse El. & Mfg. Co.	B-H-J
Electric Appliance Co.	L	Ruston Alloy Co.	S	Page Belting Co.	7	Wiles & Neufeld	S
Elec. Sel. & Sig'l Co.	S	Internat Aut. L't & P'r Co.	S	Upson & Co.	E	Wund, L. J. & Co.	S
Electric Heat Alarm Co.	Y	India Rubber Comb Co.	S	Ringler, F. A.	R	Zucker & Levett Chem. Co.	F

DO NOT BUY SOCKETS

without first getting a sample of our

ACME SOCKET

with price. You will agree with us, that it is a very superior article at a low price.

We are prepared to quote exceptionally

LOW PRICES

on our full line of

General Supplies.

ELECTRIC APPLIANCE COMPANY,

242 Madison Street, CHICAGO.

Electrical Supplies.

THE MONTHLY ISSUE FOR OCTOBER

ELECTRICAL INDUSTRIES

Should be read by everyone interested in electrical matters. In its Table of Contents is the following:

"Electric Railway Exhibit at the Fair."
"American Search Lights at the Fair."
"Duquesne Lines of Pittsburgh."
"A New Incandescent Arc Lamp." By L. B. Marks.
"The Return Circuit of Electric Railways." By Thos. J. McTighe.
"The Business End of Electricity." By H. C. Thom.
"Three Point Incandescent Switches." By Albert Scheible, M. E.
Together with illustrations of the recent applications of electricity.

The paper contains regularly

A Buyer's Directory of Manufacturers and Dealers in Electrical Supplies and Appliances.

A Complete Directory of Electric Light Stations in North America and a Complete Directory of Electric Railways in North America.

These directories are revised each issue to the date of going to press and are to be found in no other electrical journal in the World. Its articles are read carefully and its directories used constantly by all the buyers in the trade. These facts make it without a superior as an advertising medium. Sample copies and rates sent on application.

Subscription price $3 per year.

ELECTRICAL INDUSTRIES PUB. CO.,

Monadnock Block, CHICAGO.

WM. BARAGWANATH & SON,

LIST OF HEATERS

TO BE SEEN IN OPERATION AT THE WORLD'S FAIR.

Two 500 H. P. East End Boiler Gallery doing 1800 H. P. work.
One 300 H. P. heater and receiving tank, Wellington Catering Co's., plant.
One 150 H. P. heater at Hygeia plant.
One 200 H. P. Libby Glass Works.

55 WEST DIVISION STREET, CHICAGO.

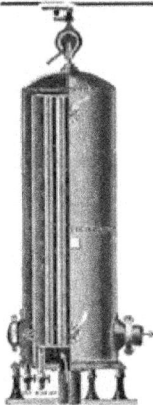

OUR IMPROVED SYSTEM

...OF...

Automatic Fire Alarm,

covered by patents recently issued, is the
embodiment of all factors contributing to the

GREATEST SAFETY,

and the MOST RELIABLE

PROTECTION FROM FIRE.

Western Electric Company,

CHICAGO and NEW YORK.

CLARK COMPANY, NEW YORK.

ELECTRIC

192 Broadway and 11 John Street.

MANUFACTURERS OF ARC LIGHTING APPARATUS FOR EVERY PURPOSE A SPECIALTY.
The CLARK ARC LAMPS for use on EVERY CURRENT, have the reputation of being the best and most durable of any ever made in the United States.

RAWHIDE PINIONS FOR ELECTRIC MOTORS
A SPECIALTY.
RAWHIDE DYNAMO BELTING

Greatest Adhesive Qualities. A Non-Conductor of Electricity.
Cannot be Chopped Thin any other Belt.

THE CHICAGO RAWHIDE MANUFACTURING CO.
THE ONLY MANUFACTURERS IN THE COUNTRY

LACE LEATHER ROPE
and OTHER RAWHIDE

GOODS
OF ALL KINDS
BY KRUEGER'S PATENT

This Belting and Lace Leather is not affected by steam or dampness, never becomes hard, is of much more durable and the most economical Belting made. The Rawhide Rope for Round Belting Transmission is superior to all others.

75 Ohio Street, CHICAGO, ILL.

STANDARD ELECTRIC COMPANY.

GENERAL OFFICES: 625 Home Insurance Building.

WORKS: So. Canal Street,

CHICAGO.

STANDARD SYSTEM

AT THE

WORLD'S FAIR.

MACHINERY HALL, Sec. Q, 2 Standard Arc Dynamos.
Sec. S, 20 " " "
ELECTRICITY BUILDING. Sec. P, Space 2, Arc Lighting Exhibit.

he Standard Lamps Light the Power Plant, Machinery Hall, Agricultural Hall, Shoe and Leather Building, and Other Buildings and Portions of the Grounds.

See our Double Service All Night Lamp Before Buying an Old Style Two Rod Lamp.

Mile after mile of
SIMPLEX WIRE

Supplied to the
FERRIS WHEEL

• • •

By...George Cutter,
The Rookery, CHICAGO.

SIMPLEX WIRES

SIMPLEX
Ever Onward and Upward

INSURE
HIGH
INSULATION

Simplex Electrical Co.
620 Atlantic Ave.,

George Cutter, Chicago.   BOSTON, MASS.

XNTRIC

"That's the Switch"

And we control that movement.

H. T. PAISTE,

10 South 16th St.,
PHILADELPHIA,
PA.

Made 5 amp. S. P.
10 mp. S. P.
5 amp. 3 way.
10 amp. 3 way.

China Window Tube (Patented).
Made only by PASS & SEYMOUR,
George Cutter,   SYRACUSE, N. Y.
CHICAGO.

Enterprise
Electric
Company

GENERAL WESTERN AGENTS

N. I. R.

307 Dearborn Street,
Chicago....

Manufacturers' Agents and Mill Representatives for

Electric Railway,
Telegraph, Telephone and
Electric Light

SUPPLIES OF EVERY DESCRIPTION

Agents for   Cedar Poles,
Cypress Poles, Oak Pins,
Locust Pins, Cross Arms, Glass
Feeder Wire,   Insulators,

WIRES, CABLES, TAPE and TUBING

BEAR IN MIND

that the regular monthly issue of ELECTRICAL IN-
DUSTRIES contains the most complete and correct
directories published of the electric light central stations
and the electric railways in North America.

World's Fair Headquarters Y 27 Electricity Building.

CITY OFFICES, Monadnock Block.

Consolidated Electric Co.

Manufacturers and Dealers in all kinds of

ELECTRICAL . SUPPLIES,

115 Franklin Street,

CHICAGO.

WAGNER ELECTRIC FAN MOTORS

For Direct or Alternating Currents.

These motors give a greater power with less consumption of current than
any other fan motor on the market. They are full 1 3 horse power. Six bladed
blade fan. Self-oiling. Furnished with or without guards.

IT WILL PAY YOU TO SEE THE WAGNER BEFORE BUYING ELSEWHERE.

TAYLOR, GOODHUE & AMES,
346 Dearborn Street, CHICAGO.

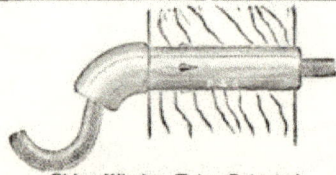

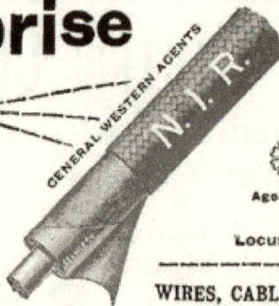

WEEKLY WORLD'S FAIR

ELECTRICAL INDUSTRIES

DEVOTED TO THE ELECTRICAL AND ALLIED INTERESTS OF THE WORLD'S FAIR, ITS VISITORS AND EXHIBITORS.

Vol. I, No. 18.   **CHICAGO, OCTOBER 12, 1893.**   FIVE MONTHS $1.00   TEN CENTS A COPY

Exhibit of the Phœnix Glass Company.

The many forms that glass will assume in the skillful hands of the glassblower, as well as the colors and delicate shading it will retain, has made it a most useful article to the decorator. Since the introduction of the electric light glass has been used more than ever for decoration, and the beautiful forms and creations in the line of electroliers, pendant and bracket lights in the modern house and hotel would fill with wonder even the more recent workers in this art that ranks as one of the oldest.

The center space in Electricity Building is occupied by the exhibit of the Phœnix Glass Company, which gives the company, as it deserves, a very prominent position. The exhibit shows a great variety of designs in glass goods for gas and electric lights. The pavilion, as shown in the

EXHIBIT OF THE PHŒNIX GLASS COMPANY.

accompanying illustration, is circular in form, with large columns supporting the roof.

A cylinder of glass mirrors rise to the ceiling in the center of the pavilion, forming a background to the shelves on which are displayed many of the glass goods. The shelves and supports are finished in white and gold, pre-seating a very handsome appearance. Above the shelves panels of mirrors alternate with panels of plush in the center of which cut-glass bowls of different colors are lighted up by incandescent lights concealed behind them. From the ceiling are suspended globes, stalactites and other forms of pendants enclosing incandescent lamps. The effect is enchanting, reminding one of a crystal cave. Some of the pendants are quite valuable, being of cut glass. The designs are quite intricate, requiring great skill in their execution.

On the shelves are displayed a great variety of the standard forms and colors of shades for gas and electric lights and a great many new designs for newel posts and hall lights. A great variety of tints are shown in the spot, wart, fluted and etched styles of glass. Flower-shaped shades for electroliers in many beautiful and delicate tints are shown. Some very neat bracket globes are displayed, on which the scenes and figures are hand painted and executed with great skill. The coloring of the paintings is especially interesting.

The display has been arranged with great care and shows excellent taste. But few duplicates are shown, the exhibit being made up of separate designs and shows an extensive line of goods. Most of the new shades and globes are shown on fixtures making the exhibit doubly attractive. These goods are used about the grounds and buildings. The shades in the decorations about the power plant in Machinery Hall were manufactured by this company, also a great deal of the glassware in Electricity Building and the globes on Wooded Island.

Great credit is due to Mr. E. H. Fox for the neat arrangement and the excellence of the exhibit in many ways. Mr. Fox has been at the exhibit a large part of the summer and has made many new friends and acquaintances among the electrical fraternity.

Complaint of the lack of public appreciation has often been heard from firms who have prepared elaborate displays in the electricity and other buildings. There is little doubt but that many visitors simply take an outside view of the Fair and, if they enter the buildings at all, bestow but a passing glance upon exhibits which have been carefully prepared and are full of interest. One reason of this, no doubt, has been the attractiveness of the grounds and the special efforts made by the Fair management to entertain the crowds with band concerts, swimming contests, Midway processions, etc. But as the cooler days of fall approach it may be expected to find the attendance in the buildings considerably increased and a greater interest taken by visitors in the various exhibits. A great improvement in this direction has been noticed already.

Exhibit of the Eureka Tempered Copper Company.

In a pavilion designed after a Moorish mosque the Eureka Tempered Copper Company of North East, Pa., has arranged an elaborate display of rolled and cast copper goods. This display is located in the southwestern part of the gallery, close to the front, where it can be seen from all parts of the main floor.

The process used by this company was discovered by an operator in the Pennsylvania oil fields. The value of tempered copper was little understood by him at that time, and the success that has attended the company far exceeded the hopes of the capitalists who formed this company. The number of uses found for this metal have gradually increased, and, consequently, the output of the metal, in the various forms. As an anti-friction metal its value was first appreciated, and large quantities were early used for journal bearings. With the rapid growth of electric lighting, and the general adoption of electric motors and other electrical machines, its special advantages for commutator bars and other parts of the machines brought it at once into great demand. Being one of the best conductors of electricity, and having a great value as an anti-friction metal, it at once filled a place in the new industry for which nothing has been found to take its place.

This company makes a specialty of commutator bars which are cast, rolled and hardened without the use of alloys. The metal when finished still has its natural toughness and pliability,

EXHIBIT OF THE EUREKA TEMPERED COPPER COMPANY.

which insures its wearing qualities and superior conductivity. The exhibits are artistically arranged about the interior of the mosque and show a great variety in the products of the company's factories. Commutators are shown in sizes ranging from ⅛ to 750 horse power; commutator bars in a number of styles, trolley wheels, gear pinions, bearings to show the extensive use of tempered copper for that purpose, and an extensive line of other goods, in the manufacture of which copper is used to a more or less extent, are shown.

The commutators of the Westinghouse machines at the Fair are made of Eureka tempered copper, which demonstrates the excellence of the metal more than the most elaborate array of samples. A very neat souvenir of the company and the Fair is given away at their exhibit. On one side it is a fac-simile of the John Scott medal given to the company for its improvements in casting and hardening copper by the Franklin Institute. On the other appears the name of the company, the World's Fair, and a commutator, brush, trolley wheel, etc., representing the goods exhibited. It is of copper—a most appropriate metal. Mr. W. H. Grissom, manager of the Chicago office, is looking after the exhibit and the interests of the many visitors.

Power Plant the Mather Electric Company.

In Machinery Hall, Section C, is located the power plant of the Mather Electric Company. It immediately attracts the attention of the visitor on account of its light color, the machines being finished in white and gold. To the street railway man it possesses many points of interest, as it is especially adapted for furnishing power for that class of work. And for operating motors for factory purposes, it is also adapted on account of its close regulation and high efficiency.

In this plant are four large generators, of which two are of 225 K. W. capacity, each working at a potential of 550 volts at 150 revolutions per minute. The other two genera-

which is constructed of slate slabs set in steel frame, all of which is enclosed in a neat and ornamental frame, on the upper panel of which appears the name of the company. The four generators are connected to one set of bus bars on this switch board, so that all the machines may be operated in multiple if so desired. The switchboard contains four rheostats, four triple pole switches, through which the generators are connected to the bus bars; four ammeters, which record the quantity of current generated by each machine; one volt meter which is so connected by means of the company's improved switch that readings may be taken from either generator or from the bus bars by making the proper connection on the switch; one differential indicator which may be connected by means of the company's indicator switch to anyone of the generators when it is desired to throw one or another of the generators into circuit while one or more are in operation, and this instrument indicates exactly when the machine to be thrown into circuit attains the proper potential. There are also on the switchboard four double pole single throw switches, four ammeters,

THE POWER PLANT OF THE MATHER ELECTRIC COMPANY.

tors which are compound wound like the preceding have a capacity of 120 K. W. and are working at a potential of 550 volts at 525 revolutions per minute. The former generators are six pole machines, with field magnets and armature supports cast in one piece without joint, except between the upper and lower half of the field.

The winding of the armature is such that there are only two paths for the current through the armature, and yet opposite bars of the commutator are connected and so maintain the same potential. The commutators are cross connected, the opposite bars being maintained always at the same potential, so that it is possible to use either all the sets of brushes or any two sets, as may be desired. The generators are mounted on sliding iron bases fitted with four screws connected together in pairs by means of a chain, and operated by ratchets, so that the generator can be kept in perfect alignment and the tension of the belt regulated. These base frames are firmly bolted to the foundations.

Near the center of the space is placed the switch board.

which show the quantity of current used on each circuit. These circuits are protected by means of double pole safety cut outs which are provided with fusible strips of suitable capacity and are also further protected by means of automatic circuit breakers mounted on separate frames behind the switchboard. The bus bars and all the connections between the switches and instruments are of finely finished copper symmetrically arranged on the back of switchboard and present a very neat appearance. These generators have been supplying the current for operating the motors in the Mines and Mining Building, Agricultural Building, for the movable sidewalk on the pier and also for the Mather Electric Company's exhibit in Electricity Building. The plant presents a neat and substantial appearance and is worthy of the attention of all power users.

The engines which operate these generators also form an interesting part of the plant. A 600 horse power Woodbury engine, manufactured by the Stearns Manufacturing Co. of Erie, Pa., operates the two larger generators and a 375 horse power the two smaller ones. These engines are of the condensing, tandem compound type and are belted directly to the generators.

The lighting plant of the Libbey Glass Works on Midway Plaisance which is an interesting feature of that exhibit was installed by this company and may well be considered a model lighting plant. The plant consists of three 55 kilowatt continuous current dynamos, wound for 125 volts, and operated by two Russel engines of 125 and 200 horse power respectively.

The current furnished by this station operates about 1,000 incandescent lamps and from 35 to 40 arc lamps. These lights are lavishly arranged about the works both inside and out, and give them a very bright and attractive appearance at night. The continuous service demanded of this plant puts the machines under a severe test that they have stood since they were installed with very creditable results. The dynamos are of the Mather ring type which is as much admired for its neat mechanical appearance as its efficiency in operation.

A machine has recently been devised by Mr. C. S. Bradley by which currents of high potential may be generated without the injurious effects to the commutator on account of the great difference of potential of the neighboring brushes. The Electrical Engineer describes the machine as follows: "The inventor has adopted the ingenious plan of mounting upon the same shaft an armature wound with a number of independent circuits, all of which contribute to the resultant effect through the medium of a number of commutators, the brushes of which are coupled together in such a way that the several windings will be in series relation to one another. The dielectric strain upon the wire insulation is reduced by connecting some intermediate point of the internal circuit with the frame of the machine. The windings are highly insulated from each other and so arranged that parts of the windings which lie in close juxtaposition upon the armature will have as low a difference of potential as is compatible with a symmetrical system of winding. Each commutator has a large number of segments in order that the potential difference between any two adjacent ones may be as small as possible and yet the aggregate difference at the brushes may be high. Each has a pair of brushes. Two brushes of opposite sign are connected together and the remaining two are connected to the terminals of the machine. There will thus be thrown upon the line a current having an e. m. f. behind it equal to the sum of the e. m. f.'s developed by the two windings, and the dielectric strain between the frame of the machine and the windings would be equal to the aggregate e. m. f., a condition which it is very necessary to avoid, inasmuch as the insulation of that portion of the winding where the potential is highest is liable to rupture. In order to avoid this difficulty Mr. Bradley connects the frame of the machine with some portion of the circuit between the line brushes, so as to raise the potential of the machine and thus lower the dielectric strain by connecting the two brushes which interlink the two windings by a good, firm connection with the base of the machine. Thus if each winding develops an e. m. f. of a thousand volts, the frame of the machine will be brought to a potential of 1,000 volts and the strain upon the insulation will only be the difference between the final e. m. f., or 2,000 volts, and 1,000 volts, or just one-half of what it would be without the base connection. In mounting the machine care is taken to highly insulate it from the earth in order to prevent leakage from the base, which, under the system of connections described, will have a considerable potential."

ELECTRICAL INDUSTRIES.

PUBLISHED EVERY THURSDAY BY THE

ELECTRICAL INDUSTRIES PUBLISHING COMPANY,

INCORPORATED 1893.

MONADNOCK BLOCK, CHICAGO.

TELEPHONE HARRISON 450.

E. L. POWERS, Pres. and Treas.   R. E. WOOD, Secretary.

E. L. POWERS, - Editor.
W. A. REMINGTON, - - - - - - - - - - - - - - - Associate Editor.
E. E. WOOD, - - - - - - - - - - - - - - - - - Eastern Manager.
FLOYD T. SHORT, - - - - - - - - - - Advertising Department.

EASTERN OFFICE, WORLD BUILDING, NEW YORK.

World's Fair Headquarters, Y 27 Electricity Building.

SUBSCRIPTION:

FIVE MONTHS, ... $1.00
SINGLE COPY, ... 10

Advertising Rates Upon Application.

News items, notes or communications of interest to World's Fair Visitors are earnestly desired for publication in these columns and will be heartily appreciated. We especially invite all visitors to call upon us or send address at once upon their arrival in city or at the grounds.
ELECTRICAL INDUSTRIES PUBLISHING CO.,
Monadnock Block, Chicago

The convention of the American Street Railway Association at Milwaukee next week should be attended by every one in any way interested in street railways, electric traction or rapid transit. Milwaukee is but a few hours ride from the World's Fair and an excursion, either by train or boat, could be easily made up. The profit derived from such a trip would more than compensate for the expense connected with it.

To mass at one point three quarters of a million people in a few hours last Monday morning was the unparalleled task which Chicago railways were called upon to perform. That it was accomplished is shown by the fact that that many people went to Jackson Park and back on the long to be remembered Chicago Day. Every piece of rolling stock possessed by the railways was called into service, and carried such loads, inside and out, on the roof as well as on the sides, as are rarely carried. The crowds for blocks around the down-town terminal stations were a remarkable sight. That such crowds were carried with so few accidents speaks well for the railways and their numerous employes. The experience of the preceding months of the Fair was worth a great deal in handling this immense traffic.

At the next meeting of the Jury of Awards, to be held on Saturday next, it is expected that the balance of the awards will be announced except those requiring considerable time for tests. The jury has been very prompt thus far in making the announcements, and has given sufficient time for adjusting any differences that may arise. Although there is dissatisfaction on the part of some at not receiving awards or at not receiving the award desired, in general there has been but little complaint in the Department of Electricity in comparison with some of the other departments. The department should certainly feel pleased with the work so far accomplished. The task is a difficult one although the awards are numerous and broad in the field covered. It would be hard to fix a value to an award, and yet to many exhibitors they will represent some return for the time and expense required to make their exhibit.

WORLD'S FAIR NOTES.

The Calumet Electric Street Railway Company has recently erected a large sign, illuminated with incandescent lamps, at the 60th street entrance to the Fair grounds, to guide the visitor on his homeward way to the proper place to take its lines for South Chicago.

Over a million paid admissions in two days is the remarkable record for Monday and Tuesday of this week. This brings the total paid admissions up to 16,814,535, and that we may see the 20 million mark reached before the Fair closes is possible.

The spray from the electric fountain, south of the Mac Monnies fountain, cut a wider pathway in five minutes in the crowd on Chicago Day than a battalion of Columbian guards was able to make in half an hour. This is not a reflection on the much abused guards.

Among the visitors at the Fair last week was Miss Sarah J. Farmer, of Eliot, Me., daughter of Prof. Moses G. Farmer a model of whose electric street car, made in 1847, is part of the exhibit of the Western Electric Company, a description of which was published in WORLD'S FAIR ELECTRICAL INDUSTRIES No. 1.

An interesting entry on the register of the American Institute of Electrical Engineers at the World's Fair is that of Mr. W. C. Wilkinson, of 5835 Drexel avenue, Chicago. It reads: "I knew Thomas Davenport in Brandon, Vt. I remember riding, when a boy, in his circular electrical railway, I cannot recall the year—perhaps in 1846. W. C. W."

If the floats are used again on Manhattan Day, as is now contemplated, it is to be hoped that those in charge will profit by former experience and that everything will move smoothly. It is hardly necessary to say that the electrical lighting of the grounds made possible the night pageant, and the electrical features contributed very materially to its success.

We are pleased to see that the impressions which our foreign visitors received on their visit to Chicago during the electrical congress were of such a kind that, on their return home, they can speak of the cordiality and hospitality of their American friends. The reports that have appeared in the foreign papers speak of the pleasure, profit and success of the congress.

The Reliance Gauge Company, Cleveland, Ohio, received the only award made by the judges in the Department of Machinery on Reliance safety alarm water columns and patent solderless floats. The exhibit of this company in the Electricity Building was described in No. 7 of the WORLD'S FAIR ELECTRICAL INDUSTRIES. The awards to his company are the bright spots in Mr. George B. Clarke's life on those days when the other boys come nearer the attendance figures than his own estimate.

One of the daily papers announced last week that a book had been opened in Electricity Building for bets on the attendance on Chicago Day. The paper gave the alleged bookmaker's name, the odds offered, etc., etc. We find there is nothing in it. The Chicago daily papers have become so accustomed to record the bookmaker as part of the attractions which draw crowds that it is hard for them to comprehend the Fair without bookmakers, but neither Electricity Building nor the Fair need bookmakers to bring it visitors.

It is to be regretted that, owing to a delay in the latter part of the procession of floats, many of the visitors on Chicago Day left the grounds before the Commerce float, lighted by many hundreds of incandescent lights, made its way around the grounds, for it was conceded to be the most beautiful of all the floats. Long after the rest of the already delayed procession had made the tour of the grounds the Edison dragon was got under way, but it was not possible to follow the original line of march, and many who had the patience to wait for it were finally disappointed. Some difficulty was also experienced with the lights on this float.

If one company more than another, of the exhibitors in the Department of Electricity, is remembered by the casual visitor it will be due to the spectacular effects produced by that company. The constant additions made to the display of the Western Electric Company have been mentioned from time to time in this column. The latest novelty in this exhibit is in the shape of a border of red, white and blue incandescent lamps, arranged in consecutive groups of five or six lamps of each color around the large sign on the north side of the Scenic Theater. The effect is produced by two strips of light of about six feet in length, apparently being to chase each other through the lamps around the border. While no display of this sort is needed to attract visitors to the Scenic Theater, the device might be used to advantage by some of the Midway Plaisance or even down town attractions.

There has been a demand on the part of exhibitors who have been awarded medals for some sort of distinguishing mark to be placed on their exhibits, and to meet this demand the authorities have decided to issue a so-called "official ribbon" to those entitled to medals and diplomas. The first ones to make their appearance in Electricity Building adorn W. R. Brixey's exhibit of Day's Kerite. In design the ribbon is neat and at the same time rich. It is of the traditional blue, about three inches wide and twelve inches long, with gold braid at top and tassels at the bottom. Printed in silver on the ribbon is a fac simile of the design of the medal together with the signatures of Gen. Geo. R. Davis, John Boyd Thatcher and the chief of the Department in which the award is made. Then follows the name of the medal winner and the article or articles on which the award was made.

Awards Department of Electricity.

The Jury of Awards announces the following additional awards in the Department of Electricity.

UNITED STATES.

Brush Electric Co.
1. Carbons for Arc Lamps.

General Electric Co.
1. The Thomson Lightning Arrester.

Queen & Co.
1. Cable testing set.
2. Conductivity apparatus.

Westinghouse Electric & Mfg. Co.
1. Automatic carbon shunt circuit breaker.

GERMANY.

Von Poppenberg.
1. Dry Batteries.

Schuckert & Co.
Station Ammeters and Voltmeters.

RUSSIA.

1. N. Wladimiroff.
Portable Storage Battery.

PERSONAL.

Mr. Wm. Hochhausen, E.E., Brooklyn, N.Y., is at the Exposition for a few days.

Mr. Fred W. Royce, E.E., Washington, D.C., is in the city and in attendance at the Fair.

Mr. Wm. S. Lintner, E.E., Gloversville, N.Y., has been doing the Fair for a few days past.

Mr. E. A. Ross, Montreal, Que., registered at the American Institute headquarters Tuesday.

Mr. George Francis Myers, Pittsburgh, Pa., was seen about the Electricity Building recently.

Mr. J. C. Van Buren, Albany, N.Y., has been viewing the electrical exhibits the past few days.

Mr. C. S. Van Nuis, New York, of "Ajax" fame, is visiting the Fair and calling on his many electrical friends.

Mr. Arthur Venning, Brooklyn, was one of the three-quarters of a million people who visited the Fair on October 9th.

Mr. C. D. Jenney of the Jenney Electric Motor Co., Indianapolis, is spending a few days in Chicago and at the Fair.

Prof. H. A. Storrs of the University of Vermont, is registered at the office of the American Institute of Electrical Engineers.

Mr. M. I. Bench, superintendent of the Malone Light, Heat, Power & Coal Company of Malone, N.Y., is in the city visiting the Fair.

Mr. H. A. Olmsted, manager of the Monterey Electric Light Co., paid a visit to ELECTRICAL INDUSTRIES World's Fair headquarters this week.

Mr. Charles Bayliss, supt. Massillon Electric Railway Co., Massillon, Ohio, called at the World's Fair office of ELECTRICAL INDUSTRIES this week.

Mr. F. L. Freeman, Washington, D.C.,assisted in the celebration of Chicago Day to the extent of getting as far as the office of the American Institute.

Mr. Edward H. Fox, of the Phoenix Glass Co., has just returned from a trip to Toledo, Cincinnati, etc., and speaks encouragingly of the outlook for the fall trade.

Mr. Thomas A. Edison, in company with Mr. John I. Beggs, formerly district manager for the Edison Company in this city, now of New York, registered at the World's Fair office of ELECTRICAL INDUSTRIES Tuesday, October 10th.

Mr. W. H. Flemming, E.E., general manager of the International Trading & Electric Co., New York, called at the World's Fair office of ELECTRICAL INDUSTRIES Tuesday evening, and expressed himself as enjoying the Exposition thoroughly.

BUSINESS NOTES.

THE CENTRAL ELECTRIC COMPANY desires to call particular attention to the Eichburg Patent Tree Insulator, listed in its new general catalogue. This is a new device which it has recently put upon the market, and which is intended to overcome the trouble brought about by the destruction of the insulation on line wire, caused from constant rubbing against branches of trees. The insulator perfectly insulates the line and gives it plenty of play when attached to a branch, permitting it to move in any direction with the swaying of the limb. This device is listed and described fully in its new catalogue.

THE ELECTRIC APPLIANCE COMPANY has just received the order for the wire and cable to be used in wiring the new west side tunnel. It will be wired throughout with celebrated Paranite High Grade Rubber Covered Wire. The selection was made entirely on the merits of the insulation, after a series of very exhaustive tests. It was one of those deals, which are altogether too rare, where quality and not price was made the sine qua non, and the Electric Appliance Company are considerably elated over their success. The conditions in tunnel service of this kind are of course very severe, and satisfactory results can be obtained only with the highest grade of insulation.

Mr. C. R. Huntley of Buffalo, has purchased for the New Lighting Co., at Niagara, N.Y., one of the No. 8 C. light, 7,000 C.P., arc dynamos comprising the service plant of the Brush company in Machinery Hall, to be delivered at the close of the Exposition. This leaves but one of the sixteen dynamos unsold. There will also accompany the dynamos 100 of the double carbon Brush-Adams lamps which are now doing duty on the "all night" circuit at the Fair. The Indianapolis Light & Power Co., have placed an additional order for 100 double carbon lamps which makes 856 which they have bought within eight months. These, taken in conjunction with the 15.65 light dynamos which they have purchased during the same period shows their appreciation of the Brush system.

DEPARTMENT OF ELECTRICITY.

OFFICES: SECTION M, ELECTRICITY BUILDING.

Chief, JOHN P. BARRETT.
Assistant Chief, J. ALLEN HORNSBY.
General Superintendent, J. W. BLAISDELL.
Electrical Engineer, W. W. PRIBE.

DEPARTMENT OF MECHANICAL AND ELECTRICAL ENGINEERING.

OFFICES SOUTH OF MACHINERY HALL.

Mechanical Engineer, C. F. FOSTER.
Electrical Engineer, R. H. PIERCE.
First Asst. Mechanical Engineer, JOHN MEADEN.
First Asst. Electrical Engineer, S. G. NEILER.

AMERICAN INSTITUTE OF ELECTRICAL ENGINEERS.

World's Fair Headquarters,
SECTION S, ELECTRICITY BUILDING.

RALPH W. POPE, Secretary

Open from 9 a.m. to 5 p.m.

Amusements.

HOOLEY'S THEATER—M. Coquelin and Mme. Jane Hading, repertoire. 149 Randolph street.

COLUMBIA THEATER—Mr. Henry Irving and Miss Ellen Terry, "The Merchant of Venice." Saturday, "The Bells." Monday "Becket." 108 Monroe street.

GRAND OPERA HOUSE—Hoyt's "A Trip to China Town." 87 Clark street.

AUDITORIUM—Imre Kiralfy's Spectacle "America." Congress street and Wabash avenue.

McVICKER'S THEATER—Wm. H. Crane, in "Brother John," 82 Madison street.

CHICAGO OPERA HOUSE—American Extravaganza Company in "Sinbad." Washington and Clark streets.

SCHILLER THEATER—Felix Morris, in "Kerry and the Major." Randolph, near Dearborn.

HAVERLY'S CASINO—Haverly's United Minstrels. Wabash avenue, near Jackson street.

TROCADERO—Vaudeville. Michigan avenue near Monroe street.

THE GROTTO—Vaudeville. Michigan avenue near Monroe street.

Buffalo Bill's "Wild West." 63d street. Daily at 3 and 8.30 p.m.

THE STODDARD LECTURES—Central Music Hall, commencing Oct. 2d. Seven courses exactly alike six evenings and matinee each week. Week of October 2, Eastern Japan; week of October 9, Western Japan; week of October 16, China; week of October 23, Farther India; week of October 30, Nearer India.

"Sinbad" which had such a remarkable run when presented a few years ago, is revived tonight at the Chicago Opera House. The cast is announced as follows: Sinbad, Louise Royce; Ninetta, Frankie M. Raymond; Maraschina, Ada Deaves; Salamagundi, Lillie Laurel; Cupid, Nellie Lynch; Rafael, Edith Rice; Angelo, Bessie Lynch; Fiametta, Edna Thornton; Zerlina, May Lowrey; Count Maladetto Spaghetti, William Armstrong; Smarleyow, Henry Norman; Old Man of the Sea, Joseph Doner; Nicilo, Jack Gailmette; and Fresco, Edwin Foy. The premiere dancers will be Martha Irmler, premiere danseuse assoluta, Madeline Morando and Hulda Irmler, premiere danseuse, and Signor Nicola Guerra, principal male dancer. Gerard Coventry will be the stage manager; W. H. Batchelor, the musical director; Signor Filiberti Marchetti, maitre de ballet, and Martin Krueger, principal electrician. All the scenery will be from the brush of Frederick Dangerfield. The entire force of the American Extravaganza Company will be engaged in the production, and the piece on its travels and in Chicago will be under the personal direction of David Henderson, while the business staff will be headed by George Bowles, business manager, and Daniel McCullough, acting manager.

ELECTRICAL INDUSTRIES

THE
FERRIS WHEEL

When you visit the World's Fair, you will
naturally take a ride on the FERRIS WHEEL
and be interested in the ELECTRIC LIGHT
INSTALLATION, which is wired throughout
with

OKONITE WIRE

FURNISHED BY THE

CENTRAL ELECTRIC COMPANY,
116-118 Franklin Street,
CHICAGO, ILLS.

ELECTRICITY BUILDING—EXHIBITORS AND THEIR LOCATION.

GALLERY.

MAIN FLOOR.

LINE-WIRE		CROSS-ARMS
PINS	**INSULATORS**	BRACKETS
LAMPS	General Construction Material	SOCKETS
SWITCHES	**CUT-OUTS**	LAMP-CORDS
CARBONS	Incandescent Lighting Supplies	GLOBES
ARC CUT-OUTS	**SPARK-ARRESTERS**	CARBON HOLDERS
	Arc Lighting Supplies.	

ELECTRIC APPLIANCE COMPANY,

242 Madison Street, CHICAGO.

NEW YORK	PITTSBURGH	CHICAGO
42 Murray Street.	43 Sixth Avenue.	19 & 21 Wabash Avenue.

The Phœnix Glass Co.

Manufacture the largest line of

ELECTRIC, GAS AND OIL

GLOBES AND SHADES

IN THE WORLD.

ORIGINAL DESIGNS. NO COPIES.

An inspection will prove our lines to be desirable—and profitable to the dealer purchasing them.

PHŒNIX GLASS CO.

42 Murray Street, NEW YORK.   43 Sixth Ave., PITTSBURGH, PA.   19 & 21 Wabash Ave., CHICAGO, ILL.

Phœnix Glass Co.'s Exhibit can be found in the Center of Electricity Building.

THE REGULAR MONTHLY EDITION OF

ELECTRICAL INDUSTRIES

Is the most complete Electrical Journal published.

Every issue containing descriptions of all the new applications of electricity, complete directories of the Manufacturers and Dealers, the Electric Lighting and Railway Companies in North America, revised and corrected to the date of going to press. These special features are found in no other Electrical Journal in the world, and consequently it is read by more actual buyers than any other publication, which fact makes it without a superior as an advertising medium.

ELECTRICAL INDUSTRIES PUBLISHING CO., Monadnock Block, CHICAGO.

OUR IMPROVED SYSTEM

... OF ...

Automatic Fire Alarm,

covered by patents recently issued, is the
embodiment of all factors contributing to the

GREATEST SAFETY,
and the MOST RELIABLE
PROTECTION FROM FIRE.

———

Western Electric Company,
CHICAGO and NEW YORK.

CLARK ELECTRIC COMPANY, NEW YORK.

192 Broadway and 11 John Street.

MANUFACTURERS OF ARC LIGHTING APPARATUS FOR EVERY PURPOSE A SPECIALTY.
The CLARK ARC LAMPS for use on EVERY CURRENT, have the reputation of being
the best and most durable of any ever made in the United States.

RAWHIDE PINIONS FOR ELECTRIC MOTORS

A SPECIALTY.

RAWHIDE DYNAMO BELTING

THE CHICAGO RAWHIDE MANUFACTURING CO.

THE ONLY MANUFACTURERS IN THE COUNTRY

LACE LEATHER ROPE
AND OTHER RAWHIDE
GOODS
OF ALL KINDS
BY KRUEGER'S PATENT

75 Ohio Street. CHICAGO, ILL.

STANDARD ELECTRIC COMPANY.

GENERAL OFFICES: 625 Home Insurance Building,

WORKS: So. Canal Street,

CHICAGO.

STANDARD SYSTEM

AT THE

WORLD'S FAIR.

MACHINERY HALL, Sec. Q, 2 Standard Arc Dynamos.

Sec. S, 20 " " "

ELECTRICITY BUILDING, Sec. P, Space 2, Arc Lighting Exhibit.

The Standard Lamps Light the Power Plant, Machinery Hall, Agricultural Hall, Shoe and Leather Building, and
Other Buildings and Portions of the Grounds.

See our Double Service All Night Lamp Before Buying an Old Style Two Rod Lamp.

12   ELECTRICAL INDUSTRIES.

Mile after mile of
SIMPLEX WIRE
Supplied to the
FERRIS WHEEL
...
By..George Cutter,
The Rookery, CHICAGO.

SIMPLEX WIRES

INSURE
HIGH
INSULATION

SIMPLEX

Ever Onward and Upward!

Simplex Electrical Co.
620 Atlantic Ave.,
George Cutter, Chicago.   BOSTON, MASS.

XNTRIC

"That's the Switch"

And we control that movement.

H. T. PAISTE,

10 South 18th St.,
PHILADELPHIA,
PA.

Made 5 amp. S. P.
      10 mp. S. P.
      5 amp. 3 way.
      10 amp. 3 way.

China Window Tube (Patented).

Made only by **PASS & SEYMOUR,**
George Cutter,   SYRACUSE, N. Y.
CHICAGO.

Enterprise
Electric
Company

GENERAL WESTERN AGENTS

N. I. R.

307 Dearborn Street,
Chicago....

Manufacturers' Agents and Mill Representatives for

Electric Railway,
Telegraph, Telephone and
Electric Light

SUPPLIES OF EVERY DESCRIPTION

Agents for   Cedar Poles,
              Cypress Poles, Oak Pins,
Locust Pins, Cross Arms, Class
———— Feeder Wire,   Insulators,

WIRES, CABLES, TAPE and TUBING

BEAR IN MIND

that the regular monthly issue of ELECTRICAL IN-
DUSTRIES contains the most complete and correct
directories published of the electric light central stations
and the electric railways in North America.

World's Fair Headquarters Y 27 Electricity Building.

CITY OFFICES, Monadnock Block.

Consolidated Electric Co.

Manufacturers and Dealers in all kinds of

ELECTRICAL . SUPPLIES,

115 Franklin Street,

CHICAGO.

WAGNER ELECTRIC FAN MOTORS

For Direct or Alternating Currents.

These motors give a stronger breeze with less consumption of current than
any other fan motor on the market. They are full 1⁄8 horse power. Six bladed
16-inch fan, self-oiling. Furnished with or without cords.

IT WILL PAY YOU TO SEE THE WAGNER BEFORE BUYING ELSEWHERE

TAYLOR, GOODHUE & AMES,
348 Dearborn Street, CHICAGO.

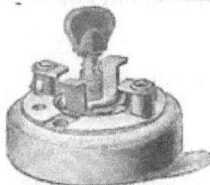

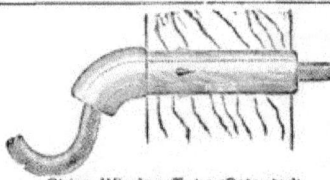

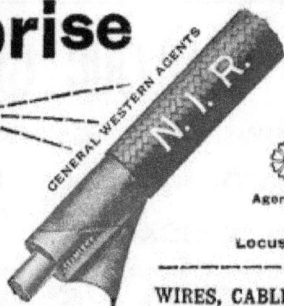

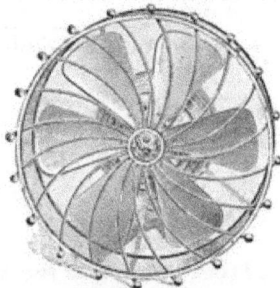

WEEKLY WORLD'S FAIR

ELECTRICAL INDUSTRIES

DEVOTED TO THE ELECTRICAL AND ALLIED INTERESTS OF THE WORLD'S FAIR, ITS VISITORS AND EXHIBITORS.

Vol. I, No. 19.　　　　CHICAGO, OCTOBER 19, 1893.　　　　FIVE MONTHS $1.00
TEN CENTS A COPY

Exhibit of Pullman Street Cars.

No department probably shows such extended and varied exhibits as are displayed in the department of Transportation. These exhibits show a gradual advancement and a complete evolution of ideas in the succeeding stages of the art. Among car builders the lines followed in the construction of street cars and coaches for steam roads have rapidly diverged since the introduction of the railway. While in cars for steam roads the comfort and safety of the passenger has been a special point aimed at, in street cars these points have been in general but a secondary consideration. However in the designs of the street cars exhibited by the Pullman company at the World's Fair these points have received special attention.

In connection with the extensive display of passenger

EXHIBIT OF PULLMAN STREET CARS.

cars made by the Pullman's Palace Car Company are a number of street cars embodying the more advanced ideas and improvements in street car construction. The accompanying engraving gives a view of the exhibits looking toward the west. Besides the features affording comfort and increased safety to the passenger, the double carrying capacity and the reduced cost per passenger of operation are worthy of attention. In the foreground is shown a double deck center vestibule motor car. The following dimensions will indicate its size. It is thirty feet eight and seven eights inches long over all; eight feet two inches wide and fifteen feet five inches in height. The center vestibule divides the car below into two compartments and dispenses with the end platform. At each end of the upper deck a pilot house is provided for the motor man.

The outside of the car is sheathed in narrow widths and painted in royal blue with special decorations in gold. The interior is finished in mahogany with decorated ceiling. The windows are of Chances' crystal sheet glass and plate glass in the doors. In the doors of each compartment are bevelled mirrors also four in the vestibule and four at each door. The windows are fitted with fringed curtains hung on Hartshorn rollers. The car is lighted by Hicks & Smith combination oil and electric lamps, one in each compartment. The seats are of Hale & Kilburn spring pattern covered with tapestry. The car is provided with eight

might these platform seats be given to passengers who wish to smoke.

The stairways going to the upper deck of this car are not of the same design as those in the center vestibule motor car above described there being but a single stairway constructed into each side of the car instead of dividing stairways. The inside finish of the cars is of Mexican mahogany of special design with ceiling handsomely decorated in gold. It is fitted with spring seats upholstered with Wilton plush carpet. The glass in the windows is Chances' crystal plate bevelled and in the doors one-quarter inch polished plate. The curtains at the windows are adjusted by Hartshorn rollers and the floor is fitted with Everett floor matting. The trimmings are all of bronze. The upper deck has slat seats and has an iron handrail and wire screen guard around the top of the car. The trucks under this car are of Pullman manufacture equipped with iron wheels and steel axles and in addition to the ordinary brakes, with air brakes.

Immediately to the rear of the center vestibule double

FIG. 1.—EXHIBITS OF ELECTRIC RAILWAY CARS AT THE FAIR—STEPHENSON AND BROWNELL CARS IN ELECTRICITY BUILDING.

electric heaters, electric signal bell and gongs on each pilot house. The trimmings are all of bronze and four folding gates are placed on the platform. The car is mounted on Pullman trucks which are equipped with 30 inch Allen wheels, two 20-horse power Westinghouse electric motors and Sessions' friction brake.

Next to this is a double deck center vestibule trail car. As will be noticed it is but a little higher than the ordinary street car, being eleven feet three and one-half inches high or three and a half inches higher than the ordinary car. The outside is finished similar to the above with gold ornamentation of a special design. There are six large windows and one entrance on each side. There is a platform at each end for the motor man, very much like the platforms on ordinary cars except that they are entirely closed by a wire screen so as not to permit anyone to get on or off except through the entrances at the sides. Seats are provided on each platform to accommodate several passengers thus further carrying out the idea of utilizing all the available space and giving the greatest comfort possible, especially

deck trail car will be seen an ordinary single truck car equipped with the double deck feature, with stairways leading to the upper deck from the platforms at each end; while to the right of the illustration can be seen the new wide vestibules which have attracted so much attention in the Pullman company's exhibit of coaches.

Exhibits of Electric Railway Cars at the Fair.

The display of street cars at the Fair does not appear as extensive to the casual observer as the importance of the subject would seem to demand. This is in a measure true, but the exhibits of cars are considerably scattered and consequently to the visitor to Transportation Building, the display looks small after following aisle after aisle of steam cars and locomotives. The exhibits display the latest ideas in the industry and are certainly models of their kind.

At the east end of street car row appear two large sketches of the first street car, John Mason, built by the John Stephenson Co., in 1831. These are interesting especially when compared with the modern cars exhibited in the same space by that company. The stage coach style of car in its clumsiness is but a crude affair when placed by the side of the trim, neat but substantial cars which the John Stephenson Company, Ltd., of New York, have placed on exhibition. These cars are but sample cars and were built as part of recent orders, one for the Broadway cable line and the other for the electric line of the Elmira and Horse-heads Street Railway of Elmira, N. Y.

The latter is mounted on a Tackaberry truck and is built to accommodate Westinghouse electric motors and equipment. The car presents a very neat external appearance and is painted a light buff color. The interior is finished in quartered oak, upholstered seats, wide windows with window shades, double doors etc. The floor is covered with rolling wooden matting; the platforms are wide and the steps low and broad. The cars show excellent workmanship and great skill in the design. The company has also a car on exhibition in the exhibit of the Westinghouse Electric Mfg. Co. in Electricity Building which is com

tions are undoubtedly the most elaborate ever attempted and the car attracts the attention of every visitor. This company also have a car on exhibition in connection with the Westinghouse company in Electricity Building.

Both open and closed cars are exhibited by J. M. Jones Sons Company of West Troy, N.Y., near the east end of Street Car Row. The closed car is designed for electric traction and is mounted on a Taylor improved electric truck. The body of the car is 18 feet long. There are six windows on each side of French plate glass and two on each end. The platforms are large and designed to carry a large number of passengers. It is fitted with Wilsons improved sand box, foot gongs, Jones ratchet brake handle, and the West End trolley. The doors are double at each end so constructed that both are opened or closed as the case may be whenever either is moved. The interior is finished in mahogany with quartered oak ceiling. Smith electric lamps and shades are provided of a very neat design. The seats and backs are upholstered in blue plush and windows equipped with curtains of blue and gold

FIG. 2.—EXHIBITS OF ELECTRIC RAILWAY CARS AT THE FAIR—LAMOKIN OPEN CAR.

pletely equipped with Westinghouse apparatus and supplied with current showing to excellent advantage the lighting of the car, the motive power and controlling apparatus.

The Brownell Car Company has on exhibition a model car which is certainly handsome in appearance both inside and out. It is finished in white, yellow and gold. The car is 28 feet over all and 7 feet 10 inches wide. It is mounted on a Brownell non oscillating electric motor car truck. The platforms are wide and have solid wrought metal steps fitted with rubber pads. On the center panel appear the Latin words "capacitas, pecuniae, dividenda, commoditas," indicating the things most sought for in a street car, i.e. capacity and commodiousness, money and dividends. The interior is finished in white and gold and is handsomely decorated. The seats and backs are old gold silk plush upholstered with curled hair. In the center panels are plate glass mirrors. The doors are of mahogany and are placed two at each end. The roof is of the improved Brownell truss form, has patented truss trolley bridge and is equipped with trolley complete. The decora-

Bronze poles with nickel hand straps are also provided for standing passengers. The trimmings of the car both inside and out are of polished bronze. The main panels on the sides and ends are finished in blue and bronze borders and gold stripes.

The open car is also constructed for electric service and is mounted on a Bemis truck. It has five reversible seats and four lamps. The reversible seats are maple while the others are paneled in cherry. At each end there are three drop sashes with French plate glass and the sides are provided with heavy curtains, which are adjustable and run in guides. The ceiling is of white birch decorated with cherry moulding. The car is also equipped with solid bronze trimmings both inside and out. It is also provided with the Smith lamps, New Haven register, foot gongs, etc.

The Lamokin Car Company, have also an attractive exhibit in Street Car Row, consisting of two street cars one open and the other closed, which show great skill in construction and design. The closed car which is equipped with two 25 horse power Westinghouse motors is mounted on a McGuire truck. The car is vestibule with large observatory windows. The steps are of the type made by the Stanwood Manufacturing Company, the trimmings are of solid bronze, the seats are of antique oak, the backs are upholstered in plush, and the windows are furnished with curtains of the roller type. The interior of the car is finely

car is mounted on an all steel Robinson truck manufactured by the Robinson Machine Company of Altoona, Pa. The fittings are all of bronze and the decorations both inside and out are neat and tasty.

Street car trucks are extensively exhibited. The exhibit made by the Peckham Motor Truck & Wheel Company, of

EXHIBIT OF THE STANDARD STREET CAR STOVES AT THE WORLD'S FAIR

finished and is equipped with the Cochrane electric heaters. The doors are of solid oak and the windows are of French plate double glass.

The open car is of a design that has met with favor with a great many companies and is especially adapted for heavy summer excursion traffic. It has vestibuled ends with large

Kingston, N.Y., comprises the latest and most improved form of truck which are built for both long and short cars. The excellence of the construction is shown in the individual parts of the truck separately exhibited. In the accompanying illustration is shown truck No. 6 A. design for closed cars up to 20 feet in length. It is of the new

FIG. 5.—EXHIBITS OF ELECTRIC RAILWAY CARS AT THE FAIR—McGUIRE BICYCLE TRUCK.

windows. The sides are provided with drop curtains, the car has seven seats and on each side of the motor man are side seats capable of holding two passengers each. The Stanwood steel step is used and the platforms are also equipped with foot gongs and headlights. It has also Cochrane sand boxes and a Cochrane trolley board. The

oscillating type and has spiral springs on the underside of the frame.

The McGuire Manufacturing Company displays a number of styles of trucks in its space in the Transportation Building as well as in connection with other exhibitors in Electricity Building and in other parts of the grounds.

The new adjustable bicycle truck which possesses many points of excellence has undoubtedly attracted the most attention. It is the latest improvement made by the com-

THE PECKHAM TRUCK.

pany in trucks for long cars. It overcomes several of the objectionable features of long cars on swivel trucks. The truck has large drive wheels with small guide or auxiliary

known Columbian four wheeled truck which is familiar to most street car men. It is also shown in connection with several exhibitors in Electricity Building and elsewhere.

Exhibit of Stirling Boilers.

The accompanying illustration gives an excellent view of The Stirling Company's exhibit in the main boiler room at the World's Columbian Exposition. It comprises two batteries of 800 horse power each of the well known Stirling type of water tube boilers, the construction of which is familiar to most of our readers. A complete description appeared in the June number of ELECTRICAL INDUSTRIES. These boilers have received a good deal of attention since they were installed and the fact that they received a medal and diploma from the jury of awards is sufficient evidence

EXHIBIT OF STIRLING BOILERS AT THE FAIR.

wheels. The car may be mounted as low as a four wheel car, the truck being open for the motor. The entire load is carried, it is said, on the eight spiral springs, located over the two driving axles, on a straight track while on curves the proportion transferred to the guide wheels varies, but is always sufficient to prevent derailment.

This is effected by a device consisting of a half elliptical spring located nearly over the drive wheel and a double cam attached to the car body directly over the spring. Mounted on the spring is a roller which fits into a recess in the center of the double cam thereby performing two functions. While on a straight track it serves as a centering device for the truck, at the instant a curve is encountered the roller acts upon the cam depressing the spring and thereby assuming part of the load and supporting it on the guide wheels. Among the other trucks shown is the well

of their excellence. The company has made a good fight for recognition at the Fair and is to be congratulated on the success it has attained.

Exhibit of the Standard Street Car Stoves.

In the Transportation Annex, street car row, the Standard Railway Supply Company, Monadnock Block, Chicago, has a comprehensive exhibit of the Standard stoves. The general arrangement of the exhibit is shown in the accompanying cut. Car seats are arranged about the space of both the upholstered and veneered patterns with the Standard stove attached showing how it can be placed in any car without cutting the seat, no matter what the finish or the shape of the seat may be.

The old street car manager as he inspects this display of

neat and compactly designed heaters recalls his varied experience in heating his cars. He remembers the cars where there was no stove and in which the passenger poked his feet in the straw on the bottom to keep them warm; then the introduction of the common house-heating stove which was placed at one end of the car, affording some heat and great inconvenience. Then he recalls how some enterprising inventor persuaded him to cut out the seat at the center of the car and install a stove which was protected by heavy boxing. But even this had many objectionable features. As winter approached the car was run into the shop and equipped for the winter and when spring came it made another trip to the shop to be prepared for the summer traffic.

In the "Standard" stoves here displayed the acme of perfection seems to have been reached. It requires but a few minutes to place it in the car and but a few minutes to remove it when it is no longer required. The cushioned seats are uninjured and the car is neither disfigured nor marred when it is taken out. The stoves are constructed entirely of cast iron in such a manner as to make them durable and efficient.

The boxing or casing of the stove is constructed of hard wood which is finished to match the interior of the car. It is heavily insulated with asbestos, thus protecting the car and shielding passengers sitting next to the stove. The stove has a large ash pan which, with the opportunity for a coal box beneath the seat, all the necessary adjuncts are conveniently furnished. The neat design and elegant nickel finish of the metal work make it an ornament to the car.

American Battery Company's Exhibit.

The exhibit of the American Battery Company at the World's Fair shows its storage batteries in operation for lighting and for power. A plant of 52,450 ampere hour cells connected in two series of 26 each is shown, furnishing a current for 30 50-volt 16 candle power lamps. The use of this battery for power is shown in the electric carriage. This carriage has a seating capacity for six people and is equipped with 24 150-ampere hour cells.

A two and one-half horse power motor wound for 50 volts is geared to the back axle of the carriage. The controlling apparatus consists of a contact board and a reversing switch, both fastened to the under side of the carriage floor. A spring plug and foot switch is placed in the floor of the carriage directly in front of the driver, who has only to remove his foot from this spring switch and allow the small contact plug to raise from the cup of mercury. This action breaks the circuit between the battery and motor. One of the most important features of the carriage is the arrangement of the front wheels to facilitate steering. Each of the two front wheels turns upon a bearing that is swivelled to swing horizontally and independently of the axle.

The front axle is rigidly fixed in a position parallel to the back axle. A yoke bar carrying a rack couples the two front wheels and the gears working in this rack and operating through the steering wheel shaft, permits of the two wheels being swung to the right or left so as to guide the carriage.

These batteries are what might be called an improved form of the Planté cell. No lead oxides or active material is mechanically applied. The plates are formed by oxidization in an electro-chemical bath. The plates consist of a series of horizontal strips of rolled lead crimped on both sides three-eighths and one-half inch wide and of such thickness and separated by such distances as by tests and experiments gives the greatest capacity combined with the longest life. These strips are solidly united at the ends by a process which leaves the sides of the plate a solid bar of metal of special composition, not affected by chemicals. The plate is then supplied with foot and terminal pieces very firmly connected of special composition. The plate is then ready for formation and the plate is submitted to the forming process which oxidizes the surface of the lead strips, but attacking but slightly the surrounding support. The oxide taking more space, expands and the process is continued until the interstices between the strips are filled solidly, thus leaving the alternate strips of lead and active material horizontally across the plate and surrounded by a solid frame of metal. This makes the plate very solid and strong. The lead strips being not

(Continued on page 8.)

ELECTRICAL INDUSTRIES.

PUBLISHED EVERY THURSDAY BY THE

ELECTRICAL INDUSTRIES PUBLISHING COMPANY,

INCORPORATED 1893.

MONADNOCK BLOCK, CHICAGO.

TELEPHONE HARRISON 396.

R. L. POWERS, PRES. AND TREAS.   K. E. WOOD, SECRETARY.

E. L. POWERS, - - - - - - - - - - - - - - - - Editor.
W. A. REMINGTON, - - - - - - - - - Associate Editor.
K. E. WOOD, - - - - - - - - - - - - Eastern Manager.
FLOYD T. SHORT, - - - - Advertising Department.

EASTERN OFFICE, WORLD BUILDING, NEW YORK.

World's Fair Headquarters, Y 27 Electricity Building.

SUBSCRIPTIONS.
FIVE MONTHS, - - - - - - - - - - - - - - $1.00
SINGLE COPY, - - - - - - - - - - - - - - - - 10
Advertising Rates Upon Application.

News items, notes or communications of interest to World's Fair
Visitors are earnestly desired for publication in these columns and will
be heartily appreciated. We especially invite all visitors to call upon us
or send address at once upon their arrival in city or at the grounds.
ELECTRICAL INDUSTRIES PUBLISHING CO.,
Monadnock Block, Chicago.

The fire last Thursday night at the 39th St. barns of the Chicago City Railway resulted in a heavy loss to the barns and contents. Nearly four hundred horses were lost, a number of cars, feed, etc. The damage was more severe, occurring as it did just at the time when the traffic was heaviest on account of the World's Fair. The resources of the company, however, were such as to enable it to provide an immediate service. It is the intention of the company to equip the entire system of cross town lines with electric traction, a part of which has already been equipped and the system will gradually be extended until it is completed.

A most disastrous and probably fatal accident occurred at the power house of the Atlantic Avenue Railroad Company, of Brooklyn, on last Wednesday, October 11. The power house is located at the corner of Third Ave. and Second St., and supplies power for the extensive Atlantic Ave. system. One of the large fly wheels, 18 feet in diameter and weighing some 20 tons, burst and flew in every direction, badly wrecking the building and seriously injuring the engineer and several persons in adjoining buildings. The accident necessarily stopped the cars, but power was secured from the Brooklyn City Railway's power house, and the cars were again running.

Manhattan Day, for which Oct. 21st has been designated, will undoubtedly be a memorable occasion. Mayor Gilroy has issued a proclamation full of wise counsel for the citizens of the metropolis and the New York papers urge an attendance on that day of every citizen of New York City who can possibly attend. The World says: "New York owes it to herself to make Oct. 21st an occasion effectively and forever contradicting all ridiculous suggestions of jealousy on our part toward the sister city which has rendered so notable a service to the whole country in making this the greatest of World's Fairs." Everything is favorable for a large attendance on that day. Railroad rates are extremely low, the railroads are prepared to handle a large traffic, and Chicago is ready to welcome any number of the citizens of the eastern seaport.

The Fair is certainly an educator. Prof. H. B. Smith, of Perdue University is spending the week with his class in electrical engineering, numbering nearly thirty, inspecting the exhibits in the Department of Electricity. It is to be regretted that other colleges, not located as near to Chicago as the Perdue University, cannot well follow this most excellent example set by Prof. Smith.

Mr. Willard A. Smith, chief of the Department of Transportation, has expressed his willingness to co-operate in any arrangements made for the convenience of the railway people who may come from the convention to the World's Fair. Should the convention desire to visit the Transportation Building in a body, he will have the street railway exhibits roped off and the general public excluded during the pleasure of the street railway visitors.

The Milwaukee Convention.

The conditions are in every way favorable for an unusually large attendance at the street railway convention held in Milwaukee this week. No doubt the nearness of the meeting to the World's Fair has prevented many, and especially the larger exhibitors, from making the elaborate display they otherwise would were it not for their costly exhibits at the Fair, but the street railway men will be present just the same. The number and variety of practical papers that are to be presented and the discussions that will arise will be of great value and will well repay all who hear them. The membership of the American Street Railway Association has steadily grown to large proportions and its importance is fully recognized. The number of members, however, should be larger and we hope to see a very material increase in the membership during the coming year. Street railway men, as a rule, are progressive as is shown by the rapid adoption of electricity as a method of traction over horse and other motive powers and great benefit to the association will arise from the meeting.

Supply men will of course be present in large numbers and what would a convention be without them? Their aid in developing this new field has been of incalculable value.

The arrangements made by the committees in charge for the accommodation and pleasure of the delegates and their friends during their stay in the Cream City are all that could be desired. Milwaukee has a well deserved reputation that is known far and wide for hospitality and generous welcome to strangers. We are sure that no one will go away without the best impressions of the city and her street railway men. Much good will come of the meeting. It is to be hoped that all of the delegates will avail themselves of the opportunity afforded to visit the World's Fair and examine the street railway exhibits in the Electrical and Transportation buildings. The chiefs of both departments have extended a cordial invitation to them to come and we trust that the exhibits made there by the representative manufacturers will be inspected by every street railway manager. We are sure they will be well repaid.

The band stand just south of the Electricity Building was occupied by the Elgin Band last week, while this week Innes' 13th Regiment Band of New York City caters to the public ear.

(Continued from page 6.)

over one-tenth of an inch apart, allow a large as well as intimate connection with all parts of the active material thus enabling them to obtain heavy charge and discharge rates. Corrugations of the lead strips effectively key the active material in place preventing its falling out, while the strips being placed edgewise to the plate solidly filled with active material, gives the plate rigidity sidewise, nearly equal to a solid lead plate of light thickness thereby preventing buckling or bending.

After the plates are formed, they are assembled together, the number of plates being governed by the size of cell desired commencing with the outside plate, and are separated by rubber insulators called combs. Every alternate one is negative while those between form the positive, there being one more negative than positive plates. The plates are then firmly united by fusing a T shaped piece of rolled lead, positive to positive and negative to negative plates. This piece having a projecting strip for conveniently connecting the cells. The elements are then placed in a containing jar of glass or hard rubber which is then filled with a solution of one part sulphuric acid to five parts water. The cell when charged is ready for work.

WORLD'S FAIR NOTES.

The display of Christoffet Cie in the French section shows the progress which has been achieved by the aid of electricity in art of galvano plastic and of electro chimic. Many of the reproductions in metal of ancient works of art are more than ordinarily instructing.

"Can you tell me where I will find the exhibit of aerolites," inquired a Vassar girl of a Columbian guard. "In the Electricity Building," calmly answered the blue-coated and tin-sworded dignitary with a graceful movement of his white gloved hand towards the statue of Ben Franklin, and added, "All the lights are there."

The exhibit of A. C. Mather, so long in an unfinished condition, has at last, during the past week, been put in running order. It shows a proposed solution of the problem occupying the attention of Gov. Flower, of New York, as to utilizing electricity on the Erie canal. An electric railway system is also shown, together with a model of a gigantic water wheel for furnishing power from Niagara.

This is children's week at the Fair, and from 30 to 60 thousand Chicago school children have visited the Fair daily. The Electricity Building has many exhibits which interest them, but none more so than the Electrical Door. Mr. Troy has given up trying to "keep order" in his exhibit, but lets the boys and girls go through as many times as they wish, much to their delight.

CONVENTION NOTES.

The McGuire Manufacturing Company, Chicago, makes a nice exhibit of trucks. The company is well represented by Messrs. McGuire, Cooke and Hubbard.

Taylor, Goodhue & Ames, Chicago, are to be represented at the convention by Messrs. Wm. Taylor and Dee, who have with them a number of their electric railway specialties.

The Electrical Installation Company, Chicago, is represented by its general manager, Mr. L. E. Myers. This company make no exhibit but point with pride to the work that it has accomplished during the past few months in the

way of railway and lighting construction. Among other contracts it has just finished 54 miles of overhead electric railway construction in Chicago and built complete 3½ miles of road for the Belle City Street Railway Company of Racine, Wis. The company has also recently installed an arc and incandescent light plant for the West Chicago Street Railway. It is now installing a large arc plant for the Chicago & West Indiana Railroad for lighting the yards.

The Western Electric Company, Chicago, is making a very fine exhibit of wires and cables at the convention, which is well worth the careful inspection of electric street railway men. The exhibit is in charge of Mr. H. C. Eddy.

Pullman's Palace Car Company, Chicago, is well represented at the convention by Mr. W. S. Loutit, assistant contracting agent, who with just pride points to the 100 handsome new closed and open cars running on the Milwaukee street railway lines manufactured by his company. It is an exhibit worth seeing.

The Siemens & Halske Electric Company, Chicago, has as a practical exhibit a 750 K. W. Generator in operation connected to a Buckeye engine. This dynamo is sold to the Consolidated Street Railway Company of Cincinnati, where it will be shipped after the convention. Mr. A. W. Wright and other members of the company are in attendance.

The Westinghouse Company is fully represented at the convention. Among those present are Mr. Lemuel Bannister, vice-president and general manager, Messrs. W. C. Clark, B. F. Stewart (Chicago), W. F. Zimmerman (New York), W. S. Brown (Boston), E. C. Bragg (Phila.), J. A. Rutherford, W. T. E. Gray and several others. The company is distributing large quantities of literature.

The Miamisburg Electric Company, Miamisburg, O., is represented at the Street Railway Convention by Mr. D. H. Allen, general manager of the company, and Mr. A. L. Daniels, salesman of the company, has on exhibition finished street railway commutators and segments of tempered copper in Section A Exposition Building. In the exhibit are also Imperial dry and Burnley cartridge batteries.

The Western Electric Heating Company, Chicago, is represented at the convention by its manager, Mr. Geo Cutter, who has on exhibition a full line of electric street car heaters besides office heaters for street railway offices. The season is now at hand when managers will be looking after the best means of heating their cars in the coming winter season and they will no doubt examine the exhibit with much interest and profit.

The Railway Equipment Company of Chicago is represented by its manager Mr. W. R. Mason. Its headquarters are in Parlor 92, Plankinton Hotel. Among the specialties there on exhibition is a new rail bond spring bushing which is of special interest in view of the recent discussion of rail bonding. It is simple and effective, and in its use on several roads, it is said, has proved very satisfactory.

The headquarters of the General Electric company are at the Pfister Hotel. The interests of the General Electric company at the convention are under the care of Mr. Theo. P. Bailey, of the Chicago office, assisted by Mr. G. K. Wheeler, Mr. G. Atterbury, and others from the same office. Mr. O. T. Crosby, Mr. Theodore Stebbins and Mr. R. H. Beach, are present from the main office in New York, Mr. W. J. Clerk, from the Cincinnati office, Dr. T. Addison, from the San Francisco office, Mr. W. H. Knight, Mr. J. B. Blood and Mr. A. K. Baylor represent the engineering department.

GALLERY.

MAIN FLOOR.

Exhibitor.	Section	Exhibitor.	Section	Exhibitor.	Section	Exhibitor.	Section
Austria	Y	Electrical Review	T	Jaeger, Chas. L.	T	H-bauer Gauge Co.	T
Amonds Electric Co.	X	Electricity	T	Johns Mfg. Co., H. W.	T	Roessler & Hasslacher Chem. Co.	S
Am. Inst. of Elec. Eng.	S	Electric Gas Co.	R	Jewell Belting Co.	V	Street Railway Journal	Y
American Battery Co.	R	Electrical Engineer	T	Jenney Elec. Motor Co.	S	Sinswger Aut. Telph. Co.	T
Axford, H. M	S	Electrical World	T	Knapp Electrical Works	T	Standard Paint Co.	T
Allg. Elec. Gesellschaft	S	Eddy Electric Motor Co.	B	K. A. P. Elec. Novelty Co.	C	Sprackels, S. L.	S
Bates Mfg. Co.	R	Excelsior Electric Co.	B	Knapp & Buckley	S	Sias Iron Tower Co.	W
Bryant Electric Co.	R	Electrical Forging Co.	D	Kennedy Electric Co.	L	Spain	S
Billinge & Spencer	R	Equitable Dynamo Co.	G	Lawton, H. A	Y	Schering, Chas. A. & Co.	D
Binley, W. R	T	McAleton Mfg. Co.	P	LeClanche Battery Co.	S	Schomburg & Sohn	E
Belknap Motor Co.	E	Electrical Conduit Co.	P	McNeil Toner Elec. Co.	L	Siemens & Halske	E
Bell Telephone Co.	E ? G	England	G	Marcus, W. N	S	Schuchart & Co.	R
Brush Electric Co.	L	Empore China Works	S	Meeker, Dr. G	S	Shirl Electric Co.	S
Caldwell El. Cloth Cut Mch. Co.	Y	Franklin Elec. Appliances	S	McIntosh Bat. & Opt Co.	W	Sperry Elec. Railway Co.	A
Consol. Elec. Storage Co.	A	French Piano Exhibit	Y	Manson, C. Belting Co.	D	Standard Undwg. Cable Co.	L
Cutter George		Felton & Guilleaume	F	Mather, E.	F	Standard Electric Co.	Z
Clinton Elec. Co.	T	France	K F U	Mather Electric Co.	M	Samson Battery Co.	S
Chicago Elec. Wire Co.	R	Ft. Wayne Elec. Co.	M	Newman Clock Co.	Y	Tate Aut. El. Signal Co.	Y
Copenhagen Fire Alarm Co.	S	Gault & Co., N. C	Y	Non-Magnetic Watch Co.	B	Todd, Applegate Co.	S
Central Electric Co.		Caldwell Fire Alarm Co.	Y	N. Y. Insulated Wire Co.	T	Taylor, Goodhue & Ames	S
Continental Cable Co.	Y	General Electric Co.	B H N C A J	National Carbon Co.	T	Thomson Elec. Welding Co.	O
C & C Elec. Motor Co.	A	General Incand't Arc Lt Co.	L	Norwich Elec. Wire Co.	T	T-lautograph, Elisha Gray	W
Cleveland Elec. & Mfg. Co.	A	Greeley, E. S. & Co.	T	North Am. Phonograph Co.	X	Union Electric Co.	A
Chicago Belting Co.	D	Germany	D	N. Y. A. I. E. A.	S	Veder, E.	W
Industry Clock Co.	R	Hubbard, Wm. & Co.	T	Nat. Aut. Fire Alarm Co.	A	Webb, G. F	Y
Department of Electricity	R	Hjalman, J. & Co.	S	Nat. Engraving Machine Co.	O	Weston El. Instrument Co.	R
Electrical Insurance	V	Hart & Hegeman Mfg. Co.	T	Owen, D. A	I	Washburn & Moen Mfg. Co.	R
Elec. Enoch & Nat. Co.	R	Hoje Elec. Appliance Co.	S	Phoenix Glass Co.	I	Western Union Tel. Co.	T
Electric Separator	T	Hall, Chas. F		Paint, R. T	Y	Waite & Bartlett Mfg. Co.	L
Edgerton, E. M	T	Holmes, N. S	W	Fabermacher Gulb Co.	Y	White, S. S., Dental Mfg. Co.	T
Elgin Telephone Co.	T	Hartman & Braun		Panoply, J. K	T	Western Electrician	T
Edison Elec. Mfg Co.	L	Hanson & Van Winkle	L	Pratt El. Med. Aut. Sup. Co.	S	Wilde Aut. Regular Lt. Co.	A
Enterprise Elec. Co.	S	Hirsh, J. M		Powell, Wm. & Co.	Y	Western Electric Co.	A
Eureka Temp. Copper Co.	C	Hardemuth, F. & Co.	P	Phelps, A. H	S	Westinghouse El. & Mfg. Co.	B H J
Electric Appliance Co.	S	Bibholz Alloy Co.		Pope Belting Co.	D	Wiley & Stiehle	P
Elec. Sel. & Sig'l Co.	A	Internat. Aut. L't & P'r Co.	S	Queen & Co.	R	Wing, L. J. & Co.	E
Electric Heat Alarm Co.	Y	India Rubber Comb Co.	S	Ringler, F. A	R	Zucker & Levett Chem. Co.	F

PERSONAL.

Mr. E. E. Bartlett, New York City, is in the city and at the Fair this week.

Mr. A. L. Parcelle, E. E., Boston, was viewing the electrical exhibits last week.

Mr. Arthur Churchill, E. E., Schenectady, N. Y., visited the Electricity Building Monday.

Mr. G. Herbert Condict, Germantown, Philadelphia, was in Chicago and at the Fair last week.

Mr. Wm. S. Turner, E. E., New York City, is registered at the A. I. E. E. World's Fair office.

Mr. Wm. C. Cuntz, of the Pennsylvania Steel Co., Steelton, Pa., is visiting the Fair this week.

Mr. George Westinghouse and family are registered at the Hotel Wildemere and are visiting the Fair.

Mr. C. G. Young, New York City, registered last week at the office of the A. I. E. E., Electricity Building.

Mr. Harry Alexander of the Alexander-Chamberlain Electric Co., New York City, is at present in the city.

Mr. E. E. Davis, E. E., Newburn, Tenn., called on ELECTRICAL INDUSTRIES during his stay at the Fair last week.

Mr. M. M. Kimble, of Claflin & Kimble (Inc.), Boston, selling agents of the "Novak" lamp, is spending a few days at the Fair.

Mr. Felika Rycerski, C. E., Warsaw, Poland, was last week devoting considerable time to the exhibits in the Electricity Building.

Mr. H. Ward Leonard, E. E., New York City, is registered at the World's Fair office of the American Institute of Electrical Engineers.

Mr. J. C. Boyd, supt. of the Jewett Car Co., Jewett, Ohio, is visiting the Fair this week on his way to the convention at Milwaukee.

Mr. J. H. McGraw of the Street Railway Journal, New York, called at the ELECTRICAL INDUSTRIES World's Fair office Monday of this week.

Mr. Geo. Manson who has been spending two weeks at the Fair has just returned east. Mr. Manson expresses himself as much pleased with his visit.

Mr. T. Ahearn, of Ahearn & Soper, the Canadian agents of the Westinghouse Electric & Mfg. Co., at Ottawa, Ont., is in the city and at the Fair this week.

Mr. J. J. Carty of the Metropolitan Telephone Co., New York City, is visiting the Fair and making his headquarters at the Western Electric Company's exhibits.

Mr. B. E. Greene, publisher of Electricity, accompanied by Mrs. Greene, spent the forepart of the week at the Exposition on his way from New York to the Milwaukee convention.

Mr. W. P. D. Crane, M. E., manager electrical department of the H. W. Johns Mfg. Co., New York, arrived in Chicago the latter part of last week, and is putting in his time sight seeing at the Fair.

Mr. Edward H. Chapin, formerly with the Street Railway Journal, and at present secretary and treasurer of the Shipman Engine Mfg. Co., Rochester, N. Y., called at the World's Fair office of ELECTRICAL INDUSTRIES Tuesday.

Mr. Wm. H. McKinlock, of the Metropolitan Electric Company feels that the outlook for business is encouraging, and sees the promise of liberal support for a good, strong company conservatively managed. Evidence of this fact is fast multiplying.

Prof. George Forbes, F. R. S., the widely known electrical engineer of 34 Great George St., London, Eng., has opened an American office in the Mills Bldg., corner of Wall St., and Broadway, New York. Mr. Horatio A. Foster is associated with him as chief assistant.

Mr. J. A. Corby finding that his other business interests require his closer attention, has resigned the presidency of the Railway Equipment Company, and Mr. W. R. Mason has been elected to that office. The company is doing a large business in electric railway supplies, which will no doubt be greatly increased in the near future.

Electric Railway Power Plant Equipment.

The transportation lines of Chicago were tested on Chicago Day as no systems of local transit were ever tested before and the companies acquitted themselves in a most satisfactory manner. The Chicago daily papers are the severest critics when the matter of transportation is brought up, but in this case they are all united in their praises for the magnificent manner in which the crowds were handled by the companies. Although each and every company performed their duties to the public in an admirable manner, we believe the cross-town electric lines of the Chicago City Railway are at least entitled to be placed at the head of the entire list. Every possible car and man was placed on this line, and although the equipment is one of the finest in the country, they were utterly unable to meet the requirements of the crowds.

The cars were run as close as it was deemed practical and they were so crowded that every conceivable place was taken up, even so far as the windows and the tops of the cars. These lines moved more passengers per mile of track than any other line, and how they ever ran trains in such a crowd without injuring a person or a breakdown is a mystery. This is without doubt due to the carefulness and thoughtfulness of Supt. Bowen and Chief Engineer Hill in the selection of the equipment and the superintendence of the system.

As in every system of electric traction much depends on the successful operation of the power equipment. The generators and boilers were heavily taxed and the results obtained from the engines were everything that could be desired. These engines are in pair form with cylinders 24 inches in diameter by 48-inch stroke, running 100 revolutions per minute.

These engines, which are of the improved Wheelock type, are equipped with Hill's patented valve system, which system allows the engine to be operated at a greater speed than the regular Corliss practice and also allows one to take steam to full stroke if necessity demands, thus obtaining a greater maximum out of a certain sized cylinder than is the usual practice. Power is transmitted direct to the driven shaft, on each end of which is a Westinghouse generator, by means of the Hoadley system of compound wind rope transmission. The operation of the ropes is absolutely noiseless, and it was interesting to see how easily they performed their heavy duty on that day.

The above engines were manufactured and installed by the California Engineering Company, Hoadley Bros., engineers, who also installed the system of rope transmission. This company have also the exclusive right for manufacturing the improved Greene engine which embodies the latest patents of Mr. N. G. Greene. A 650-horse power tandem compound engine of this type is being installed by this company for the Siemens & Halske Electric Company in this city. The engine is directly connected to the armature shaft. The armature, which is 20 feet in diameter, runs 100 revolutions per minute and acts as a fly wheel.

BUSINESS NOTES.

THE FILER & STOWELL COMPANY, Milwaukee, has recently gone into the manufacture of the improved Greene engine, improved Wheelock engine and Corliss engine together with a full line of power transmitting machinery. The company also manufactures friction clutches, pulleys, etc. A new catalogue has just been issued.

THE METROPOLITAN ELECTRIC COMPANY, offices 319-320 Manhattan Building, and salesroom 367 Dearborn St., recently organized, will make an announcement shortly as to the scope and general plan of the company. In the meantime they are receiving orders for general supplies and N. I. R. wire with which they are generously supplied.

THE CENTRAL ELECTRIC COMPANY, Chicago, has listed among the new specialties recently placed on the market the Hammond cleat. This new device, it is said, is meeting with considerable favor with the trade and this company feels confident that it is a specialty that has come to stay as it is very practical. Special circulars bearing on this subject are sent out on request by the company.

THE ELECTRIC APPLIANCE COMPANY report that the opening of the fall season is causing a brisk demand among other things for their celebrated Acme Expanding wire guard which they first placed on the market about a year ago and which, though comparatively an insignificant specialty, has proven to be a most successful seller. Its improvement over other wire guards for the protection of incandescent lamps lies mainly in the fact that it is easily and firmly fastened to a socket without any auxiliary fastenings or clips and is particularly light and symmetrical in appearance.

W. L. ADAMS & COMPANY, 84 Adams St., Chicago, have succeeded to the business of W. L. Adams. This change was made in order to secure additional facilities for handling the increased business. The firm is to do a business of manufacturing and selling everything necessary for the equipment, repairs and maintenance of electric railways. Owing to the long and practical experience of the members of the firm in the electric railway field together with every facility for filling orders with dispatch the new concern feel confident that they are able to anticipate the wants of the trade and by supplying goods of the highest character meet the most exacting requirements.

THE ATTENTION

OF THE

DELEGATES

and others who attend the Convention of the American Street Railway Association in Milwaukee this week is called to the

100 NEW OPEN and CLOSED CARS

RECENTLY FURNISHED BY THE

PULLMAN COMPANY,

and now in operation on the Milwaukee Street Railway.

YOUR ATTENTION IS ALSO INVITED TO THE

PULLMAN EXHIBIT

at the World's Columbian Exposition as illustrated and described in this issue.

Address all correspondence to

CONTRACTING AGENT,

PULLMAN CO., Chicago, Ill.

Map of Chicago.

Showing Location of its Electrical and Allied Business Interests, Principal Hotels, Theatres, Depots and Transportation Lines to the World's Fair Grounds. (Index numbers refer to the black squares.)

EXCLUSIVELY
HIGH GRADE
ELECTRICAL SUPPLIES

Our stock is complete in every particular. Our prices are as low as consistent with first-class material. Our facilities are such that Our careful personal attention to all orders is a fact. Our dealings are always satisfactory to our customers.

ELECTRICAL SUPPLIES

ELECTRIC APPLIANCE COMPANY,
242 Madison Street, CHICAGO.

WM. BARAGWANATH & SON,

LIST OF HEATERS
TO BE SEEN IN OPERATION AT THE WORLD'S FAIR.

Two 500 H. P. East End Boiler Gallery doing 1800 H. P. work.
One 300 H. P. heater and receiving tank, Wellington Catering Co's., plant.
One 150 H. P. heater at Hygeia plant.
One 200 H. P. Libby Glass Works.

55 WEST DIVISION STREET
CHICAGO.

EVERY STREET RAILWAY MANAGER, SUPERINTENDENT
AND ELECTRICIAN should read regularly the Monthly Issue of

ELECTRICAL INDUSTRIES

IT WILL GIVE YOU THE NEWS
IT WILL INFORM YOU HOW TO RUN YOUR PLANT ECONOMICALLY AND THUS INCREASE DIVIDENDS

Subscribe Now
AND TAKE ADVANTAGE OF OUR SPECIAL OFFER TO STREET RAILWAY MEN

All Delegates and their Friends are cordially invited to call and register at our Headquarters.
Section Y 27, Electricity Building when they visit the World's Fair.

Electrical Industries, CHICAGO.

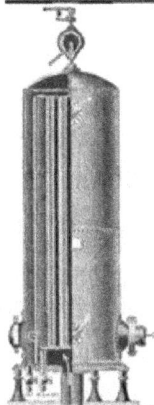

CENTRAL ELECTRIC COMPANY,

CHICAGO.

STREET RAILWAY SUPPLIES.

Construction Tools,   **Line Materials,**

PINS, BRACKETS, INSULATORS, ETC.

OKONITE
FEED
WIRES,

CAR WIRES,

OKONITE
and.. MUNSON
TAPES,

ARE
THE
LEADERS.

INCANDESCENT
LAMPS,
TIED
AND
FREE
FILAMENTS.

LUNDELL
MOTORS,
FOR
STREET
RAILWAY
CIRCUITS.

THE FERRIS WHEEL...WIRED WITH OKONITE.

Interior Conduit System of Electric Wiring.

THE ONLY PERFECT AND COMPLETE SYSTEM FOR INTERIOR WIRING.

STANDARD PAINT COMPANY'S Celebrated P. & B. Electrical Paints and Compounds carried in Stock.

ELECTRIC LIGHT CARBONS FOR ALL SYSTEMS.

Central Electric Company...CHICAGO.

SPECIAL AGENTS:

SOUTHERN ELECTRICAL SUPPLY CO., St. Louis. - GATE CITY ELECTRIC CO., Kansas City, Mo.

THE ONLY STORAGE BATTERY

Manufactured in the U. S. to receive

A DIPLOMA

AT THE WORLD'S COLUMBIAN EXPOSITION, CHICAGO.

The following is a copy of the Award given by the Judges of Storage Batteries at the World's Columbian Exposition, Chicago, 1893, to the only storage battery made in this country deemed worthy of any notice whatever:

"We affirm that the 'American' battery has been examined and tested by us, and found worthy of an award for its excellency of design and construction, and for its efficiency and indications of durability."

Signed WILBUR M. STINE.
W. LOBACH.

THE "AMERICAN" STORAGE BATTERY IS EFFICIENT AND DURABLE.
ITS ACTIVE MATERIAL IS FORMED FROM ITS OWN SUPPORT.
ITS PLATES WILL NOT BUCKLE UNDER ANY CIRCUMSTANCES.

GUARANTEED TO BE SUPERIOR TO ANY OTHER MADE.

Estimates furnished on application for complete storage battery.

EQUIPMENTS FOR CENTRAL STATIONS, ISOLATED PLANTS, DRAIN LIGHTING, TRACTION AND BOAT WORK.

AMERICAN BATTERY CO.,

Office: Room 709 Security Building. CHICAGO.
188 Madison Street,

W. L. ADAMS & CO.

Manufacturers and Dealers in

RAILWAY SUPPLIES

CAN FURNISH YOU WITH ANYTHING FROM A SPIKE, TO THE COMPLETE EQUIPMENT FOR AN

ELECTRIC RAILWAY,

AT PRICES CONSISTENT with FIRST CLASS MATERIAL.

Let us make you quotations; You will save money.

GENERAL OFFICES, 84 Adams Street, CHICAGO.

ELECTRIC CAR HEATERS

Under the patents of the
American Electric Heating
Corporation.

See Exhibit at the
Milwaukee Convention.

Western Electric Heating Co.

851-853-855 The Rookery,
CHICAGO.

Westinghouse
Electric and Manufacturing
Company
Pittsburg, Pa.

■ ■ ■ ■ ■

USERS OF OUR

Electric Railway Apparatus

unanimously attest its

ELECTRICAL and MECHANICAL SUPERIORITY.

Get Our New Pamphlet of Testimonials,

Many statements which fully emphasize the fact, that the

WESTINGHOUSE SINGLE REDUCTION MOTOR, GENERATOR,

and all other

Electric Railway Apparatus

is the **BEST** in point of

CONSTRUCTION, DURABILITY, EFFICIENCY.

and low cost of maintenance.

Westinghouse Electric and Manufacturing Co.
PITTSBURGH, PA.

THE GREAT VARIETY OF

WIRES

...AND...

CABLES

EXHIBITED BY THE

WESTERN ELECTRIC COMPANY

AT THE...CONVENTION...OF THE

AMERICAN STREET RAILWAY ASSOCIATION

...AT...

Milwaukee

ARE WORTHY THE CAREFUL INSPECTION OF EVERY

Electric Street Railway Man.

Western Electric Company

CHICAGO...AND...NEW YORK

CLARK ELECTRIC COMPANY, NEW YORK.

192 Broadway and 11 John Street.

MANUFACTURERS OF ARC LIGHTING APPARATUS FOR EVERY PURPOSE A SPECIALTY.
The CLARK ARC LAMPS for use on EVERY CURRENT, have the reputation of being
the best and most durable of any ever made in the United States.

RAWHIDE PINIONS FOR ELECTRIC MOTORS

A SPECIALTY.

RAWHIDE DYNAMO BELTING

Greatest Adhesive Qualities.   A Non-conductor of Electricity.

THE CHICAGO RAWHIDE MANUFACTURING CO.

THE ONLY MANUFACTURERS IN THE COUNTRY

LACE LEATHER ROPE
AND OTHER RAWHIDE

GOODS
OF ALL KINDS
BY KRUEGER'S PATENT

75 Ohio Street,   CHICAGO, ILL.

STANDARD ELECTRIC COMPANY.

GENERAL OFFICES: 625 Home Insurance Building.

WORKS: So. Canal Street,

CHICAGO.

STANDARD SYSTEM

AT THE

WORLD'S FAIR.

MACHINERY HALL, Sec. Q, 2 Standard Arc Dynamos.

Sec. S, 20   "   "   "

ELECTRICITY BUILDING. Sec. P, Space 2, Arc Lighting Exhibit.

The Standard Lamps Light the Power Plant, Machinery Hall, Agricultural Hall, Shoe and Leather Building, and
Other Buildings and Portions of the Grounds.

See our Double Service All Night Lamp Before Buying an Old Style Two Rod Lamp.

Mile after mile of
SIMPLEX WIRE
Supplied to the
FERRIS WHEEL
• • •
By...George Cutter,
The Rookery, CHICAGO.

SIMPLEX WIRES

**INSURE
HIGH
INSULATION**

Simplex Electrical Co.
620 Atlantic Ave.,
George Cutter, Chicago.   BOSTON, MASS.

Made 5 amp. S. P.
10 amp. S. P.
5 amp. 3 way.
10 amp. 3 way.

XNTRIC

"That's the Switch"

And we control that movement.

H. T. PAISTE,
10 South 18th St.,
**PHILADELPHIA,
PA.**

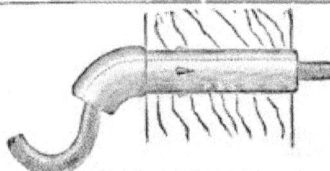

China Window Tube (Patented).
Made only by **PASS & SEYMOUR,**
George Cutter,   SYRACUSE, N. Y.
CHICAGO.

Enterprise Electric Company

307 Dearborn Street.
Chicago....

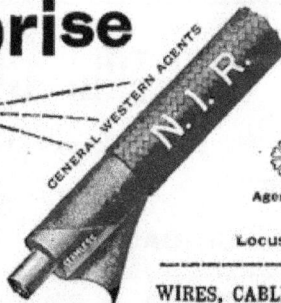

GENERAL WESTERN AGENTS

N. I. R.

Manufacturers' Agents and Mill Representatives for

Electric Railway,
Telegraph, Telephone and
Electric Light

SUPPLIES OF EVERY DESCRIPTION

Agents for   Cedar Poles,
Cypress Poles, Oak Pins,
Locust Pins, Cross Arms, Glass
Feeder Wire,   Insulators,

WIRES, CABLES, TAPE and TUBING

Miamisburg Electric Co.

MANUFACTURERS OF

COMMUTATORS
AND
COMMUTATOR SEGMENTS
OF TEMPERED COPPER
FOR STREET RAILWAYS

Also Brushes and Brush Copper, Burnley's
Cartridge Batteries and Imperial Dry Batteries.
See our Exhibit at the Street Railway Convention, Milwaukee. Section A, Exposition Building.

MIAMISBURG ELECTRIC CO., Miamisburg, O.

BEAR IN MIND

that the regular monthly issue of ELECTRICAL IN-
DUSTRIES contains the most complete and correct
directories published of the electric light central stations
and the electric railways in North America.

World's Fair Headquarters Y 27 Electricity Building.

CITY OFFICES, Monadnock Block.

Consolidated Electric Co.

Manufacturers and Dealers in all kinds of

ELECTRICAL . SUPPLIES,

115 Franklin Street,

CHICAGO.

WEEKLY WORLD'S FAIR

ELECTRICAL INDUSTRIES

DEVOTED TO THE ELECTRICAL AND ALLIED INTERESTS OF THE WORLD'S FAIR, ITS VISITORS AND EXHIBITORS.

Vol. I, No. 20.   **CHICAGO, OCTOBER 26, 1893.**   FIVE MONTHS $1.00 / TEN CENTS A COPY

The Exhibit of the Mather Electric Company in the Electricity Building.

In the southwestern part of Electricity Building, Section M, is located the exhibit of the Mather Electric Company. It covers nearly 4,000 square feet of floor space which is filled with sample machines and devices from the exten-

merous illuminated signs and attractive devices, drawing the attention of the visitor to the company and its products. One of the most conspicuous and ornamental features of the exhibit is a large horse-shoe shaped magnet representing the Mather patent ring type field which is a trade mark of the company. The magnet is 12 feet in diameter, mounted on a base 10 feet long. The ring field

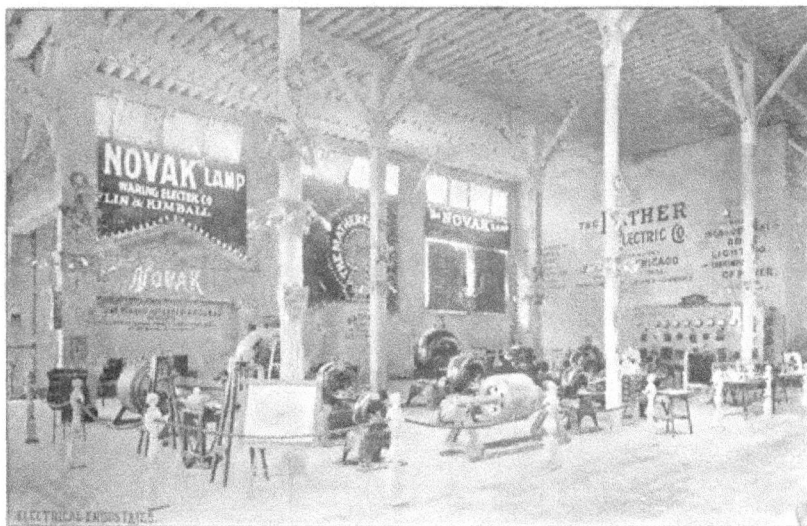

EXHIBIT OF THE MATHER ELECTRIC COMPANY.

sive works at Manchester, Conn. The space is floored with matched pine flooring, a portion of which is raised for an office. The machines and apparatus are symmetrically arranged about the space. The special fittings and parts of machines showing the construction of commutator journal boxes, journal bearings, self oiling boxes, etc., are neatly arranged on long polished oak tables.

The walls back of the space have been concealed by nu-

and pole pieces are delineated by blue and red lamps, while the words the Mather Electric Company are delineated in white opal lamps. By a special device different colors are shown alternately, then together, producing a pleasing and artistic effect.

On the north wall above the switch board is a large sign covering some 600 square feet, giving the name of the company, its principal offices, the principal products manufactured, and their applications. Distributed about the space are easels holding views of the works of the company at Manchester, Conn., and also some of the prominent buildings which are using the Mather system. Among the other attractive features of the exhibit is the handsome and interesting display of Novak lamps made by Chillin & Kimball, (inc), the general sales agent for the Novak lamp. This display has attracted a great deal of attention and has aroused more than usual interest among the many visitors.

For lighting the exhibit there are 80 incandescent Novak lamps of 16 candle power fed from a 110 volt circuit. The lamps are distributed on brackets and chandeliers of plain but very neat design. The space is marked by a very neat railing made with turned posts, painted in white and gold, through which a worsted cord of Venetian red color is gracefully looped, forming the rail. The dynamos exhibited are of the Mather patent ring type of constant potential, continuous current, compound wound pattern of sizes varying from three and a half to 55 kilowatts. These machines are all wound for a potential of 125 volts and are complete with bases, rheostats, etc.

The 55 and 10 kilowatt machines are in operation, furnishing the current for all the incandescent lights used for lighting and decoration about the exhibit. The current furnished has a potential of 110 volts from a speed of 750 revolutions. The seven kilowatt machine is also connected so as to be used when a portion of the incandescent lamps are desired for lighting purposes. There is also one multipolar slow speed, compound wound, continuous current, dynamo wound for a potential of 125 volts and having a capacity of 17 kilowatts, shown complete with base and rheostat.

An electric power generator of the multipolar, compound wound, constant potential type with a capacity of 125 kilowatts is shown. It is wound for a potential of 550 volts at a speed of 525 revolutions per minute. These generators are said to have a commercial efficiency of over 90 per cent, and to regulate absolutely from no load to full load. The mechanical part of their construction and equipment is of the latest design, the generator is set on an iron base, fitted with four screws, connected together in pairs by means of a chain and worked by ratchets so as to control the alignment of the machine and the tension of the belt. These machines have also a rheostat of suitable capacity and a commutator turning device adjustable on the base of the machine.

Electric motors are shown of a variety of sizes and of different types. There are two multipolar self-regulating motors having a capacity of 50-horse power each and wound for a potential of 250 volts. These motors have been operated in series on the 500 volt circuit supplied by the Mather Electric Company generators in Machinery Hall and are belted to run the 55 kilowatt dynamo mentioned above. The motors are connected through an automatic starting box which is wound for 500 volts. A 10-horse power motor is also run from this power circuit and is belted to the seven kilowatt dynamo.

There is also shown in the exhibit a six-horse power wound for 220 volts and a one-horse power motor wound for 110 volts. The latter runs a commutator device used for operating the changes in the lighting of the illuminated signs.

Among the other machines exhibited is the company's latest style of armature for electric power generators. This is of the ventilated type and is designed for a 225 kilowatt

machine. This armature has a slotted core and is built of copper bars. The winding is such that there are only two paths for the current through the armature which are always maintained at the same potential. The commutator is cross connected, and it is possible to use either all the sets of brushes or any two sets as may be desired. The generators for this armature are built with six poles and carries six sets of brushes. The armature for an electric power generator of 120 kilowatts capacity is shown wound with wire on the same principles as the one above described. There is also shown armatures for smaller generators and a number of field coils.

At the north end of the exhibit is placed a switchboard on which are placed the instruments and devices for

FIG. 1.—RECORDING INSTRUMENTS EXHIBITED BY THE BRISTOL CO.

controlling the current used in the exhibit. The board is of white Tennessee marble from the marble works of Davidson & Sons, Chicago. The frame in which it is mounted is neatly ornamented and finished. The instruments, switches and fittings displayed on this board are all of the Mather company's manufacture. There are used on the board a number of magnetic vane ampere meters both large and small, also volt meters, a differential indicator, three volt meter switches, both single and double pole switches and cut outs, etc.

The rheostats for the dynamos are also mounted on the board. For protecting the circuits an automatic circuit breaker is used, adjustable from 100 to 600 amperes and suitable for any voltage up to 600 volts. They are said to act within 10 amperes of the quantity for which they are set, breaking off current absolutely and thus affording complete protection to the circuits. This exhibit which the Mather company has installed shows in a comprehensive way the various electrical machines and appliances which the company manufacture. The line of goods for power and lighting stations is complete, and are all of a commercial character. The machines show skill in their construction and careful attention to detail and are worthy of the attention of electric light and power users.

Recording Instruments Exhibited by the Bristol Manufacturing Co.

The new recording volt meters exhibited by the Bristol Manufacturing Company are attracting a great deal of attention, notably those used in connection with the incan descent lamp tests made by the Jury of Awards. The main exhibit made by the company is located in Machinery Hall Section 25 and contains not only recording volt meters but also instruments for recording the pressure of air, gas, steam, water, and any liquid. The range of pressure runs

attracted to each other by a current passing in series. The current is conducted to the moveable coil A through the supporting springs D and E, which together with the special feature of the moving coil B mounted on friction less spring knife edges renders the instrument extremely sensitive to the smallest change of voltage.

The marking arm F is attached directly to the spring E and partakes of its motion recording the changes of voltage on a uniformly revolving chart. This chart shown in Fig. 2 is intended for 110 volt circuit and as will be noticed the divisions on the chart are on an increased scale in the vicinity of the voltage to be maintained. The variation of one volt is possible to be in this way noted. The coil C is an auxiliary resistance with the alternating current volt meter; the auxiliary resistance is furnished in a separate rheostat which may be adjusted to suit the rate of alterna tions of the current to be measured. One of the specialties

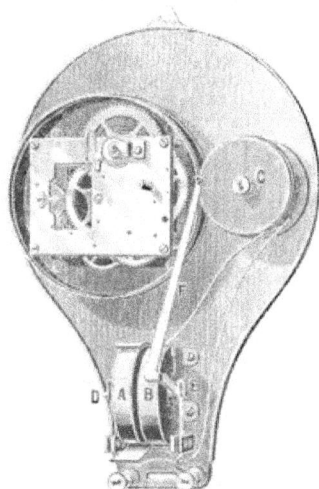

FIG. 2.—RECORDING INSTRUMENTS EXHIBITED BY THE BRISTOL CO.

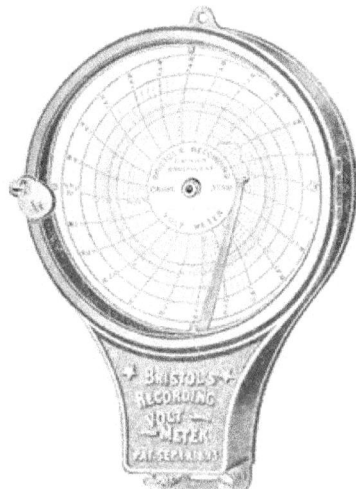

FIG. 3.—RECORDING INSTRUMENTS EXHIBITED BY THE BRISTOL CO.

from zero to fifteen hundred pounds per square inch and adapted to record continuously day and night.

Erected diagonally across the space is an arch in imita tion of cut stone; on the face of each is attached one of their gold plated recording gauges, every alternate one being provided with an electric light. On one of the pillars supporting the arch is a gauge which records the pressure of the steam used in the building. On the other pillar is placed one of their new recording volt meters which records the voltage of the current used by the lamps in the exhibit. Within the arch is a semi circular grille of wrought iron of an ornamental design which bears the name of the company and its specialties.

About the space there has been arranged on tables so that visitors may examine them, sample recording instru ments among which is shown the new recording volt meter illustrated in the accompanying cuts which show the in ternal construction and also the exterior appearance of the instrument. The coil A of Fig. 2 is mounted on the spring knife edge supports D and E and is free to move toward the parallel and stationary coil B, when they are mutually

shown in this exhibit which is also manufactured by the Bristol Manufacturing Company of Waterbury, Conn., is Bristol's steel belt lacing which is shown in a variety of ways. This company since its establishment in 1889 has developed an extensive business in these lines of goods.

The people of Chicago are getting used to the blockades incident to the increased number of trains on the street railways these World's Fair days, but, even to those of us who retain our seats in the cars for ten or fifteen minutes "waiting for them to start," the sight of crowded cars that never will start looks strange. One can see this sight almost any evening in the Westinghouse exhibit in the Electricity Building, where (a Stephenson and a Brownell) cars are filled with tired sightseers glad of a place to rest for a few minutes.

The Jury of Awards met last Saturday, but as only four members were present and the rules call for a quorum of five, no business was transacted. The next meeting will be held on Saturday of this week. This will probably be the last regular meeting of the jury.

ELECTRICAL INDUSTRIES.

PUBLISHED EVERY SECOND DAY BY THE

ELECTRICAL INDUSTRIES PUBLISHING COMPANY,

INCORPORATED 1889.

MONADNOCK BLOCK, CHICAGO.

TELEPHONE HARRISON 518.

E. L. POWERS, PRES. AND TREAS.　　F. E. WOOD, SECRETARY.

E. L. POWERS　　　　　　　　　　　　　　　　EDITOR.
W. A. REMINGTON　　　　　　　　　　　ASSISTANT EDITOR.
F. E. WOOD　　　　　　　　　　　　　EASTERN MANAGER.
FLOYD T. SHORT　　　　　　ADVERTISING DEPARTMENT.

EASTERN OFFICE, WORLD BUILDING, NEW YORK.

World's Fair Headquarters, Y 27 Electricity Building.

SUBSCRIPTIONS
FIVE MONTHS　　　　　　　　　　　　　　$1.00
SINGLE COPY　　　　　　　　　　　　　　10
Advertising Rates Upon Application.

News items, notes of communications of interest to World's Fair visitors are earnestly desired for publication in these columns and will be heartily appreciated. We especially invite all visitors to call upon us or send address at once upon their arrival in city or at the grounds.
ELECTRICAL INDUSTRIES PUBLISHING CO.,
Monadnock Block, Chicago.

After carefully considering the matter, it has been officially decided that the Fair will close promptly Oct. 30th, the time appointed by Congress. It was thought that the attendance might be so great that it would be necessary to hold open a short time in November, but when it was found that this could not be done legally it was decided to officially close on the day prescribed; consequently the work of taking away exhibits will begin early next week, although the public will be admitted to the grounds and buildings during the day for some time yet. In the Department of Electricity exhibitors are beginning to get in a hurry to pack up exhibits at the earliest possible moment. No one can look at the exhibits, the grounds and the buildings without feeling deep regret that within a few days the work of demolition will begin. The fact, however, that the Fair has been the greatest success of anything attempted in modern times is a thought that brings some consolation.

The California Mid Winter Fair will receive many exhibitors and concessioners from the World's Columbian Exposition. Many of the concessioners are preparing to go to San Francisco as soon as Jackson Park is closed. The Fair at San Francisco will be visited largely by those who have not visited the World's Fair at Chicago. The great distance has allowed but few from that section of the country to visit Chicago and should the winter be favorable their holiday will be spent at Golden Gate Park. The annual meeting of the National Electric Light Association will bring many of those interested in electric lighting together who have not previously attended the meetings of the association. To a large number of manufacturers this is an entirely new field which will become more and more valuable as the country is developed. No more favorable opportunity will present itself than the one offered the coming winter. Many manufacturers have exhibits of such a kind as are easily moved and many constructed especially for exhibition purposes which could be transferred at little expense to San Francisco, while if they were preparing new exhibits the expense would be many times as great.

The Street Railway Convention.

The 12th annual meeting of the American Street Railway Association, which was held at Milwaukee last week, was well attended. The preparations for the convention had been carefully made, and to the Milwaukee Street Railway Company and the citizens of Milwaukee the association is indebted for its pleasant reception and many marks of attention during its stay in the cream city. The papers read were fully up to the standard, and several called forth lively and interesting discussions. This is specially true of the paper read by Mr. E. G. Connette on Power House Engines. The great attractions of the World's Fair greatly lessened the attendance toward the close, many leaving for Chicago before the convention ended. Owing to the enormous passenger traffic of the railroads, the railroad yards and side tracks are full of freight waiting to be moved, and on this account a great many manufacturers of railway apparatus were unable to exhibit their goods as they had intended. Many although greatly disappointed at not receiving their goods made the most of the situation and established headquarters in the exposition building at their hotel where they could meet their friends. The tendency toward consolidation is noticeable in the report of the executive committee, which shows that 10 roads formerly members of the association have been absorbed by other companies. The consolidation of the roads of a city has its advantages to the roads themselves and to the public in general, still it often has objectionable features. The convention closed Friday. The next meeting is to be held at Atlanta, Ga.

The banquet at Milwaukee Thursday evening at the Pfister Hotel was attended by 256 ladies and gentlemen. At the table of honor sat the newly elected president, Mr. H. C. Payne and the new executive committee of the association. At Mr. Payne's right sat Mrs. Payne while on his left sat the Rev. Judson Titsworth. The banquet hall was on the seventh floor of the hotel. The floral decorations were elaborate and of handsome and appropriate designs. The banquet was well planned and all voted it a great success.

Ex-president Longstreet acted as toastmaster and to the first toast, "Transportation and Civilization," Mr. J. C. Flanders responded. In his remarks he called attention to the facilities for rapid transit and said that the annual saving to the four billion passengers carried annually, by the introduction of electricity was a million hours, which if calculated in dollars and cents would show a large amount of money which the public owes to street railway managers. "Street Railway Employes and the Public," responded to by Mr. Ogden H. Fethers was heartily appreciated and received great applause. Mr. Winfield Smith responded to "The Early Days of Street Railways" and Mr. H. H. Windsor to "The Street Railway Man of the Past, Present and Future." Mr. Horace Rublee upheld "The Press" and Mr. J. Harvey Stedman read an original poem.

One of the mementos of the banquet was a souvenir spoon for each guest of sterling silver. In the bowl of the spoon was a representation of President Payne's private car, to whom the delegates were indebted for this further evidence of Milwaukee's hospitality.

The tests of incandescent lamps will probably be completed on Saturday. The results of the tests will be announced about the first of next year.

CONVENTION NOTES.

The Hale & Kilburn Mfg. Co., Philadelphia, made a display of car seats of several styles.

Mr. J. H. Stedman, Rochester, N. Y., was on hand with several new styles of transfer tickets.

The Garton-Daniels Electric Co., Keokuk, Iowa, had several of their lightning arresters on exhibition.

The Curtis Electric Motor Co., Jersey City, had a truck equipped with the Curtis motors on exhibition.

The Graham Equipment Company, Boston, Providence and Chicago, were represented by Mr. J. H. Graham.

The Pawtucket Brass Foundry, Pawtucket, R. I., showed motor bearings of its manufacture of several sizes.

The Davis Car Shade Company, Portland, Me., had a number of shades on model frames showing their operation.

Mr. C. S. Van Nuis, of New York, was present and distributed a great many fac simile Ajax lightning arresters.

The Pinkham Car-Track Sander Co., of Boston, Mass., displayed in the exposition building its electric sand box.

The Edison Mfg. Co., of New York, had on exhibition a number of Kennelly standard volt meters showing the construction and points of excellence.

The Gibbs Electric Company, Milwaukee, had several trolleys complete, trolley wheels and line supplies on exhibition.

The Taylor Electric Truck Co., Troy, N. Y., had on exhibition a Taylor truck and was represented by Mr. Taylor.

The Burrowes Car Shade Co., Portland, Me., was represented by Mr. John W. Baker, who had a number of shades on exhibition.

L. Katzenstein & Co., New York, had on exhibition a number of journal boxes, showing the use of Katzenstein's metallic packing.

The Walker-Marshall Automatic Switch Co., Milwaukee, made an exhibit consisting of the Walker-Marshall automatic switches.

The American Arch, Iron and Brass Works, Chicago, displayed Robinson trolley wheels of which this company are the manufacturers.

Mr. W. E. Ludlow, Cleveland, Ohio, displayed a new combined step and gate. When not used for a step it acts as a gate and vice versa.

The Cutter Electrical & Mfg. Co., Philadelphia, had an exhibition board erected on which were displayed magnetic cutouts, rheostats, etc.

The R. D. Nuttall Co., Allegheny, made an exhibit of gears, pinions and trolleys of the design which has become well known to street railway men.

The Empire State Radial Truck Company, Cleveland, Ohio, had trucks on exhibition. Mr. Pfetsch of the company was present at the convention.

Harrison & Carey, Chicago, agents for the W. T. C. Macullen Co., of Boston, had on exhibition circuit breakers, cross overs and general line supplies.

The Tousley Trolley Company, Chicago, had on exhibition its self lubricating trolley which is claimed to possess several new features of excellence.

Washburn & Moen Mfg. Co., Worcester, Mass., made through its Chicago representative, Mr. H. B. Cragin, quite an elaborate exhibit of trolley and feeder wires.

The Link Belt Machinery Company, Chicago, had on exhibition besides the regular line of goods a new friction clutch which attracted a good deal of attention.

The Baltimore Sand Box Co., Baltimore had sample sand boxes on exhibition, placed on a section of car platform showing their operation both by hand and foot levers.

The St. Louis Car Co., St. Louis had several cars on exhibition. The two placed opposite the 5th St., entrance of the exposition building attracting considerable attention.

The Railway Equipment Company, Chicago, had very pleasant headquarters at the Plankinton Hotel. Mr. W. B. Mason, manager of the company was in attendance.

The F. Wall Mfg. Supply Co., Alleghaney City, Pa., made an extensive exhibit of street car gongs. Mounted on a display board were arranged gongs in various sizes and kinds.

The Hartford Woven Wire Mattress Co., Hartford, Conn., had a number of car seats displayed showing the construction of its spring seats. Mr. H. F. Evans western agent with headquarters at St. Louis was looking after the interests of his company.

The Consolidated Car Heating Co., Albany, N. Y., was represented by Mr. McElroy. The exhibit made of car heaters was of the practical order and received considerable attention.

Fairbanks, Morse & Co., Chicago, displayed railway supplies. The exhibit of lifting jacks was quite extensive, showing jacks of several styles together with a number of models.

The National Time Recorder Co., Milwaukee, had a Bolte automatic time keeper on exhibition showing its operation as used in connection with street railway systems, factories, etc.

Chas. Munson Belting Co., Chicago, was well represented by Mr. W. C. Groetzinger and Mr. J. H. Shay. The display of belting comprised samples of the various kinds manufactured.

The Superior Machine Company, Cleveland, Ohio, had a number of gears and pinions on exhibition which the company manufactures in standard sizes for the different kinds of motors.

The Vose Spring Co., New York, was represented by Mr. Gus. Suckow, sales agent, who was on the lookout for new orders. A number of trucks displayed were fitted with the Vose springs.

The St. Louis Register Company, St. Louis, was represented by Mr. Gardner McKnight. The exhibit of registers in the exposition building was attractive and showed to advantage their general use.

Wm. Wharton Jr. & Co., Baltimore, engineers, founders and machinists, made an exhibit of track material and gave out neat pamphlets calling attention to their extensive exhibit at the World's Fair.

The New Haven Car Register Co., New Haven, Conn., was represented by Mr. A. N. Soper who had a display of a number of the New Haven registers which possess several novel and interesting features.

Wadhams Oil & Grease Co., Milwaukee, Wis., called the attention of the delegates in many ways to Wadham's graphite curve grease which is particularly intended for use on electric and cable railways.

The Stirling Supply Company, New York, had on exhibition car heaters, sand boxes, etc. Mr. John Kennelly, chief engineer of the company and Mr. Howard Wheeler representative were looking after the interests of the company.

H. W. Johns Mfg. Co., New York, made a display of line materials and supplies. Mr. H. A. Reeves, manager of the electrical department, Chicago, and Mr. W. P. D. Crane general manager of the electrical department, New York, were present.

Mr. F. E. Carlston, representing the lubricating department of the Standard Oil Co., at Milwaukee, had an exhibit of samples of armature and curve greases, cylinder oils and lubricants of several kinds, also paraffine for insulation.

The Cudahy Lubricating Company, Milwaukee, had samples of Kent's motor grease in several forms. Mr. H. W. Kent, the general manager, was looking after the interest of his company and the street railway men have undoubtedly become familiar with another kind of lubricant.

The Western Electric Heating Co., Chicago had a number of electric heaters in operation about its booth. Mr. Geo. Cutter the manager was present and the interests of the company were well cared for. The exhibit attracted a great deal of attention.

Jones & Laughlin, Lim., Pittsburgh and Chicago, had on exhibition a number of steel axles of which the company makes a specialty. Many of the axles had been tested in various ways and showed the excellent quality of the material used in their construction.

The Peckham Motor Truck & Wheel Co., New York, was represented by its president, Mr. Edgar Peckham. The extra long 6 D truck exhibited by this company attracted considerable attention. It was equipped with G. E. 800 motors and stood ready for the car body.

The Eureka Tempered Copper Co., North East, Pa., had a very fine line of copper commutator bars, pinions, etc., on exhibition which showed both the character and quality of the goods manufactured. Mr. Alfred Short, president of the company, was in attendance at the convention.

The Miamisburg Electric Company, Miamisburg, O., was well represented by Mr. D. H. Allen, general manager and A. L. Daniels, salesman of the company. The exhibit made at the exposition building was comprehensive, showing to good advantage the specialties made by the company, consisting of pure tempered copper commutator bars and segments. Burnley dry batteries, etc.

THE ROCHESTER CAR WHEEL WORKS, Rochester. N. Y., made an extensive exhibit of car wheels of all kinds. Mr. F. D. Russell, general manager of the street railway department, and Mr. Eldridge Packer, general sales agent were present looking after the interests of their company.

BAGLETT & COMPANY, New York, distributed some very handsome souvenirs in the shape of a little pamphlet entitled a "Modern Triumvirate" containing some very fine engravings. The press work is excellent and the embossed cover handsomely designed. The triumvirate is represented on this cover in the heads, representing the artist, the engraver and the printer.

GENERAL ELECTRIC COMPANY, New York, made an extensive exhibit of trolley and line supplies. Mounted on a Peckham truck were exhibited G. E. 800 motors. Representatives from the main office in New York, from the Cincinnati office and also from Chicago and San Francisco offices were present. An illuminated sign was one of the attractive features of its offices in the Pfister Hotel.

THE J. G. BRILL COMPANY, Philadelphia, had a number of trucks and sections of cars, showing the construction, on exhibition in the exposition building. The double decked car that made regular trips between the exposition building and Pfister Hotel for the accommodation of delegates undoubtedly attracted the most attention. Mr. J. Randell and Mr. J. C. Brill were present at the convention.

THE SIEMENS & HALSKE ELECTRIC CO., Chicago, had one of the most practical exhibits made. In a tent at the side of the exposition building there was operated a complete power plant. A Buckeye engine was directly connected to a Siemens & Halske generator. The plant was built for Cincinnati and set up temporarily before being shipped to that point. It received a constant stream of visitors.

THE MICA INSULATOR CO., New York, had one of the most complete exhibits in the exposition building. An illuminated sign was one of the attractive features of the display. Micanite was shown in a great variety of forms; in boards, in the required shapes for insulating commutator segments, armature coils, etc. A very neat souvenir was distributed, consisting of a card on which was mounted a piece of Micanite, in the shape of a commutator bar. Mr. Chas. W. Jefferson, manager of the company, was in attendance.

THE WESTINGHOUSE EL. & MFG. CO., occupied a very prominent office on the main floor of the Pfister Hotel, accessible both from the street and the lobby of the hotel. A display of railway material was made there, consisting of the new series multiple controller for electric railways, lightning arresters, for power stations and various supplies. Representatives from the eastern and Chicago offices were present. A very neat souvenir was given away in the shape of a book of testimonials of its railway apparatus, through which are placed views of many prominent electric roads.

PERSONAL.

Mr. E. F. Peck, Brooklyn, N. Y., is at the Fair this week.

Mr. C. Schlesinger, La Salle, Ill., is in Chicago visiting the Fair.

Mr. W. C. Johnson, Fitchburg, Conn., was in Chicago last week.

Mr. Cary T. Hutchinson, New York City, is visiting the Exposition.

Mr. Chas. K. Wead. Washington, D. C., is in the city for a few days.

Mr. Joseph Sachs, New York City, was in Chicago last week viewing the Fair.

Mr. Chas. H. Bigelow, E. E., Salem, Mass., is at the Exposition for a few days.

Mr. Joseph Lindsey, Biddville, N. C., has been at the Exposition the past week.

Mr. T. Fred'k Crawford, Urbana, Ohio, called on ELECTRICAL INDUSTRIES Tuesday last.

Mr. M. De Forest Yates, Schenectady, N. Y., was one of the electrical people at the Fair last week.

Mr. E. Ward Wilkins, of Patrick & Carter, Co., Philadelphia,

Pa., was seen around the Electricity Building last week, meeting old friends.

Mr. I. Bally Craig, Boston. Mass., registered at the Exposition office of the A. I. E. E. on the 17th.

Mr. Wm. J. Nieschurtz, with the Electric Light & Power Co., Canton, Ohio, visited the Fair last week.

Mr. S. H. Brownell, Essex Junction, Vt., was an interested visitor to the Electricity Building this week.

Mr. A. Wickenheiser, Pskor, Russia, called recently at the World's Fair office of ELECTRICAL INDUSTRIES.

Mr. F. B. H. Paine, St. Louis, Mo., is making his headquarters at the office of the A. I. E. E. for a few days.

Mr. D. H. Keeley, of the Government Telegraphs, Ottawa, Canada, is registered at the A. I. E. E. World's Fair office.

Mr. James Burke, Schenectady. N. Y., registered at the office of the A. I. E. E. in the Electricity Building on the 21st.

Mr. H. Loewenberg. M. E., of the Metropolitan Telegraph and Telephone Co., New York, visited the Fair last week.

Mr. J. Stanley Thornton, of the Standard Electric Co., 327 Washington St., Buffalo, N. Y., is at the Fair for a few days.

Mr. John A. Seeley, New York City, accompanied by Mrs. Seeley, was a visitor to the Department of Electricity last week.

Mr. J. D. McIntyre, of the G. E. S. Co., Syracuse, N. Y. called at the World's Fair office of ELECTRICAL INDUSTRIES recently.

Mr. Geo. F. Porter, New York City, secretary of the National Electric Light Association, called at the World's Fair office of ELECTRICAL INDUSTRIES last week.

Mr. O. G. C. Hahn, 136 Liberty St., New York, called on ELECTRICAL INDUSTRIES while making the round of the electrical exhibit a few days since.

Amusements.

HOOLEY'S THEATER—M. Coquelin and Mme. Jane Hading, repertoire. 149 Randolph street.

COLUMBIA THEATER—Mr. Henry Irving and Miss Ellen Terry, "Becket." Saturday, "The Bells." 108 Monroe street.

GRAND OPERA HOUSE Hoyt's "A Trip to China Town." 87 Clark street.

AUDITORIUM—Imre Kiralfy's Spectacle "America." Congress street and Wabash avenue.

McVICKER'S THEATER—Wm. H. Crane, in "Brother John," 82 Madison street.

CHICAGO OPERA HOUSE—American Extravaganza Company in "Sinbad." Washington and Clark streets.

SCHILLER THEATER—"Lady Windemere's Fan" Randolph, near Dearborn.

HAVERLY'S CASINO Haverly's United Minstrels. Wabash avenue, near Jackson street.

TROCADERO—Vaudeville. Michigan avenue near Monroe street.

THE GROTTO—Vaudeville. Michigan avenue near Monroe street.

Buffalo Bill's "Wild West." 63d street. Daily at 3 and 8.30 p.m.

At the Auditorium the same crowded houses are seen. Visitors coming to the Fair are securing seats a long time in advance in order to see the spectacle while in Chicago.

Mr Crane and his excellent company are receiving the encouragement of full houses in the presentation of the fourth week of "Brother John" at McVicker's. The engagement continues until November 11th.

At the Grand Opera House "A trip to Chinatown" draws crowded houses. This comedy of Hoyt's as presented by the talented company seems to meet with the approval of World's Fair visitors, and as the manager says the only dissatisfied people are those who cannot get admission. The dancing of the pretty Bessie Clayton is a special feature that is admired by all.

WE HAVE AT THE PRESENT TIME

A STOCK OF ABOUT

4,000 INSULITE SOCKETS

OF THE

SAWYER-MAN or WESTINGHOUSE STYLE.

THESE SOCKETS ARE WELL ADAPTED FOR USE IN

DRY PLACES,

and we are prepared to offer the entire lot, or part of same, at ten cents each, net cash.

WESTERN ELECTRIC CO.,

NEW YORK--CHICAGO.

MAP OF ELECTRICITY BUILDING—EXHIBITORS AND THEIR LOCATION.

GALLERY.

MAIN FLOOR.

Exhibitor.	Section	Exhibitor.	Section	Exhibitor.	Section	Exhibitor.	Section

AFTER THE FAIR

It is presumed that the energies of the country will again be devoted more strictly to business interests.

We have fully prepared ourselves to meet the large increase of business which will naturally result and our stock is

LARGE AND COMPLETE

in every line with prices lower than ever before on high-grade specialties and first-class material.

ELECTRIC APPLIANCE COMPANY,
242 Madison St., CHICAGO.

ELECTRICAL SUPPLIES

WM. BARAGWANATH & SON,

LIST OF HEATERS

TO BE SEEN IN OPERATION AT THE WORLD'S FAIR.

Two 500 H. P. East End Boiler Gallery doing 1800 H. P. work.

One 300 H. P. heater and receiving tank, Wellington Catering Co's., plant.

One 150 H. P. heater at Hygeia plant.

One 200 H. P. Libby Glass Works.

55 WEST DIVISION STREET
CHICAGO.

THE MONTHLY ISSUE FOR OCTOBER

ELECTRICAL INDUSTRIES

Should be read by everyone interested in electrical matters. In its Table of Contents are the following:

"Electric Railway Exhibit at the Fair."

"American Search Lights at the Fair."

"Duquesne Lines of Pittsburgh."

"A New Incandescent Arc Lamp." By I. B. Marks.

"The Return Circuit of Electric Railways." By Thos. J. McTighe.

"The Business End of Electricity." By H. C. Thom.

"Three Point Incandescent Switches." By Albert Scheible, M. E.

Together with illustrations of the recent applications of electricity.

The paper contains regularly

A Buyer's Directory of Manufacturers and Dealers in Electrical Supplies and Appliances.

A Complete Directory of Electric Light Stations in North America and a Complete Directory of Electric Railways in North America.

These directories are revised each issue to the date of going to press and are to be found in no other electrical journal in the World. Its articles are read carefully and its directories used constantly by all the buyers in the trade. These facts make it without a superior as an advertising medium. Sample copies and rates sent on application.

Subscription price $5 per year.

ELECTRICAL INDUSTRIES PUB. CO.,
Monadnock Block, CHICAGO.

CENTRAL ELECTRIC COMPANY,
CHICAGO.

═══════

STREET RAILWAY SUPPLIES.

Construction Tools,   Line Materials,
PINS, BRACKETS, INSULATORS, ETC.

OKONITE FEED WIRES,

CAR WIRES,

OKONITE and..MANSON TAPES,

ARE THE LEADERS.

INCANDESCENT LAMPS, TIED AND FREE FILAMENTS.

LUNDELL MOTORS, FOR STREET RAILWAY CIRCUITS.

THE FERRIS WHEEL...WIRED WITH OKONITE.

Interior Conduit System of Electric Wiring.
THE ONLY PERFECT AND COMPLETE SYSTEM FOR INTERIOR WIRING.

STANDARD PAINT COMPANY'S Celebrated P. & B. Electrical Paints and Compounds carried in Stock.

ELECTRIC LIGHT CARBONS FOR ALL SYSTEMS.

✻✻✻✻✻✻✻✻✻✻✻✻

Central Electric Company...CHICAGO.
SPECIAL AGENTS:
SOUTHERN ELECTRICAL SUPPLY CO., St. Louis. - GATE CITY ELECTRIC CO., Kansas City, Mo.

CLARK ELECTRIC COMPANY, NEW YORK.

192 Broadway and 11 John Street.

MANUFACTURERS OF ARC LIGHTING APPARATUS FOR EVERY PURPOSE A SPECIALTY.
The CLARK ARC LAMPS for use on EVERY CURRENT, have the reputation of being
the best and most durable of any ever made in the United States.

RAWHIDE PINIONS FOR ELECTRIC MOTORS
A SPECIALTY.
RAWHIDE DYNAMO BELTING

Greatest Adhesive Qualities.   A Non-Conductor of Electricity.
          Causes Less Friction than any other Belt.

THE CHICAGO RAWHIDE MANUFACTURING CO.
THE ONLY MANUFACTURERS IN THE COUNTRY.

| LACE LEATHER ROPE AND OTHER RAWHIDE GOODS OF ALL KINDS BY KRUEGER'S PATENT | This Belting and Lace Leather is not affected by steam or dampness, never becomes hard; is stronger, more durable and the most economical Belting made. The Rawhide Rope for Round Belting Transmission is superior to all others. |

75 Ohio Street,   CHICAGO, ILL.

STANDARD ELECTRIC COMPANY.

GENERAL OFFICES: 625 Home Insurance Building.

WORKS: So. Canal Street,

CHICAGO.

STANDARD SYSTEM

AT THE

WORLD'S FAIR.

MACHINERY HALL, Sec. Q, 2 Standard Arc Dynamos.

Sec. S, 20   "   "   "

ELECTRICITY BUILDING. Sec. P, Space 2, Arc Lighting Exhibit.

The Standard Lamps Light the Power Plant, Machinery Hall, Agricultural Hall, Shoe and Leather Building, and
Other Buildings and Portions of the Grounds.

See our Double Service All Night Lamp Before Buying an Old Style Two Rod Lamp.

Mile after mile of
SIMPLEX WIRE
Supplied to the
FERRIS WHEEL
•••
By...George Cutter,
The Rookery, CHICAGO.

SIMPLEX WIRES
INSURE HIGH INSULATION

SIMPLEX

Ever Onward and Upward

Simplex Electrical Co.
620 Atlantic Ave.,
George Cutter, Chicago.   BOSTON, MASS.

XNTR C
"That's the Switch"
And we control that movement.
◆
H. T. PAISTE,
10 South 16th St.,
PHILADELPHIA, PA.

Made   5 amp. S. P.
      10 amp. S. P.
       5 amp. 3 way.
      10 amp. 3 way.

P. & S. CHINA CLEAT.

George Cutter,
CHICAGO.

PASS & SEYMOUR,
SYRACUSE, N. Y.

Enterprise Electric Company
307 Dearborn Street, Chicago....

GENERAL WESTERN AGENTS

N. I. R.

Manufacturers' Agents and Mill Representatives for

Electric Railway,
Telegraph, Telephone and
Electric Light
SUPPLIES OF EVERY DESCRIPTION

Agents for   Cedar Poles,
Cypress Poles, Oak Pins,
Locust Pins, Cross Arms, Glass
—— Feeder Wire,   Insulators,
WIRES, CABLES, TAPE and TUBING

Miamisburg Electric Co.
MANUFACTURERS OF
COMMUTATORS
AND
COMMUTATOR SEGMENTS
OF TEMPERED COPPER
FOR STREET RAILWAYS
Also Brushes and Brush Copper,
**BURNLEY'S CARTRIDGE BATTERIES
and IMPERIAL DRY BATTERIES.**

MIAMISBURG ELECTRIC CO., Miamisburg, O.

BEAR IN MIND
that the regular monthly issue of ELECTRICAL INDUSTRIES contains the most complete and correct directories published of the electric light central stations and the electric railways in North America.

World's Fair Headquarters Y 27 Electricity Building.
CITY OFFICES, Monadnock Block.

Consolidated Electric Co.
Manufacturers and Dealers in all kinds of
ELECTRICAL . SUPPLIES,
115 Franklin Street,
CHICAGO.

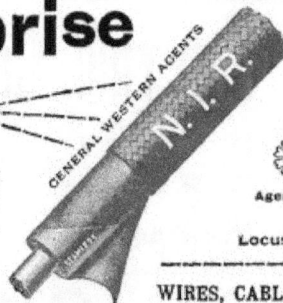

WEEKLY WORLD'S FAIR

ELECTRICAL INDUSTRIES

DEVOTED TO THE ELECTRICAL AND ALLIED INTERESTS OF THE WORLD'S FAIR,
ITS VISITORS AND EXHIBITORS.

Vol. I, No. 21.   CHICAGO, NOVEMBER 2, 1893.   FIVE MONTHS $1.00
TEN CENTS A COPY

The Department of Electricity.

In January, 1891, Prof. J. P. Barrett was selected chief of the Department of Electricity, and at once gathered

ever apparent. Difficulties were met in the work which at the time seemed almost insurmountable, and which were overcome only by the persistent efforts of the chief and his staff. The formation and construction of an exposition of

STACEY BUTLER.   ORLANDO SHEPARD.   GEO. W. MOORE.   GEO. J. HENRY, JR.
WILLIS HAWLEY.   JOHN W. BEARSDELL.   PROF. J. P. BARRETT.   DR. J. ALLEN HORNSBY.   M. W. CRUMB.

about him an efficient corps of assistants. The wisdom with which he selected his staff has never been questioned, but in the struggle to create a department that would be representative of the industry his wisdom was

this character was something that had never before been accomplished, and consequently experience was out of the question in this department. But in order to profit as much as possible by former expositions, Dr. J. Allen Hornsby, who had been selected as first assistant, was dispatched to Europe to visit the Frankfort Exposition and gain as much information as possible from foreign expositions which would be valuable in aiding the department to do the work that was to contribute probably more to the success of the Fair than any other feature. Thus, under the direction of the chief, to whom Dr. Hornsby, the acting head of the department, submitted his plans, the Department of Electricity gradually came to maturity. In the accompanying engraving, taken from a photograph by our photographer, we present to our many readers the chief of the department and his assistants. In the February number, 1891, of ELECTRICAL INDUSTRIES, appeared a sketch with portrait of Prof. Barrett, and in the April number of the present year appeared a portrait and sketch of Dr. Hornsby, so that our readers are in a measure familiar with their history. Mr. John W. Blaisdell, general superintendent, has occupied a position which to a less competent person would have been extremely difficult, but his genial manner, especially in handling the allotment of passes, has preserved most friendly feelings. To Mr. W. W. Primm, who, for the past two years has been with the department, occupying the position of electrical engineer, much credit is due to the perfection of the details of the work. Mr. Willis

EXHIBIT OF THE BELL TELEPHONE COMPANY.

Hawley, the assistant, had previous to his connection with the department an extended experience in electrical work. Commencing with the Western Electric Company, he became familiar with different branches of electrical construction, installation and maintenance.

The inspectors of the department are Mr. Geo. J. Henry and Mr. Orlando Shepard, and to their efforts and watchfulness was due the magnificent fire protection of the building. Mr. Frank J. Sullivan and Mr. Stacy Butler have attended to the receipt of the exhibits, and have kept a record of every piece of apparatus installed. Between them they have divided the twenty-four hours of the day so that one of them has been in constant attendance. Mr. Geo. E. Moore has fulfilled the duties of stenographer, which has by no means been a light task. As the department disbands and the members return to their former occupations the goodwill and best wishes of the many friends and acquaintances which they have made in connection with the discharge of their duties will go with them.

Exhibit of the Bell Telephone Company.

In the pavilion of the American Bell Telephone Company which occupies a prominent place at the south end the visitor to the Fair could gain a very good idea of the telephone system now so extensively used. In the west peristyle is located a switchboard of the latest pattern to

which are connected all the telephone subscribers at the World's Fair and which forms what is called the World's Fair station. An iron railing separates the visitors from the space occupied by the operators. An attendant is nearly always present to explain the details in the operation.

To the south of the board through the grating in the floor the visitor can see the cables as they emerge from the conduit and rise to the frames where the wires of the cable are separated and run to the different places on the switchboard. In the east peristyle are a number of cases which contain the historical exhibits of the company and which show the first telephone in practical use and the succeeding ones up to the latest pattern of the long distance telephone. There is also shown a telephone adapted

EXHIBIT OF THE ELECTRIC SELECTOR AND SIGNAL COMPANY.

to the requirements of a diver and shown attached to a divers suit.

In the center of this pavilion are maps showing the extent to which the telephone system has been introduced in the United States, the amount of territory covered by the line and statistics giving the number of subscribers and stations at the present time.

Exhibit of the Electric Selector & Signal Company.

The exhibit of the Electric Selector & Signal Company of 45 Broadway, New York, comprises a practical demonstration of its system of electric selection and signals. Its block system is fully shown in its method of transmitting and operating its signals. Along the back part of the space are arranged 10 tables representing 10 railroad stations. At the end is another table representing the train despatchers office. Each table is supplied with the necessary instruments and equipment for an office such as it represents.

But one line is used for this work and over this one line any signal may be transmitted to any station or operated at any one station without disturbing the others, from the train despatchers office. The system involves the use of the same batteries, relays and keys used in the ordinary telegraph system. By the use of small discs, notched on their edges forming the different combinations, the train despatchers office communicates with the signals having the same combination as that of the disc which the train dispatcher uses. This same method of selection is used for cutting out or in lights, motors, or any kind of machinery or apparatus operated by the electric current. A number of arc and incandescent lights, motors, etc., are shown in the exhibit illustrating the operation of the company's system. This exhibit was one of the earlier installed and at the exhibit some one has always been in attendance to explain the apparatus shown. Mr. J. H. Raymond who has been at the exhibit most of the summer has been very diligent in

arrangement of strain insulators made of moulded mica, of which it required nearly 1,000 to complete the figure. Just beyond this are some large panels on which are displayed trolley appliances and line appliances in which moulded mica is used as an insulator.

In the case at the end of the exhibit is shown asbestos in the form of gaskets, washers, various forms of packing, etc. In another case are displayed incandescent lamp sockets, rosettes, etc., of moulded mica. These different appliances are shown in another case cut in two, so that the relative

EXHIBIT OF THE NEW YORK INSULATED WIRE COMPANY.

extending the knowledge of this system and its various applications.

The Exhibit of the H. W. Johns Manufacturing Company.

One of the most important parts of every electrical machine and apparatus is its insulation and consequently to electrical manufacturers and users of manufactured articles the exhibit of the H. W. Johns Mfg Company has been very interesting. It is located in the west gallery of the Electricity Building and is quite conspicuous on account of the

position of the insulation and metal is shown. Armature rings and a great variety of forms and shapes of vulcabeston are shown as it is used in connection with numerous electrical devices. Near the center of the exhibit are shown insulations for armature and field coils, magnet spools, etc.

Exhibit of the New York Insulated Wire Company.

Among the exhibits of insulated wire in Electricity Building that of the New York Insulated Wire Company occupies a prominent place. About the space, which is near the

THE EXHIBIT OF THE H. W. JOHNS MANUFACTURING COMPANY.

tricity Building and is quite conspicuous on account of the large white balls mounted on posts and placed at the corners of the space and also the large pavilion finished in white and gold in the center of the space.

The illuminated sign which flashes out the name of the company and then the word "insulation" in red lamps attracts the attention of visitors on the ground floor. Placed about the exhibit are a number of unique objects constructed of various pieces of apparatus manufactured by the company. On an oak stand just south of the pavilion is an urn-shaped

south end of the east gallery, are arranged a number of pyramids composed of reels of wire of different sizes. The log cabin built of vulca duct has probably attracted the most attention as it was designed after a primitive log cabin. The method of construction, and the chimney that rises a foot or two above the ridge are each suggestive of the old-time cabin. A number of cases supporting signs are prominent objects about the exhibit. These signs, which are of neat and attractive design, call attention to the Grimshaw wire, tapes and compounds.

Exhibit of J. Lang & Co.

In the west end of the long pavilion erected under the direction of the Ansonia Electric Co., between the exhibit of Stanley transformers and the dynamo built by Mr. Wm. Wallace, in 1875, is located a handsome display of switches made by J. Lang & Co of 44 Michigan St. The switches, which comprise single, double and

EXHIBIT OF H. A. LAWTON & CO.

triple pole switches, are all finely polished and skillfully made.

They are nearly all of the knife pattern, which closes between substantial contact pieces. The double pole clipper switch and the single pole quarter throw switch which have more recently been introduced are shown. In all these switches the current passes through the blade of the switch and in no case through the arm.

Exhibit of the C. & C. Electric Motor Co.

At the south end of Electricity Building near the scenic

EXHIBIT OF J. LANG & COMPANY.

theatre of the Western Electric Company is located the exhibit of the C. & C. Electric Motor Co., of New York. The dazzling brightness of the illuminated sign placed on the wall above the exhibit immediately attracts the visitors attention. In this sign that spells out the company's name are 300 16-candlepower incandescent lamps. Another of the bright attractions of the display is the large clusters of incandescent lamps at each side of the entrance to the space.

The standard that supports these clusters are bronze of

a very neat design as is also the bronze railing that surrounds the exhibit. Current is supplied to the space by the 500-volt power circuit from Machinery Hall. A 50 kilowatt motor drives an 800 light dynamo which is of the standard 110 volt type. Small motors are shown in operation illustrating some of the different uses to which these motors are applied. A five-horse power motor is connected to a Sturdevant blower; a 10-horse power motor to a mine hoist and another 10-horse power motor to a 60 inch exhaust fan.

On the switch board which is of white marble are shown an excellent display of instruments, switches and connections. This board is placed immediately beneath the illuminated sign. The exhibit has been in charge of Mr. Arthur Capen who has been painstaking in explaining to the many visitors the machines and appliances on exhibition.

Exhibit of the Jewell Belting Company.

The space occupied by the Jewell Belting Co., of

EXHIBIT OF THE C. & C. ELECTRIC MOTOR COMPANY.

Hartford, Conn., in the north-eastern part of Electricity Building is filled with a variety of belt goods of interest to power users. Probably the most prominent feature of the display is the immense star suspended above the exhibit. It can be seen from the distant parts of the building, and draws the visitors in that direction as they make their way through the building.

The frame work of this star is made up of Jewell belting and shows in the method of joining the different pieces the different methods of joining the lengths of leather in the manufacture of belts. The surface of the star points is made of oak bark, while around the large cut glass jewell that forms the center is the name of the company.

Belts of different widths from four inches to 52 inches are shown. The latter represents the width of a steer hide.

EXHIBIT OF THE JEWELL BELTING COMPANY.

A brand of belting known as the diamond is shown, which is used extensively for driving electrical machinery. It is short lapped and tanned by the oak process, making an exceedingly durable belt. In a case at the right is shown a belt that has seen continual service for 36 years.

piece of belting represents a considerable, the practical use of a belt will indicate more exactly its value for transmitting power. In Machinery Hall there may be seen in use Jewell belts of several sizes; for driving the Eddy generators in Machinery Hall four 24 inch Jewell belts have been in constant use; for transmitting the power from the Atlas engine to the larger Westinghouse generator, a 72-inch three ply Jewell belt is used. These belts demonstrate in the only practical way the efficiency of these belts. Numerous views about the space in Electricity Building

EXHIBIT OF THE NEW ENGLAND BELT Co.

show belts manufactured by this company in use for driving different kinds of machinery.

Exhibit of the Hope Electric Appliance Co.

In the south end of the west gallery is located the ex-

EXHIBIT OF THE HOPE ELECTRIC APPLIANCE COMPANY.

having been taken off the pulleys and brought direct to the World's Fair. There is also shown in this space belt dressing and cement, bearing the well-known Jewell trade mark.

Although to the experienced eye the appearance of a

hibit of the Hope Electric Appliance Co. of Providence, R. I. The appliances displayed are all practical and useful. They are so arranged as to show their operation, current being supplied to the space.

On posts at each side of the large arch that spans the

space are shown mast arms and cut out boxes for both arc and incandescent circuits. To one of the posts Wright's automatic mast arm is shown, which is of great value for use in crowded streets where it is dangerous to lower a lamp in the center of the street. This arm consists of one large piece of wrought iron pipe securely attached and braced to the post. Beneath this pipe is a swinging arm hinged at the post. To the outer end of this swinging arm an arc lamp is attached. A thin tape of strong bronze metal, ⅜ of an inch wide, is attached to the lamp, runs over a pulley at the end of the stationary arm and through it to post, where it passes over another pulley and down the post to a windlass. By turning the handle attached to the windlass the tape is unwound and the lamp lowered until it hangs at the end of the swinging arm beside the post, at the side of the street out of the way of passing teams.

The wires are carried to a series cut-out on the post from whence they are carried through the movable arm to the lamp. When the lamp is lowered the lamp is cut out of the circuit, thus protecting the trimmer while handling the lamp. The windlass is covered and the tape entirely concealed in a cleat on the pole and within the fixed arm so that it is entirely protected from injury from the weather.

The other mast arm shown is designed for use on streets where trolley lines prevent the lamp being lowered from the center of the street. In this case the lamp is drawn into the pole. From the arch are suspended numerous incandescent lamps which are fed from a transformer, which shows the operation of the double pole primary switch which opens the primary circuit. The cut-out boxes shown are perfectly weather proof and are durable in construction. The special feature of these cut-outs is the quick break. The construction is such that the instant the lever reaches a certain point the spring is released and the circuit is instantly opened or closed. A number of styles of line insulator brackets are shown. Mr. Wright, of the company, has been at the exhibit most of the season.

Exhibit of the New England Butt Co.

Under the direction of the Knapp Electrical Works of Chicago, the western agent of the New England Butt Co., a number of braiding machines have been shown in operation in Electricity Building. The machines are located in the east gallery, and power is furnished by an electric motor. Single, double and triple braiding machines are shown in operation. The exhibit has been surrounded by many interested visitors who were unfamiliar with the process of manufacturing insulating wire.

The incandescent lamp tests were brought to a close on the 30th ult. The preparation of the report of the judges, which will be very exhaustive, will be begun immediately. This report will not only show the exact cost of maintaining a 110, as well as 50 volt lamp, for a given time, but will give the cost of the candle power of the different makes of lamps tested, including the first cost of the lamp. In other words, the report will show which of the lamps tested is the cheapest for the public to use.

The Western Electric Co., shut off all current from their Machinery Hall plant, except that used to light and heat their offices at midnight on the 30th.

ELECTRICAL INDUSTRIES.

PUBLISHED EVERY THURSDAY BY THE

ELECTRICAL INDUSTRIES PUBLISHING COMPANY,

INCORPORATED 1899.

MONADNOCK BLOCK, CHICAGO.

TELEPHONE HARRISON 380.

E. L. POWERS, PRES. AND TREAS.   E. S. WOOD, SECRETARY.

E. L. POWERS,	EDITOR.
W. A. REMINGTON,	ASSOCIATE EDITOR.
E. S. WOOD,	EASTERN MANAGER.
FLOYD T. SHORT,	ADVERTISING DEPARTMENT.

EASTERN OFFICE, WORLD BUILDING, NEW YORK.

World's Fair Headquarters, Y 27 Electricity Building.

SUBSCRIPTION:

FIVE MONTHS,	$1.00
SINGLE COPY,	10

Advertising Rates Upon Application.

News items, notes or communications of interest to World's Fair Visitors are earnestly desired for publication in these columns and will be heartily appreciated. We especially invite all visitors to call upon us or send address at once upon their arrival in city or at the grounds.
ELECTRICAL INDUSTRIES PUBLISHING CO.,
Monadnock Block, Chicago

The World's Columbian Exposition having closed, the publication of the WEEKLY WORLD'S FAIR ELECTRICAL INDUSTRIES will be discontinued with this issue. From the assurances received we believe that our readers will regret that this is the last number they will receive, but the time having expired that it was to run, and its field having been taken away, we must bid our friends adieu. When the Exposition opened it was apparent that there was an important and useful field for a weekly paper devoted exclusively to the electrical and allied interests of the World's Fair, and ELECTRICAL INDUSTRIES with its characteristic enterprise promptly undertook to cover this field. The enterprise met with favor from the start, and the great success of the paper has been due to the support accorded it by both subscribers and advertisers, for which we are grateful. The many original illustrations that have appeared in these columns have cost much labor, time and expense, but we have the satisfaction of knowing that our efforts to give our readers the fullest descriptions of exhibits, the freshest and most timely notes on current events of interest at the Fair have been fully appreciated. We have endeavored to give our subscribers and every visitor, so far as possible, desirous of seeing the different features of the electrical exhibits, information not otherwise available, and the demand for copies assures us that the maps published for their information have been used very extensively. Our efforts in fact, we feel have been fully rewarded. We again thank you and make our final bow.

THE World's Columbian Exposition is now at an end. Its close was marked by no great demonstration, but in sorrow over the death of Chicago's Mayor, the flags were

7

lowered, a salute fired and the Fair was officially at an end. Festival Hall that was to have witnessed a scene of joy over the success of the Fair, witnessed instead, memorial exercises over the death of Mayor Carter H. Harrison. Thus closed the greatest exposition ever held. The whole world united in producing this Exposition, but to Chicago's wonderful enterprise we are indebted for the magnitude and success of the undertaking. When the success is considered in connection with this year of business depression what the success would have been in an ordinary year can only be conjectured. As a civilizing agency the Fair has been far reaching in its influence and effects. Nations but little acquainted with the United States have become familiar with its people, its country and products. Countries in all quarters of the globe have through their exhibits and representatives made the people of the United States and other countries acquainted with their manufactures and products. But to the electrical industry it has been a triumph. Of but recent growth, its displays at former expositions have been small and have attracted but little attention. Electricity was wisely adopted as the means of transmitting power from the central station throughout the grounds. It made it possible to operate in any building in any part of the grounds machinery without disfiguring in any way the light colored interior or exterior of the buildings. It furnished the power for different forms of intramural transportation both on land and on the canals and lagoons. It carried the visitor to the roof of the highest building. It made trips to the galleries of the different buildings easy. It illuminated the grounds and turned night into day. By different combinations and devices, as a means of decoration, places otherwise unattractive were made points of interest and amusement. The power plant and system of distribution together with the displays made by the different manufacturers of electrical goods familiarized the general public with this new industry. Under the able direction of the Department of Electricity, the Electricity Building was made one of the most attractive buildings on the grounds, and in it both day and night there was always a crowd. The displays made by the larger companies were certainly a credit to them and the manner in which they placed attendants in charge and the pains with which they explained the exhibits and systems is commendable. Aided by the information and ideas gathered from the machines, plans and literature of foreign and domestic manufacturers which have been thoroughly discussed during the past months the American manufacturers will be able to produce even better machines and the public to purchase and use them more intelligently. After the exhibits are removed and the buildings demolished, there will still remain vivid recollections of the White City, its buildings, grounds, the displays and festive occasions, and the friendships and acquaintances that have arisen from the close relations which have existed during the six months of the city's life.

The Antwerp Exposition which is to open at Antwerp, Belgium, May 5th, 1894 has brought into existence the American Propaganda which has for its object, the presentation of American industry at foreign expositions. It also acts as commercial agent for exhibitors and will collect and install exhibits and statistics such as will present the country's resources and products in a creditable manner. A building is to be erected called the American Propaganda Building for these exhibits and statistics, and which will also afford a headquarters for American visitors.

WORLD'S FAIR NOTES.

From October 31st, the Electricity Building will be open to the public only from 8 A.M. till 4 P.M. All current, whether for lighting or power was cut off at midnight on October 30th.

There was such a crowd of people demanding admittance to the Scenic Theater of the Western Electric Co., Monday, that at one time it was thought the theater would have to be closed before the day was over, not to be opened in the evening, for fear of a disturbance on the part of those who were not fortunate enough to gain admission.

The E. S. Greeley & Co. exhibit was the first to be torn down in the Electricity Building. Mr. Auerbacher having had his packing cases delivered at his exhibit Monday. The tearing down and packing of the exhibit was begun early Tuesday morning, and Mr. Auerbacher hopes to have the space entirely cleared and the goods shipped to New York by Saturday of this week.

It is rumored that some of the manufacturers of insulated wire are not desirous of having a "soak" test, as was originally contemplated by the jury, and this part of the test may be abandoned. This is to be regretted, for a wire which may show excellently in a "break-down" test may be the first to show defects when subjected to the "soak" test. The sooner the faults in a particular brand of wire are known, the sooner does the manufacturer set about remedying them.

Additional Awards in the Department of Electricity.

UNITED STATES.

Brush Electric Co.
Historical exhibit.
General Electric Co.
Ornamental lamp posts.
Greves Arc Lamp Co.
Arc lamps.
N. S. Keith.
Constant current motors.
The Mather Electric Co.
Automatic adjustable circuit breakers.
National Engraving Machine Co.
Jewelers engraving machine.
Standard Paint Co.
Insulating compound, liquid.
S. S. White Dental Co.
Application of electricity as a motive power for electric drills.
Western Electric Co.
Automatic fire alarm apparatus.
Westinghouse Electric & Mfg. Co.
Electric Meter, "Shallenberger."
Constant potential, alternating current arc lamps.
Ft. Wayne Electric Co.
Constant potential dynamos and motors.
Mather Electric Co.
Constant potential dynamos and motors latest "ring" type.
Queen & Co.
Hot wire voltmeter.
Washburn & Moen.
Bare copper wire.
Westinghouse Electric & Mfg. Co.
(1) Regulator, Stillwell.
(2) Continuous current, constant E.M.F. "Letter" type dynamos and motors.

GERMANY.

Siemens Bros & Co.
Cored carbons.
Schuckert & Co.
Registering watt meters.

ENGLAND.

General Electric Co., Ltd.
    Carbons.

ITALY.

Societa Ceramina Richard, Milan.
    Insulators.

JAPAN.

Lizaomoau Tukagawa, Saga.
    China insulator.

PERSONAL.

Mr. S. Dana Green, of the General Electric Co., left for New York on Thursday last.

Mr. E. E. Winters, general manager of the Consolidated Street Railway, Macon, Ga., was a visitor at the Fair last week.

Mr. Edw. H. Chapin, secretary and treasurer of the Shipman Engine Company, Rochester, N. Y., was a recent visitor to the Fair.

Mr. W. R. Scott, of the Buffalo Street Railway, was a visitor at the World's Fair headquarters of ELECTRICAL INDUSTRIES recently.

Mr. C. C. Benson, electrical engineer of Atlanta, Ga., was one of the interested exhibitors in the Department of Electricity last week.

Mr. J. T. Myhan, supt. of the Macon & Indian River Railroad, Macon, Ga., spent several days at the Fair and in Chicago last week.

Mr. C. J. Purdy, general superintendent of the Canandaigua Electric Light & Railway Co., made an extended visit to Chicago and the Fair recently.

Mr. W. Worth Bean, president of the St. Joseph & Benton Harbor Electric Railway & Light Co., has been in Chicago and at the Fair for several days.

Mr. J. C. Liggett has been appointed to the position of electrical engineer of the Citizen's Street Railway Company, Detroit, and is too enter upon his duties at once.

Mr. P. E. Elevier, electrical superintendent for the Superior Water, Light & Power Co., of West Superior, Wis., has spent several days at the Exposition and returns home this week.

Mr. G. W. Johnson, treasurer of the Western Electrical Supply Company, of Omaha, Neb., has been in the city for several days visiting the Fair and looking after the business interests of his company.

Mr. Chas. A Schieren, of Chas. A. Schieren & Company, New York, has received the nomination for mayor of the city of Brooklyn on the republican ticket. His many friends and business acquaintances will unite with us in hoping that he be elected.

Mr. W. D. Packard, secretary and treasurer of the New York & Ohio Company, Warren, O., has been in the city for several days past doing the Fair and looking after certain business interests. Mr. Packard is very enthusiastic over the new lamps which he has just brought out.

Mr. Ralph W. Pope, secretary of the American Institute of Electrical Engineers, is at the Fair for a few days closing up the World's Fair headquarters of the institute. The headquarters of the institute has been a most convenient meeting place during the Fair and the information and curiosities to be had there have been greatly appreciated by thousands of visitors. Under Mr. Pope's skillful management to members of the institute it has been a means of extending their acquaintance and of great benefit to the institute in extending its field of usefulness.

BUSINESS NOTES.

THE LACLEDE CARBON & ELECTRIC COMPANY has recently removed its plant to Kokomo, Ind., where it will continue the manufacture of electric supplies and apparatus, including glass and porcelain insulators, cut-outs, switches, sockets and Laclede and Hercules batteries.

THE WESTERN ELECTRIC COMPANY, Chicago, has recently closed several large contracts for nearly everything in the electrical line. The business done by this representative company during the dull times has as a whole been very satisfactory. Among the recent contracts closed is one for three dynamo for lighting the Columbia Building at Louisville, Ky.

THE RELIANCE GUAGE COMPANY of Cleveland, Ohio, supplied the safety water columns for the Washington street plant of the Chicago Edison Company. These goods have become recognized as important parts of every first class boiler plant and the awards recently received by this company for its safety alarm water columns and patent solderless floats were but a just appreciation of their worth.

THE CHICAGO RAWHIDE MANUFACTURING COMPANY, 75 & 77 Ohio St., Chicago, is to be congratulated upon having worked its factory to its full capacity all the time during the late financial stringency with no reduction in the number of hands. At the present time the factory is exceptionally busy in all departments. Among recent shipments made was a 10-inch main driving belt to John Wanamaker, Philadelphia, which is to be used for operating an electric light plant. Mr. Wanamaker has used the goods of this company's manufacture for the past 10 years, a fact which speaks well for the quality of the product. The company has also recently shipped several other large dynamo belts to various points in this country. Its foreign trade is constantly increasing. Awards of the highest character were given the company on its rawhide belting, lace leather and other products.

THE CHARLES E. GREGORY Co., Chicago, is in receipt of the following letter from the Ferris Wheel Co., which speaks well for the apparatus furnished by them.

CHICAGO, Oct. 14th, 1893.

Charles E. Gregory Co., 47-49 S. Jefferson St., City.

GENTLEMEN:—We have the pleasure to say that the two 500 light Edison dynamos, which your company furnished us, have given us most excellent service up to the present time, and that we are very much pleased with the installation of these dynamos. We have the pleasure to remain very truly yours,

THE FERRIS WHEEL Co.

"Geo. W. F. Ferris, *General Manager.*

The company also reports business exceedingly brisk having sold 22 dynamos and motors during October.

Amusements.

HOOLEY'S THEATER—A. M. Palmer's Stock Company in "A Pair of Spectacles" and "Mercedes." 149 Randolph street.

COLUMBIA THEATER—Mr. Henry Irving and Miss Ellen Terry, "Becket." Saturday, "The Bells." 108 Monroe street.

GRAND OPERA HOUSE—Hoyt's "A Trip to China Town." 87 Clark street.

AUDITORIUM—Imre Kiralfy's Spectacle "America." Congress street and Wabash avenue.

McVICKER'S THEATER—Wm. H. Crane, in " Brother John," 82 Madison street.

CHICAGO OPERA HOUSE—American Extravaganza Company in "Sinbad." Washington and Clark streets.

SCHILLER THEATER—"Lady Windermere's Fan" Randolph, near Dearborn.

HAVERLY'S CASINO—Haverly's United Minstrels. Wabash avenue, near Jackson street.

TROCADERO—Vaudeville. Michigan avenue near Monroe street.

THE GROTTO—Vaudeville. Michigan avenue near Monroe street.

The excitement attending the recent revival of "Sinbad" at the Chicago Opera House has about subsided, and the big production, which is undoubtedly the superior of any of its predecessors, is moving with the briskness and smoothness of an old-timer. The houses of course have been crowded, and a noticeable change has taken place in the clientele. Musically the piece is full of bright hits, and the comedy incidents follow one another thick and fast. Dangerfield's new scenes are revelations of stage art, and the hundreds of beautiful costumes have never been surpassed in cleverness of design or richness of material. From a spectacular point of view, the new Sinbad is much better than any of its predecessors. The great ballet scene, with its gorgeous dresses, brilliant scenic environment and sparkling light effects, is entrancingly beautiful, and the immense procession in the last act, through the corridors and over the terraces of a palace of ivory and gold, is stupendous in conception and kaleidoscopic in its wealth of color and ever-changing movement.

ELECTRICAL INDUSTRIES

CENTRAL ELECTRIC COMPANY,
CHICAGO.

STREET RAILWAY SUPPLIES.

Construction Tools,   Line Materials,
PINS, BRACKETS, INSULATORS, ETC.

OKONITE
FEED
WIRES,

CAR WIRES,

OKONITE
and..MANSON
TAPES,

ARE
THE
LEADERS.

INCANDESCENT
LAMPS,
TIED
AND
FREE
FILAMENTS.

LUNDELL
MOTORS,
FOR
STREET
RAILWAY
CIRCUITS.

THE FERRIS WHEEL...WIRED WITH OKONITE.

Interior Conduit System of Electric Wiring.
THE ONLY PERFECT AND COMPLETE SYSTEM FOR INTERIOR WIRING.

STANDARD PAINT COMPANY'S Celebrated P. & B. Electrical Paints and Compounds carried in Stock.

ELECTRIC LIGHT CARBONS FOR ALL SYSTEMS.

Central Electric Company...CHICAGO.
SPECIAL AGENTS:
SOUTHERN ELECTRICAL SUPPLY CO., St. Louis. - GATE CITY ELECTRIC CO., Kansas City, Mo.

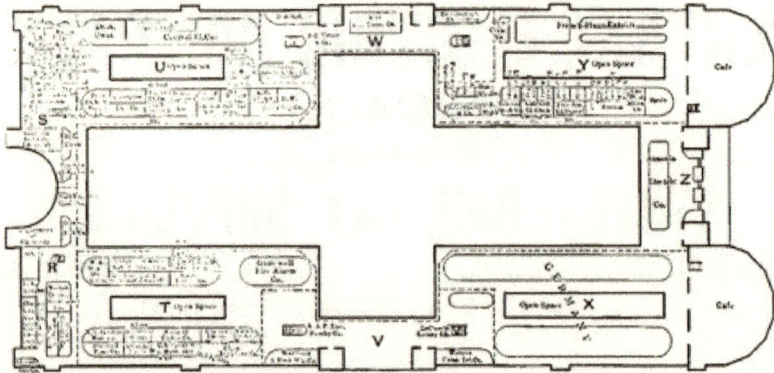

GALLERY.

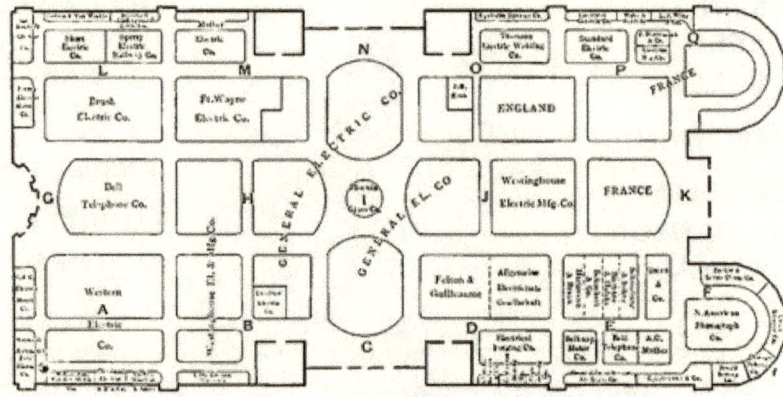

MAIN FLOOR.

Exhibitor.	Section.	Exhibitor.	Section.	Exhibitor.	Section.	Exhibitor.	Section.
Austin	Y	Electrical Review		Jaeger, Chas. L.		Reliance Gauge Co	
Ansonia Electric Co	Z	Electricity	R	Johns Mfg. Co., H. W.		Roessler & Hasslacher Chem. Co	
Am. Inst. of Elec. Eng.		Electric Gas Co		Jewell Belting Co		Street Railway Journal	
American Battery Co	T	Electrical Engineer		Jenney Elec. Motor Co		Strowger Aut. Telph. Co	
Axford, H. M.	S	Electrical World		Knapp Electrical Works		Standard Paint Co	
Allg. Elec. Gesellschaft	D	Eddy Electric Motor Co		K. & P. Elec. Novelty Co		Sponholz, C. L.	
Bagw Mfg. Co		Enterprise Electric Co	A	Knapp & Buckley		Star Iron Tower Co	
Bryant Electric Co	R	Electrical Forging Co		Kennedy Electric Co		Spain	
Billings & Spencer		Equitable Dynamo Co		Lawton, H. A.		Schuere, Chas. A. & Co	
Belsey, W. B		Elektron Mfg. Co	P	LaClaeche Battery Co		Schomberg & Sobre	
Belknap Motor Co		Electrical Conduit Co		McNeil-Tasker Elec. Co		Siemens & Halske	
Bell Telephone Co	E-G	England		Marcus, W. N		Schuckert & Co	
Brush Electric Co	Y	Rangor China Works		Merker, Dr. H.		Short Electric Co	
Caldwell El. Chain Cut. Mch. Co	Y	Franklin Elec. Appliances		Mchanan Bat. & Opt. Co	W	Sperry Elec. Railway Co	
Chrond Elec. Storage Co	H	French Piano Exhibit		Munson, C., Belting Co		Standard Underg. Cable Co	
Center, George		Felten & Guilleaume		Mather, S. C	K	Standard Electric Co	
Canton Elec. Co		France	K-P	Mather Electric Co		Samson Battery Co	
Chicago Elec. Wire Co	T	Ft. Wayne Elec. Co		Newhaus Clock Co		Tate Ant. El. Signal Co	
Copenhagen Fire Alarm Co		Gault & Co., N. C		New Magnetic Watch Co		Todd, Applegate Co	
Central Electric Co		Gatewell Fire Alarm Co		N. Y. Insulated Wire Co		Taylor, Goodhue & Ames	
Commercial Cable Co		General Electric Co	B-H-N-C-J	National Carbon Co		Thomson Elec. Welding Co	
C. A. C. Est. Motor Co		General Household Art. Lt Co		Norwich Ins. Wire Co		Telautograph, Elisha Gray	W
Cleveland Elec. & Mfg. Co		Gravely, E. S. & Co		North Am. Phonograph Co	S	Union Electric Co	
Chicago Belting Co	F	Germany		N. Y. & L. E. A		Veeder, C. J. & Co	
Delaney Clock Co		Hubbard Wm. A. Co	T	Nat. Ant. Fire Alarm Co		Webb, C. G	
Department of Electricity	R	Eggleston, C. J		Nat. Engraving Machine Co		Weston El. Instrument Co	
Eastman al. Penetrator		Holt & Hopmans Mfg. Co		Owen, N. C		Washburn & Moen Mfg. Co	
Elec. Launch & Nav. Co	H	Hope Elec. Appliance Co		Phoenix Glass Co		Western Union Tel. Co	
Electric Separator		Hall, Chas. F		Paiste, H. T		Waite & Bartlett Mfg. Co	
Edgerton, E. M		Hahne, N. A		Pulvermacher Galv. Co		White, S., Dental Mfg. Co	
Elgin Telephone Co	T	Hartmann & Braun		Sympelty, F. A		Western Electrician	
Edison Elec. Mfg. Co		Hanson & Van Winkle		Pratt Rd. Med. Sup. Co		Wilder Ant. Hugular Al. Co	
Enterprise Elec. Co	A	Hersh, J. M		Powell, Wm. & Co		Western Electric Co	
Eureka Temp. Copper Co		Hardmann, F. & Co		Pudge, A. M		Westinghouse El. & Mfg. Co	B-H-J
Electric Appliance Co	C	Illinois Alley Co	S	Page Belting Co		Wiles & Scofield	
Elec. Sel. & Sig'l Co		Internat. Ant. Lt & Pr Co	U	Queen & Co		Wing, L. J. & Co	
Electric Heat Alarm Co	Y	India Rubber Comb Co		Ringlor, F. A		Zucker & Levett Chem. Co	F

ELECTRICAL INDUSTRIES

THE PACKARD LAMP AGAIN

We are again supplying the celebrated Packard High Grade Lamp at a price below that ever quoted before on strictly high grade incandescent lamp. We are also

GUARANTEEING AGAINST INFRINGEMENT

suits, and furnishing a lamp, which for efficiency, long life, and non-discoloration has

NEVER BEEN EQUALLED

Write us about it

ELECTRIC APPLIANCE COMPANY,
242 Madison St., CHICAGO

ELECTRICAL SUPPLIES

WM. BARAGWANATH & SON,

LIST OF HEATERS

TO BE SEEN IN OPERATION AT THE WORLD'S FAIR.

Two 500 H. P. East End Boiler Gallery doing 1800 H. P. work.
One 300 H. P. heater and receiving tank, Wellington Catering Co's., plant.
One 150 H. P. heater at Hygeia plant.
One 200 H. P. Libby Glass Works.

55 WEST DIVISION STREET
CHICAGO.

THE MONTHLY ISSUE FOR NOVEMBER

ELECTRICAL INDUSTRIES

Should be read by everyone interested in electrical matters. In its Table of Contents are the following:
"The Department of Electricity."
"The Convention of the American Street Railway Association."
"Power House Engines" by E. G. Connette.
"Use of Storage Batteries in Electric Generating Stations for Utilizing and Regulating Power," by C. O. Mailloux.
"Street Car Magnetic Cut-Outs," by W. S. Harrington.
"Accident at Atlantic Avenue Power House."
"The St. Pancras Electric Lighting Station."
Together with illustrations of the recent applications of electricity.
The paper contains regularly
A Buyer's Directory of Manufacturers and Dealers in Electrical Supplies and Appliances.
A Complete Directory of Electric Light Stations in North America and a Complete Directory of Electric Railways in North America.
These directories are revised each issue to the date of going to press and are to be found in no other electrical journal in the World. Its articles are read carefully and its directories used constantly by all the buyers in the trade. These facts make it without a superior as an advertising medium. Sample copies and rates sent on application.
Subscription price $3 per year.

ELECTRICAL INDUSTRIES PUB. CO.,
Monadnock Block, CHICAGO.

Map of Chicago.

Showing Location of Its Electrical and Allied Business Interests, Principal Hotels, Theatres, Depots and Transportation Lines to the World's Fair Grounds. (Index numbers refer to the black squares.)

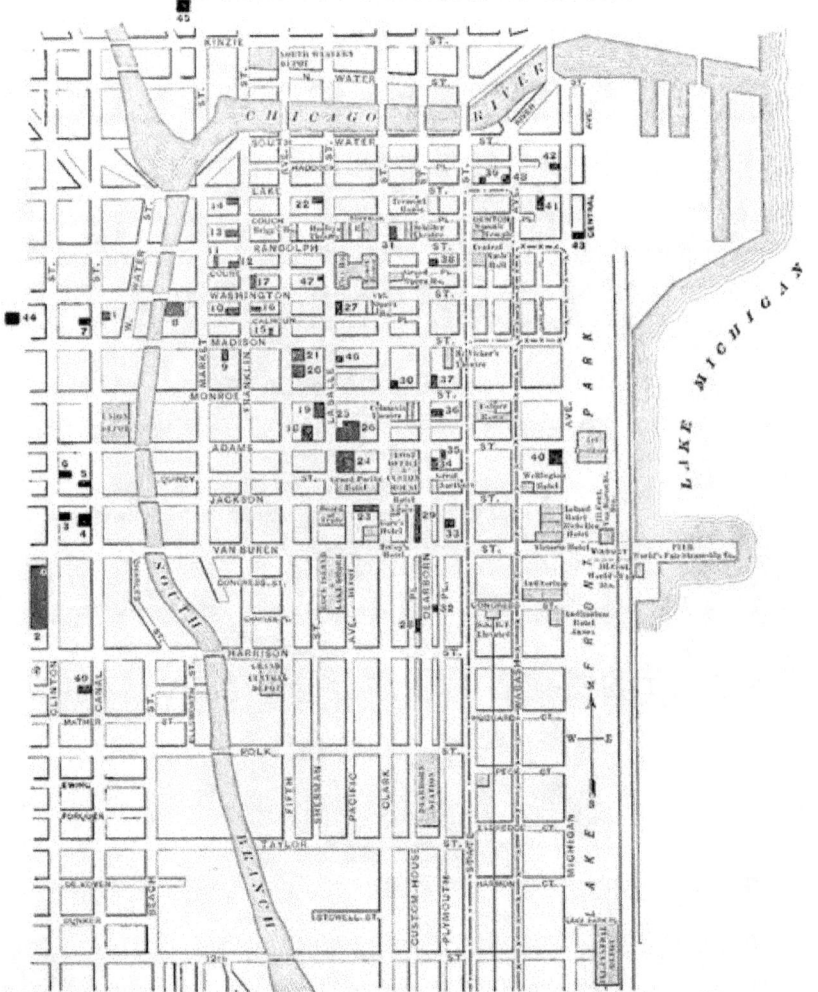

WE HAVE AT THE PRESENT TIME

A STOCK OF ABOUT

4,000 INSULITE SOCKETS

OF THE

SAWYER-MAN or WESTINGHOUSE STYLE.

THESE SOCKETS ARE WELL ADAPTED FOR USE IN

DRY PLACES,

a d we are prepared to offer the entire 'ot, or part of same, at ten cents each, net cash.

WESTERN ELECTRIC CO.,
NEW YORK--CHICAGO.

Waterhouse Arc Lamps

For Incandescent Circuits.

The best designed,
steadiest
and most
durable
arc lamps
made.

GEORGE CUTTER,

851-853-855   The Rookery,
CHICAGO.

P. & S.
China Goods
(Stock at Chicago.)

Medbery
Railway Insulators
(Stock at Chicago.)

Ferris Wheel

and the

Masonic Temple:

The tallest structures
of Chicago
of the World's Fair

Both wired with SIMPLEX

SIMPLEX ELECTRICAL CO.,
620 ATLANTIC AVENUE,
BOSTON, MASS.

GEORGE CUTTER,
851-855 The Rookery,
CHICAGO.

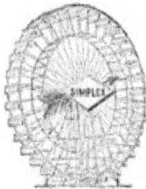

CLARK

ELECTRIC

COMPANY, NEW YORK.

192 Broadway and 11 John Street.

MANUFACTURERS OF ARC LIGHTING APPARATUS FOR EVERY PURPOSE A SPECIALTY.
The CLARK ARC LAMPS for use on EVERY CURRENT, have the reputation of being
the best and most durable of any ever made in the United States.

RAWHIDE PINIONS FOR ELECTRIC MOTORS

A SPECIALTY.

RAWHIDE DYNAMO BELTING

Greatest Adhesive Qualities   A Non-Conductor of Electricity

THE CHICAGO RAWHIDE MANUFACTURING CO.

THE ONLY MANUFACTURERS IN THE COUNTRY

LACE LEATHER ROPE
AND OTHER RAWHIDE
GOODS
OF ALL KINDS
BY KRUEGER'S PATENT

75 Ohio Street,   CHICAGO, ILL.

STANDARD ELECTRIC COMPANY.

GENERAL OFFICES: 625 Home Insurance Building.

WORKS: So. Canal Street,

CHICAGO.

STANDARD SYSTEM

AT THE

WORLD'S FAIR.

MACHINERY HALL, Sec. Q, 2 Standard Arc Dynamos.
Sec. S, 20   "   "   "
ELECTRICITY BUILDING. Sec. P, Space 2, Arc Lighting Exhibit.

The Standard Lamps Light the Power Plant, Machinery Hall, Agricultural Hall, Shoe and Leather Building, and
Other Buildings and Portions of the Grounds.

See our Double Service All Night Lamp Before Buying an Old Style Two Rod Lamp

Mile after mile of
SIMPLEX WIRE
Supplied to the
FERRIS WHEEL.

...

By...George Cutter,
The Rookery, CHICAGO.

SIMPLEX WIRES

INSURE
HICH
INSULATION

Simplex Electrical Co.
620 Atlantic Ave.,

George Cutter, Chicago.   BOSTON, MASS.

XNTRIC

"That's the Switch"

And we control that movement.

H. T. PAISTE.

10 South I th St.,
PHILADELPHIA,
PA.

Made 5 Amp. S. P.
10 mp. S. P.
5 amp. 3 way.
10 amp. 3 way.

P. & S. CHINA CLEAT.

George Cutter,
CHICAGO.

PASS & SEYMOUR,
SYRACUSE, N Y

Enterprise Electric Company

307 Dearborn Street.
Chicago . . . .

GENERAL WESTERN AGENTS

N. I. R.

Manufacturers' Agents and Mill Representatives for

Electric Railway,
Telegraph, Telephone and
Electric Light

SUPPLIES OF EVERY DESCRIPTION

Agents for   Cedar Poles,
Cypress Poles, Oak Pins,
Locust Pins, Cross Arms, Class
Feeder Wire,   Insulators,

WIRES, CABLES, TAPE and TUBING

Miamisburg Electric Co.

MANUFACTURERS OF

COMMUTATORS

AND

COMMUTATOR SEGMENTS

OF TEMPERED COPPER

FOR STREET RAILWAYS

Also Brushes and Brush Copper,
BURNLEY'S CARTRIDGE BATTERIES
and IMPERIAL DRY BATTERIES.

MIAMISBURG ELECTRIC CO., Miamisburg, O.

BEAR IN MIND

that the regular monthly issue of ELECTRICAL IN-
DUSTRIES contains the most complete and correct
directories published of the electric light central stations
and the electric railways in North America.

World's Fair Headquarters Y 27 Elec ricity Building.

CITY OFFICES, Monadnock Block.

Consolidated Electric Co.

Manufacturers and Dealers in all kinds of

ELECTRICAL . SUPPLIES,

115 Franklin Street,

CHICAGO.

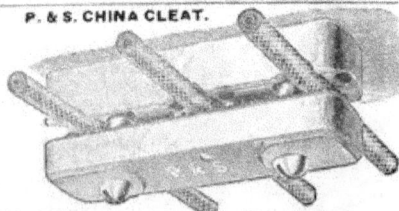

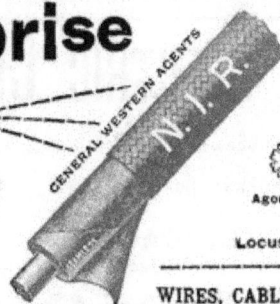